# Energy-transducing ATPases – structure and kinetics

# Energy-transducing ATPases – structure and kinetics

## Yuji Tonomura

Faculty of Science, Osaka University, Toyonaka, Osaka 560

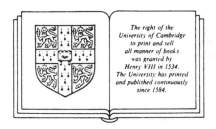

The right of the
University of Cambridge
to print and sell
all manner of books
was granted by
Henry VIII in 1534.
The University has printed
and published continuously
since 1584.

Cambridge University Press

*Cambridge*
*London   New York   New Rochelle*
*Melbourne   Sydney*

CAMBRIDGE UNIVERSITY PRESS
Cambridge, New York, Melbourne, Madrid, Cape Town, Singapore, São Paulo, Delhi

Cambridge University Press
The Edinburgh Building, Cambridge CB2 8RU, UK

Published in the United States of America by Cambridge University Press, New York

www.cambridge.org
Information on this title: www.cambridge.org/9780521104814

First published 1986
This digitally printed version 2009

*A catalogue record for this publication is available from the British Library*

*Library of Congress Cataloguing in Publication data*
Tonomura, Yuji.
    Energy-transducing ATPases – structure and kinetics.
    Includes index.
    1. Adenosine triphosphatase.    2. Actomyosin.
3. Energy metabolism.    I. Title.
QP609.A3T66    1986        599'.0133        85-26951

ISBN 978-0-521-30479-5 hardback
ISBN 978-0-521-10481-4 paperback

# Contents

# *Foreword*

In 1972 Professor Yuji Tonomura published a voluminous monograph *Muscle Proteins, Muscle Contraction and Cation Transport* (University of Tokyo Press). This book summarizes Tonomura's famous discovery of M–P–ADP complex, the intermediate of the myosin ATPase reaction. Although this volume was highly valued by the specialists, general readers claimed that it was hard to understand.

When Tonomura was invited to write a book by the Cambridge University Press, he seriously considered this criticism and decided to make the new book as readable and accessible as possible, even for undergraduate students. For this purpose he asked his young associates and graduate students (see below) to write respective chapters in Japanese. He wanted to know how the younger generation comprehended the matter. Tonomura then scrutinized the manuscripts carefully, spending several years. He did not want to make the book too bulky, so he was trying hard to shorten the length of the manuscripts to half. Unfortunately, however, Professor Tonomura succumbed to a sudden heart attack on November 28, 1982.

At that time he was preparing his lecture, entitled 'The evolution of energy-transducing ATPases' to be delivered in my laboratory at the University of Tokyo on the next day. This problem was his last adventure and, in a sense, a conclusion to his long-lasting scientific pursuit. All Japanese scientists concerned with bioenergetics were keenly looking forward to this talk. If he had lived, the last chapter of this book might have contained the essence of this lecture.

After his death, I immediately started to talk about the manuscript with Professors Koichi Yagi, Tonomura's successor at Hokkaido University, and Koscak Maruyama, Tonomura's long-time friend (Chiba University). All of us wanted Tonomura's posthumous book to be published. So we requested his colleagues responsible for each chapter to

*Foreword*

translate the Japanese version into English. Drs. Fumi Morita (Hokkaido University), Tohru Kanazawa (Asahikawa Medical College), and Taibo Yamamoto (Osaka University) cooperated with us in reviewing the English manuscripts. Thus the friendship and admiration felt for Tonomura by his friends and colleagues have made this book possible. Generous cooperation of the Cambridge University Press is also greatly appreciated, without which this book would not have been realized.

Finally, I should like to briefly refer to the late Professor Tonomura's career. He was born on February 9, 1923, in Nara, the capital of Japan in the eighteenth century. After graduation from the Department of Botany, University of Tokyo, in 1946, he worked on the kinetics of catalase under the supervision of the late Professor Hiroshi Tamiya, who profoundly influenced Tonomura personally as well as scientifically. In 1951, Tonomura started his life-long research on the actomyosin–ATP system together with Dr. Shizuo Watanabe at the Research Institute for Catalysis, Hokkaido University; in 1958 he became a Professor of Biochemistry there. He moved to Osaka University in 1963 as a Professor of Biology. He published 25 reviews and about 200 original papers (in English) in the field of contractile proteins, especially myosin ATPase, sarcoplasmic reticulum $Ca^{2+}$, $Mg^{2+}$-ATPase, $Na^+$, $K^+$-ATPase and other subjects.

I personally feel confident that the late Professor Yuji Tonomura's monumental book can now be presented to those who are interested in biological sciences. I wish to express my sincere gratitude to all those who have shared the task to publish this book. I am particularly grateful to those who have acted as moderators: Chapters 2 and 3, Akio Inoue and Toshiaki Arata; Chapter 4, Satoshi Ogihara; Chapter 5, Masami Takahashi; Chapter 6, Ichiro Matsuoka; Chapter 7, Taibo Yamamoto and Haruhiko Takisawa; Chapter 8, Motomori Yamaguchi and Junshi Sakamoto.

*February 1985 at Okazaki*                    Setsuro Ebashi

*viii*

# 1   Prologue

Living cells synthesize adenosine 5'-triphosphate (ATP) from adenosine 5'-diphosphate (ADP) and inorganic phosphate (Pi) through glycolysis and oxidation–reduction reactions. By using the energy derived from ATP hydrolysis, cells carry out their fundamental physiological functions such as mechanical work and transport of various substances as well as biosynthesis of various materials.

Progress in research to elucidate the basic principle of biological energy transduction may be divided into three major phases. It was found by Lipmann (1941) that a high-energy phosphate compound, ATP (Fig. 1.1), plays key roles in a number of cell activities. ATP is relatively stable under physiological conditions; when hydrolyzed to ADP and Pi, however, it releases enough free energy for cells to carry out one

Fig. 1.1 Structure of the ATP molecule, after Kennard *et al.* (1971).

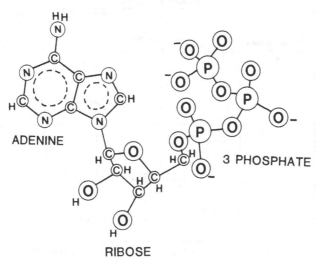

ADENINE

3 PHOSPHATE

RIBOSE

step of a chemical reaction. The free-energy change of ATP hydrolysis inside the living cell, $\Delta G'$ has been estimated to be between $-12$ and $-13$ kcal/mol (Curtin *et al.*, 1974).

The second major phase was the finding that each energy-transducing system has its own specific ATPase. In 1939, Engelhardt & Ljubimova found that the contractile protein of muscle (myosin) has an ATPase activity. Gibbons (1963) found another ATPase (dynein) involved in cell motility in cilia. In 1957, Skou found a $Na^+,K^+$-ATPase in the cell membrane, which is involved in the active transport of cations. This was followed by the discovery of the $Ca^{2+},Mg^{2+}$-ATPase of sarcoplasmic reticulum by Hasselbach & Makinose (1961). Furthermore, Racker's group identified mitochondrial membrane ATPase that plays a key role in ATP synthesis as a coupling factor ($F_1$). They also showed that a complex of $F_1$ with an oligomycin-sensitive factor ($F_0$) is actually the $H^+$-translocating ATPase (Penefsky *et al.*, 1960). Subsequently, this type of ATPase was found in chloroplasts (Avron, 1963), and also in bacteria (Abrams, McNamara & Johnson, 1960). The evolutionary relationships among the energy transducing ATPases discussed in this book are summarized in Fig. 1.2.

Fig. 1.2 The relationship among the various ATPases that appear in this book. The relations among these ATPases are discussed in Chapter 9.

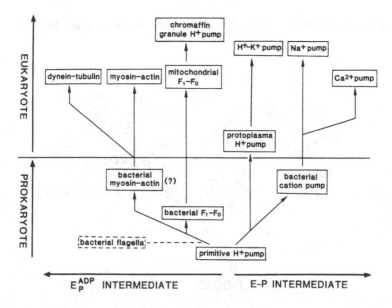

The third major phase of progress in research on biological energy transduction is characterized by a deepened understanding of the molecular aspects of energy transduction. This was achieved through studies on the dynamic and energetic properties of the elementary steps in the ATPase reaction and on the coupling mechanism between the movement of the ATPase molecule and the elementary steps in the reaction.

The object of this book is to document in detail current knowledge of the energy-transducing mechanisms in the living cell, in particular those of the contractile and transport ATPases. Accordingly, we first describe the structural aspects of these ATPases. Secondly, the dynamic properties of the elementary steps of the ATPase will be discussed in detail. Thirdly, functional states of ATPase molecules within the cell will be considered as a basis for understanding the coupling of the elementary steps of the ATPase with the movement of the enzyme molecule.

It is thought that all biological energy-transducing systems have many similarities in the mechanism of ATP hydrolysis and the molecular movement during the ATPase reaction. Therefore, the studies of one energy-transducing system at the molecular level can assist the investigation of other energy-transducing systems. The relationships among the molecular mechanisms of the various energy-transducing systems will be summarized in the final chapter and discussed from an evolutionary point of view.

This account is not intended to give an overall view of the energy-transducing systems. Although very important to living cells, the properties of various nucleotide triphosphatases (NTPases) involved in protein syntheses and structural changes of DNA are left out, since these NTPases function by mechanisms different from those of the contractile and transport ATPases discussed here. These NTPases have already been described in detail by Kornberg (1980) in his superbly written book, and their reaction mechanisms have been described lucidly by Kaziro (1978).

# References

Abrams, A., McNamara, P. & Johnson, F. B. (1960). Adenosine triphosphate in isolated bacterial cell membranes. *J. Biol. Chem.*, **235**, 3659–62.

Avron, M. (1963). A coupling factor in photophosphorylation. *Biochim. Biophys. Acta*, **77**, 699–702.

Curtin, N. A., Gilbert, C., Kretzschmar, K. M. & Wilkie, D. R. (1974). The effect of the performance of work on total energy output and metabolism during muscle contraction. *J. Physiol.*, **238**, 455–72.

Engelhardt, W. A. & Ljubimova, M. N. (1939). Myosin and adenosine triphosphatase. *Nature*, **144**, 668–9.

Gibbons, I. R. (1963). The protein components of cilia from *Tetrahymena pyriformis*. *Proc. Natl. Acad. Sci. (USA)*, **50**, 1002–6.

Hasselbach, W. & Makinose, M. (1961). The calcium pump of the granules of the muscle and their dependence on adenosine triphosphate hydrolysis. *Biochem. Z.*, **333**, 518—26.

Kaziro, Y. (1978). The role of guanosine 5′triphosphate in polypeptide chain elongation. *Biochim. Biophys. Acta*, **505**, 95–127.

Kennard, O., Isacs, N. W., Motherwell, W. D. S., Coppola, J. C., Wampler, D. L., Larson, A. C. & Watson, D. G. (1971). The crystal and molecular structure of adenosine triphosphate. *Proc. R. Soc. A.*, **325**, 401—36.

Kornberg, A. (1980). *DNA Replication*, 2nd edn. San Francisco: W. H. Freeman and Company.

Lipmann, F. (1941). Metabolic generation and utilization of phosphate bond energy. *Adv. Enzym.*, **1**, 99–162.

Penefsky, H. S., Pullman, H. E., Datle, A. & Racker, E. (1960). Partial resolution of the enzymes catalyzing oxidative phosphorylation. II. Participation of a soluble adenosine triphosphatase in oxidative phosphorylation. *J. Biol. Chem.*, **235**, 3330–6.

Skou, J. C. (1957). Influence of some cations on an adenosine triphosphatase from peripheral nerves. *Biochim. Biophys. Acta*, **23**, 394–402.

# 2 Myosin ATPase

Muscle contraction has characteristics different from those of machines. Muscle cells can transduce chemical energy directly to mechanical work with high efficiency. Also, metabolic activity in the cell increases several hundred- to several thousandfold within 100 ms after excitation. Szent-Györgyi (1949) showed that muscle contraction with these characteristics is caused by a reaction between ATP and actomyosin, a complex of two proteins, actin and myosin. ATP is hydrolyzed by myosin, and this ATPase reaction of myosin is highly activated by F-actin. This highly activated ATPase, actomyosin ATPase, is coupled with muscle contraction.

In this chapter, we introduce first the structure of the muscle cell and the sliding filament theory, the basic theory of muscle contraction, proposed from the study of the structure of muscle cell. We then discuss the structure of myosin and the elementary steps of the ATPase reaction. Many books have been published on the mechanism of muscle contraction, notably the *Cold Spring Harbor Symposium on Quantitative Biology* (1973) and the book edited by Bourne (1973) cover various fields of study on muscle contraction. Needham (1971) wrote a history of the research in this field, and Tonomura (1972) wrote a monograph on the subject.

## Muscle cells and structural proteins

### Structure of muscle cells

The structure of muscle cells varies greatly with type. In the case of vertebrate skeletal muscle, the most popular and well studied, the muscle cell, called a muscle fiber, is 20–100 μm in diameter and several centimeters in length. This huge multinucleate cell is formed by unidirectional fusing of the mononucleate myoblast during the process of development.

In living muscle fiber, components in the intracellular fluid are held stationary by the action of the plasma membrane, the sarcolemma. For the physiological study of muscle contraction, skinned fibers were prepared by mechanically removing the plasma membrane (Natori, 1954), or the membrane was made leaky by treatment with glycerol or other reagents (Szent-Györgyi, 1949).

Muscle cells have a special membrane system: the transverse (T)-system and the sarcoplasmic reticulum (SR; Porter & Palade, 1957). The T-system is a tubular structure which extends from the plasma membrane into the interior of the cell, where it joins with the SR which surrounds the contractile organ, the myofibrils. The merging of the T-system with the SR forms a structure called the triad. Excitation of the plasma membrane is transmitted to the SR through the T-system and induces the release of $Ca^{2+}$ from the SR, thus causing the contraction of myofibrils.

Muscle fibers characteristically have many myofibrils lying parallel to the main axis of the fiber. Myofibrils are about 1–2 μ in diameter and contract with the addition of $Mg^{2+}$-ATP in the presence of a low concentration of $Ca^{2+}$. As shown in Fig. 2.1, the contractile materials in the myofibrils are organized within a repeating unit called the sarcomere. A sarcomere is divided by Z-lines and is 2.3 μm in length when

Fig. 2.1 The structure of striated muscle at three levels of organization. Dimensions shown are for rabbit psoas muscle.

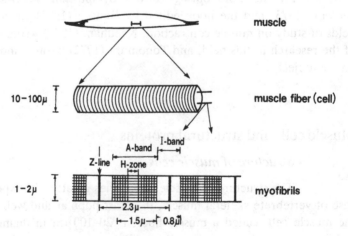

the muscle is at resting state. The central part of the sarcomere is dark under the light microscope and is called the A-band (anisotropic band). On either side of the Z-line are regions which appear light under the microscope; these together are called the I-band (isotropic band). The central part of the A-band, about $0.3\,\mu m$ in width, has lower optical density than the rest of the A-band and is called the H-zone (see review by H. E. Huxley, 1960).

### *Structure of the sarcomere and the sliding filament theory of muscle contraction*

A. F. Huxley & R. Niedergerke (1954) using living muscle and H. E. Huxley & J. Hanson (1954) using isolated myofibrils found independently that during physiological shortening both the length of A-band and the length from the Z-line to the end of the H-zone did not change, although the overall sarcomere length decreased. As shown in Fig. 2.2, they explained these results by proposing that two different protein filaments with constant lengths are distributed in the A-band and between the Z-line and the end of H-zone, respectively. Shortening is brought about by the sliding of these two filaments relative to each other. This model is generally called the 'sliding filament theory' of muscle contraction.

The two kinds of filaments proposed by these investigators were later identified by electron microscopy. The thick filaments, $1.5\,\mu m$ in length

Fig. 2.2 Schematic view of the extent of overlapping of the thin filaments with the bridges from the thick filaments, based on the sliding filament theory of muscle contraction.

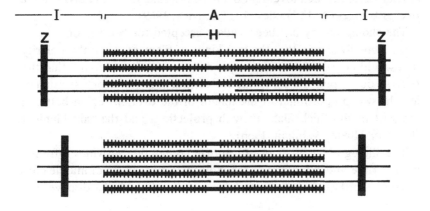

and 10 nm in width, are present in the A-band, and the thin filaments, 1.0 μm in length and 5–6 nm in width, are present running from the Z-line to the end of the H-zone (Fig. 2.2). This structure was confirmed by electron microscopic investigations of transverse sections of muscle fiber. Only thick filaments are observed in the H-zone and only thin filaments in the I-band. In the A-band, except for the H-zone, both thick and thin filaments are observed in a hexagonal arrangement (H. E. Huxley, 1957).

The A-band disappears after KCl extraction of myosin from myofibrils; the optical density of the I-band decreases markedly after extraction of actin with KI. On the basis of quantitative comparisons between the decrease in optical density and the quantity of the extracted proteins, it was demonstrated that the main component of thick filaments is myosin, and that of thin filaments is actin (Huxley, 1960). These findings agree well with findings from biochemical studies that the complex of actin and myosin (actomyosin) hydrolyzes ATP, and that muscle contraction is coupled with this ATPase reaction.

Within the framework of the sliding filament theory, the fundamental problem in the investigation of the molecular mechanism of muscle contraction is the clarification of the structure and function of the overlap region of the two filaments during the development of tension. The structure of the overlap region was investigated by Huxley and co-workers (Huxley, 1957; Huxley & Brown, 1967) using electron microscopy and X-ray diffraction. They found projections along the backbone of the thick filaments and thought that crossbridges formed between these projections and the thin filaments (Fig. 2.2). Consequently, it was proposed that muscle contraction includes at least the following three processes that are coupled with ATP hydrolysis: (1) the formation of crossbridges; (2) the movement of crossbridges; and (3) the dissociation of crossbridges (A. F. Huxley, 1957; Huxley, 1969).

The sliding theory has been widely accepted for two reasons. First, it was actually shown that the two kinds of filaments have the polarity that would be expected with this model. Second, it was found that the tension developed during isometric contraction at various sarcomere lengths was proportional to the length of the overlap region between the part of the thick filament with projections and the thin filament (Gordon, Huxley & Julian, 1966).

The sliding filament theory of muscle contraction provides a framework for the understanding of the molecular mechanism of muscle contraction. Further progress in this field was made with the discovery of the $Ca^{2+}$-regulatory system of muscle contraction. $Ca^{2+}$ is released from the sarcoplasmic reticulum as described above. $Ca^{2+}$ receptors differ according to muscle type, as will be described in the next chapter. In the case of vertebrate skeletal muscle, the thin filaments are composed of F-actin and the regulatory proteins, tropomyosin and troponin, and muscle contraction is induced by the binding of $Ca^{2+}$ to troponin.

## Structure of myosin

'Myosin', a structural protein of muscle, was discovered and named by Kuhne in 1859. Studies of energy transformation were initiated by the discovery by Engelhardt & Ljubimova (1939) that myosin has an ATPase activity. Subsequently, Banga & Szent-Györgyi (1941) discovered that the substance previously called 'myosin' is a complex of myosin and actin, another important structural protein of muscle. This complex is also called actomyosin or myosin B.

Myosin is located in the A-band of myofibrils where it forms thick filaments. It constitutes 54% of the total protein of myofibrils. Since myosin is soluble at high ionic strength (more than 0.3 M KCl) and forms the thick filaments, and precipitates at low ionic strength, it can be purified easily.

Myosin plays a central role in muscle contraction and has the following three important functions: (1) formation of thick filaments at low ionic strength; (2) hydrolysis of ATP into ADP + Pi; and (3) binding with F-actin to form crossbridges. This section describes the structure of myosin emphasizing these functions. For further information on the structure of myosin see reviews by Perry (1967), Tonomura (1972), Yagi (1975) and Margossian & Lowey (1982).

### Size and shape of the myosin molecule

Myosin has a molecular weight (m.w.) of about 480 000. It decomposes into peptides (subunits) in a solution of guanidine HCl, urea, high alkali or sodium dodecyl sulfate (SDS). The myosin molecule consists of two heavy chains (HC) of molecular weight 200 000 and four light chains (LC) of molecular weights from 15 000 to 25 000.

The size and shape of the myosin molecule have been studied by hydrodynamic methods, light scattering and precisely by electron microscopy (Slayter & Lowey, 1967; Elliott & Offer, 1978; Takahashi, 1978). As shown in Fig. 2.3, myosin is a rodlike protein with two globular tips. The total length of the myosin molecule is about 160 nm. Two heads about 10 nm in length and 5 nm in diameter are connected to a tail 140 nm in length and 3 nm in diameter. A bend is often observed around 63–73 nm from the end of the tail.

## Structure of the myosin filament

At low ionic strength myosin forms a filament essentially the same as that of the thick filament in myofibrils (Huxley, 1963). As shown

Fig. 2.3 Electron micrograph of native thick filament from rabbit psoas muscle, freeze dried and shadowed with platinum. (From Trinick & Elliott, 1979.)

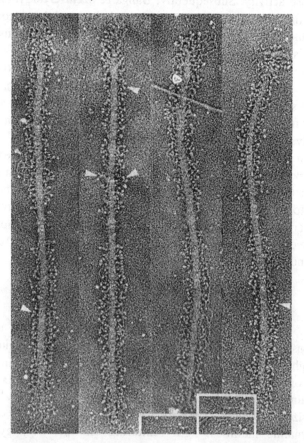

in Fig. 2.4, thick filaments have a length of 1.5 μm and a diameter of 0.1 μm. One filament contains about 300 myosin molecules. Many projections are observed on the surface of the filament except at a bare zone of about 0.15 μm in the center of the filament. The projection is composed of the head part of the myosin molecule, and the backbone of the filament is composed of the tail of the myosin molecule. The projections bind with the thin filament and form crossbridges.

The distribution of projections along the thick filament has been studied using X-ray diffraction and electron microscopy of muscle (Huxley & Brown, 1967; Squire, 1981). Fig. 2.5 shows the distribution of projections on the thick filament. The projections on the thick filament are thought to be arranged on a 6/3 helix. Three bridges project out from the backbone of the filament and the next group of three bridges occur 14.3 nm further along the filament and are rotated relative to the first three bridges by 40°. The structure as a whole is repeated at intervals

Fig. 2.4 The radial projection of thick filament in vertebrate skeletal muscle. At each 14.3 nm along the filament there are three myosin molecules. The helix has a pitch of 42.9 nm.

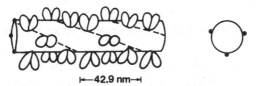

|—42.9 nm—|

Fig. 2.5 A schematic myosin molecule and electron micrographs of negatively stained myosin molecules. Magnification ×200 000. (From Takahashi, 1978).

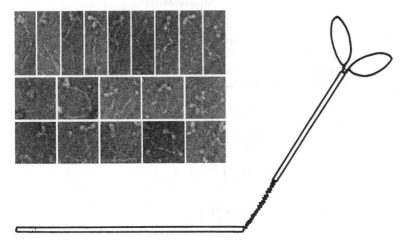

of 3 × 14.3 nm (42.9 nm). Using this model, the number of projections agrees well with the number of myosin molecules.

### Proteolytic subfragments of myosin

To study which part of the myosin molecule is responsible for the functions of myosin, various subfragments were obtained by proteolytic digestion (Lowey, 1971). Fig. 2.6 shows schematically the subfragments of myosin obtained after digestion with trypsin, chymotrypsin, or papain.

Tryptic digestion of myosin yields a heavy meromyosin (HMM) with a molecular weight of 340 000 and a light meromyosin (LMM) with a molecular weight of 120 000 and a length of 70 nm. HMM is soluble even at low ionic strength and retains the ATPase activity and the actin-binding ability. On the other hand, LMM has neither ATPase activity nor actin-binding ability, but assembles into filaments at low ionic strength. When HMM is further digested with trypsin, chymotrypsin, or papain, 1 mol of HMM yields 2 mol of subfragment-1 (S-1) and 1 mol of subfragment-2 (S-2). S-1 has a molecular weight of 120 000 and a length of 10–20 nm, while S-2 has a molecular weight of 60 000 and a length of 50 nm. ATPase activity and actin-binding ability are retained in S-1.

When myosin is digested with chymotrypsin or papain at low ionic strength, a tail devoid of heads (a rod) and S-1 are obtained. The rod

Fig. 2.6 The formation of subfragments by limited digestion of myosin by proteases.

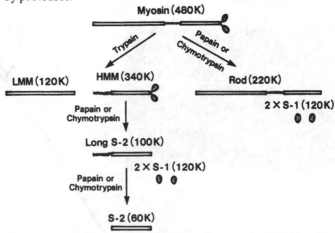

has a molecular weight of 220 000 and a length of 140 nm. It assembles into filaments at low ionic strength. When HMM or the rod is briefly digested with chymotrypsin, long S-2 with a molecular weight of 100 000 and a length of 65 nm is obtained; long S-2 is digested into S-2 with a molecular weight of 60 000. Both long S-2 and S-2 have the same N-terminal amino acid residue. Since the N- and C-terminals of myosin heavy chain exist in S-1 and LMM, respectively, it was concluded that S-2 is produced from long S-2 after digestion of a m.w. 40 000 fragment that corresponds to the S-2–LMM junction.

Hydrodynamic study of S-1 suggested that S-1, which corresponds to the myosin head, is rather spherical. The shape of S-1 was studied more precisely by measuring its rotational correlation time using fluorescence depolarization decay or saturation transfer of the EPR (electron paramagnetic resonance) spectrum. The shape of S-1 was also studied using the low-angle X-ray scattering method and by electron microscopy. These results suggest that the shape of S-1 is slightly spherical with an axial ratio of 2:4. The structure of S-1 was also studied using chemical modification of specific groups as will be discussed later in this chapter.

### Flexibility in the myosin molecule

It was shown that the S-1–S-2 junction is flexible, by measurement of rotational correlation times of S-1 and the S-1 portion of HMM and myosin (Mendelson, Putnam & Morales, 1975; Thomas, 1978). Electron microscopic observations of myosin also support the flexibility of the S-1–S-2 junction.

The flexibility of the S-2–LMM junction was suggested by the comparison of electrobirefringence of the rod, S-2 and LMM. This part of the myosin molecule often shows bending in electron microscopic observations. Thus, as shown in Fig. 2.7, the head part of myosin (S-1), which

Fig. 2.7 Structure of crossbridges based on the structure of the myosin molecule. See text for details.

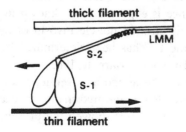

can bind with F-actin, and the tail (LMM), which forms the core of the thick filament, are linked through the S-2 part, which behaves like a flexible joint. By sliding between the thick and thin filaments, the S-2 part plays the role of a crankshaft. Furthermore, a large part (about m.w. 40 000) of the myosin light chain is digested when S-2 is formed. This large part bends even in electron microscopic observation of myosin. Because the digestion of this part by protease depends on the tension development of fibers, this large part is considered to be an elastic component of the crossbridge (Ueno & Harrington, 1981).

## The structure of subunits of myosin

As already described, myosin consists of two heavy chains (HC) of molecular weight 200 000 and four light chains (LC) of molecular weight 15 000 to 25 000. The molecular weight of the heavy chain is constant irrespective of the kind of muscle it comes from, whereas the molecular weights of the light chains differ among various kinds of muscle. Rabbit skeletal muscle myosin contains three kinds of LC with molecular weights of 25 000, 18 000 and 15 000. These LCs are called LC1, LC2 and LC3, respectively (or $g_1$, $g_2$ and $g_3$, respectively) in decreasing order of molecular weight. They are also called alkali-light-chain-1 (A-1), DTNB (5,5'-dithiobis (2-nitrobenzoic acid)) light chain and alkali-light-chain-2 (A-2), respectively, on the basis of the conditions that release each from myosin.

*Heavy chain* Since S-1 free from the LC has an ATPase activity, the ATPase active site is considered to be located on the HC. The specific chemical modification of myosin which affects the ATPase activity also occurs on HC.

Analysis of the amino acid sequence of the HC of skeletal muscle myosin was mainly performed by Elzinga (see Tong & Elzinga, 1983). The sequence of more than half of the whole HC has already been determined; the whole sequence is expected to be known in the near future. In the case of obliquely striated muscle myosin of nematode, the amino acid sequence of the HC has been determined by analysis of the DNA sequence (Karn, Brenner & Barnett, 1983).

Several points are clarified by the amino acid sequence of the HC. The amino acid sequences of two HCs in one myosin molecule are almost the same. However, Starr & Offer (1973) showed the presence of two alternative amino acid residues, Val, and Ile at the eighth position from the N-terminal. Microheterogeneity showing the presence of two alternative amino acids at a certain position has been observed through the entire HC.

In contrast to the head portion, the first-order structure of the tail part of myosin shows that it has $\alpha$-helical structure. There are repeating units of every seven amino acid residues along the entire sequence of the myosin tail, indicating the formation of a super-coil structure by two helices.

*Light chain*   Myosin has two kinds of light chains; one is difficult to remove from myosin, being retained in S-1 after proteolytic digestion. The other is rather easy to remove, can bind $Ca^{2+}$, and is phosphorylated by $Ca^{2+}$-dependent light-chain kinase. In skeletal muscle myosin, $LC_1$ and $LC_3$ belong to the former category and $LC_2$ to the latter. The amount of $LC_2$ is 2 mol/mol of myosin, while the amounts of $LC_1$ and $LC_3$ are 1.3 and 0.7 mol/mol of myosin, respectively, and the sum of both corresponds to 2 mol/mol of myosin.

The first-order structure of $LC_1$ and $LC_3$ was determined by Frank & Weeds (1974), whereas that of $LC_2$ was determined by Collins (1976) and Matsuda *et al.* (1977). Fig. 2.8 shows the first-order structure of these three kinds of LC ($LC_1$, $LC_2$ and $LC_3$) of rabbit skeletal muscle myosin. The first-order structures of $LC_1$ and $LC_3$ are identical in 141 C-terminal residues. $LC_1$ has 49 additional residues in the N-terminal region and $LC_3$ has another 11 residues. Even in these variable regions, a common sequence corresponding to 6 amino acid residues is still observed.

The amino acid sequences of $LC_1$ and $LC_3$ of chicken skeletal muscle myosin are different from those of rabbit myosin; however, the C-terminal residues of $LC_1$ and $LC_3$ are identical even in chicken skeletal muscle myosin (Matsuda, Maita & Umegane, 1981). Therefore, it is thought that C-terminal residues of $LC_1$ and $LC_3$ are coded by one common gene.

The first-order structure of $LC_2$ is rather different from that of $LC_1$ and $LC_3$. One mole of $LC_2$ can bind 1 mol of $Ca^{2+}$ or $Mg^{2+}$. The $Ca^{2+}$-binding site of $LC_2$ is estimated to be in the region of residues 35–49 from the N-terminal by comparison of its chemical structure with that of other $Ca^{2+}$-binding proteins such as parvalbumin and troponin C.

Since $LC_1$ and $LC_3$ are always included in S-1 after digestion of myosin with any proteolytic enzyme, they are considered to be present in the head portion of myosin. On the other hand, proteolytic digestion of the S-1–S-2 junction of myosin by chymotrypsin is suppressed by $Ca^{2+}$ or $Mg^{2+}$ binding to $LC_2$ (Bagshaw, 1977). Therefore, it is thought that $LC_2$ is present in the region of the S-1–S-2 junction. However, the location of $LC_2$ in myosin cannot be prescribed clearly, since it was shown by hydrodynamic studies that $LC_2$ has an elongated shape with a length of about 10 nm.

In the case of skeletal muscle myosin, the presence of LC is not necessary for the ATPase activity. Recently, it was shown that S-1 without LC showed high ATPase activity in the presence of either $EDTA(K^+)$ or $Ca^{2+}$, and its ATPase activity in the presence of $Mg^{2+}$ was activated by F-actin (Wagner & Giniger, 1981; Sivaramakrishnan & Burke, 1982). However, in some muscles, $LC_2$ plays an important role in the $Ca^{2+}$ control of muscle contraction. As will be described in Chapter 3, contraction of adductor muscle is controlled by the binding of $Ca^{2+}$ to myosin; this $Ca^{2+}$ binding requires the existence of LC. Furthermore, contraction

Fig. 2.8 Amino acid sequences of $LC_1$, $LC_2$ and $LC_3$ of rabbit skeletal muscle myosin. $LC_1$ and $LC_3$ have a common sequence following residue 50 to the C-terminal end at residue 190. The top line shows residues 1–49 of the sequence of $LC_1$; the $LC_3$ sequence (in parenthesis) is different from the corresponding positions in the $LC_1$ sequence. ($LC_1$ and $LC_3$, Frank & Weeds, 1974; $LC_2$, Collins, 1976; Matsuda *et al.*, 1977.)

$LC_1$ [$LC_3$]

① ㊿

X-PKKNVKKPAAAAAPAPAPAPAPAPAPAKPKEEKIDLSAIKIGFSKFQQDE

[Ac-SFSADQIA]E

FKGAFLLTDRTGDSKITLSQVGDVLRALGTNPTNAEVKKVLGNPSNEEMN

AKKIEFEQFLPMLQAISNNKDQGTYEDFVEGLRVFDKEGNGTVMGAELRH

VLATLGEKMKEEEVEALMAGQEDSNGCINYEAFVKHIMSI

190

$LC_2$

X-PKKAKRRAAAEGGSSNVFSMFDQTQIQEFLEAFTVIDQNRDGIIDKEDLR

DTFAAMGRLNVKNEELDAMMKEASGPINFTVFLTMFGEKLKGADPEDVIT

GAFLVLDPEGKGTIKKQFLEELLTTQCDRFSQEEIKNMWAAFPPOVGGNV

DYKNICYVITHGDAKDQE

168

of vertebrate smooth muscle is regulated by $Ca^{2+}$-dependent phosphorylation and $Ca^{2+}$-independent dephosphorylation of LC.

As already described above, the chemical structures of $LC_1$ and $LC_3$
are extremely similar. There are 1.3 and 0.7 mol of $LC_1$ and $LC_3$ per
mol of myosin, respectively, and the sum of both is 2 mol/mol of myosin.
Therefore, it is thought that $LC_1$ and $LC_3$ bind to HC randomly, and
that there are three kinds of myosin isozymes, $LC_1$–$LC_1$, $LC_1$–$LC_3$, and
$LC_3$–$LC_3$. Recently, Burke & Sivaramakrishnan (1981) showed that $LC_1$
and $LC_3$ are easily dissociated from the heavy chain of myosin under
physiological conditions (e.g., in the presence of ATP at 37 °C).

## The chemical structure of the ATPase active site

The head part of myosin, S-1, contains both the ATPase active
site and the actin-binding site. Since S-1 has not yet been crystallized,
its three-dimensional structure has mainly been studied by electron microscopic observation. However, much information on the structure of
S-1 has been obtained by the chemical modification of specific amino
acid residues in S-1.

*Chemical modification of myosin*  Chemical modification provides
an effective tool for determining the chemical structure of the active
site of an enzyme. In the case of myosin, chemical modifiers with fluorescent properties or spin-labeled probes are used to analyze the binding
between the myosin head and F-actin, as well as the movement of the
myosin head in the muscle fibers.

The myosin head contains 1 mol each of two highly reactive cysteine
(Cys) residues, $SH_1$ and $SH_2$ (Sekine, Barnett & Kielly, 1963; Sekine
& Yamaguchi, 1963). When myosin is treated with *N*-ethylmaleimide
(NEM), 1 mol of SH group ($SH_1$)/mol of myosin head is rapidly modified. This modification of SH groups leads to a severalfold increase in
the $Ca^{2+}$- and $Mg^{2+}$-ATPase activities and a decrease in the EDTA($K^+$)-
ATPase activity. F-actin-activated ATPase activity is also decreased by
this modification. Because the $SH_1$ thiol residue is extremely reactive
and is modified specifically by various SH modifiers, and because dye
bound to $SH_1$ thiol residue is very rigid, modifications of this site by
spin and fluorescent labeling are often used for the study of structural
changes in the myosin head. The ATPase activity of $SH_1$-blocked myosin
depends largely on the structure of the modifier. If $SH_1$ is modified

by a small group, the change in the ATPase activity is very small. Furthermore, it was found from an energy transfer between dye bound to $SH_1$ and an ATP analog bound to the ATPase active site that the $SH_1$ thiol residue is located about 3 nm from the ATP-binding site (Perkins, Wells & Yount, 1980; Tao & Lamkin, 1981). Therefore, the $SH_1$ thiol residue does not form an ATPase active site, but the modification of this residue alters the conformation of the ATPase active site.

When myosin is further treated with NEM in the presence of ATP or ADP, an additional 1 mol of SH group ($SH_2$) is modified. The $Ca^{2+}$- and $Mg^{2+}$-ATPase activities resulting from $SH_1$ modification decrease markedly after this modification. $SH_2$ is located near the $SH_1$ residue. The amino acid sequences around $SH_1$ and $SH_2$ have been determined; it was found that $SH_2$ is nine residues from $SH_1$ (Elzinga & Collins, 1977). Wells & Yount (1979) have reported that when $SH_1$ and $SH_2$ are S–S bonded or crosslinked with a bifunctional reagent in the presence of nucleotide, 1 mol of ADP/mol of S-1 is trapped at the ATPase active site with concomitant loss of ATPase activity.

When myosin is modified with a 20-fold molar concentration of *p*-chloromercuribenzoate (PCMB) and then treated with excess dithiothreitol (DTT) to remove the bound PCMB, 1 mol of PCMB/mol of myosin remains tightly bound (Shibata-Sekiya & Tonomura, 1975). The residue to which this PCMB binds is called $SH_r$. Modifications of $SH_1$ and $SH_2$ are inhibited by F-actin. Therefore, when myosin is modified with PCMB or NEM in the presence of F-actin, $SH_r$ is rather specifically modified. The myosin having PCMB-modified $SH_r$ (PCMB-myosin) or PCMB-HMM binds tightly with F-actin and becomes difficult to remove, even with the addition of ATP. Therefore, this modification is often used in the study of the reaction mechanism of actomyosin ATPase.

Other SH residues that can be specifically modified are those of light chains. Light chains of skeletal muscle myosin are reversibly removable, as described previously, and they contain few SH residues. Myosin with SH-modified light chains can be obtained by exchanging unmodified myosin light chains with externally added SH-modified ones.

When myosin is modified with 2,4,6-trinitrobenzene sulfonate (TNBS), 2 mol of lysine (Lys) residues/mol of myosin are rapidly modified; at the same time, the EDTA($K^+$)-ATPase activity decreases markedly (Kubo, Tokura & Tonomura, 1960; Miyanishi, Inoue & Tonomura, 1979). The amino acid sequence around this reactive Lys residue (RLR) has been studied and the position of the RLR on the HC has been determined (Miyanishi *et al.*, 1982).

S-1 is rapidly modified with phenylglyoxal, which reacts specifically with Arg at the ratio of 2 mol Arg/mol of S-1. The modification of 1 mol of this Arg is prevented by $Mg^{2+}$-ATP or $Mg^{2+}$-ADP. When these nucleotide-protectable Arg residues are modified, the $Mg^{2+}$-ATPase activity of S-1 disappears and $Mg^{2+}$-ADP becomes unable to bind (Morkin, Flink & Banerjee, 1979). It is highly likely that these Arg residues directly form the ATPase active site.

*Tertiary structure around the ATPase active site*    The positions of various amino acid residues involved in the ATPase activity of myosin have been studied by three methods.

One method determines which of the proteolytic segments of S-1, which is derived from the myosin heavy chain, contains these residues. As shown in Fig. 2.9, when S-1 is subjected to limited tryptic digestion, the m.w. 100 000 HC is digested to 22 000 and 79 000, followed by the breakdown of the heavier fragment to 51 000 and 27 000. Further digestion of the fragments of 22 000 and 27 000 yields fragments of 21 000 and 25 000, respectively (Balint *et al.*, 1975). The m.w. 25 000 segment contains the N-terminus, while the m.w. 21 000 segment is located near the C-terminal of S-1. Since a photoaffinity ATP analog binds with the m.w. 25 000 segment, the ATP-binding site is thought to reside on this segment. The reactive Lys residue (RLR) is also located on this segment. It is thought that F-actin binds to both the m.w. 51 000 and 21 000 segments. The affinity of S-1 for F-actin and the F-actin-activated ATPase activity of S-1 are decreased by limited tryptic digestion of S-1. They are induced by hydrolysis between the m.w. 51 000 and 21 000 segments.

Fig. 2.9 Degradation of heavy chains of S-1 into segments by trypsin.

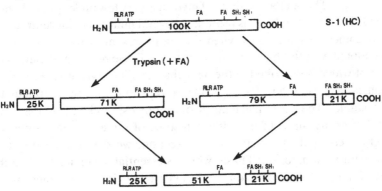

Furthermore, digestion of the linkage between the two segments is inhibited by the binding of F-actin to S-1, and they can be chemically cross-linked with F-actin. Thus, it is suggested that F-actin binds to the m.w. 51 000 and 21 000 segments. This agrees with the finding that modification of the $SH_1$ residue that is located on the 21 000 segment is inhibited by the binding of myosin heads with F-actin.

The second method of study is the measurement of the distance between reactive amino acid residues using fluorescence energy transfer (see review by Morales *et al.*, 1983). It was found that $SH_1$ and $SH_2$, which are located close to each other, are 2.5–3.8 nm from the ATP-binding site. The reactive Lys residue is also 4.3 nm from the ATP-binding site and 2.6 nm from $SH_1$. The distance between reactive amino acid residues was also studied on the acto–S-1 complex. The reactive SH group of actin at position 373 is 5.8 nm from $SH_1$ and 3.0 nm from the nucleotide-binding site of actin.

The third method of study is the direct observation of the location of the reactive amino acid residue by electron microscopy. Sutoh, Yamamoto & Wakabayashi (1984) modified the $SH_1$ residue of myosin with biotin iodoacetamide, and after the addition of avidin, a protein which binds strongly with the biotin group, they observed biotin-labeled myosin by electron microscopy. They found that $SH_1$ is located about 10 nm from the top of the head (almost at the center of the head). Since the ATP-binding site is about 3 nm from the $SH_1$ residue, the ATPase active site may also be located at the center of the head.

## Reaction mechanism of myosin ATPase

The ATPase activity of myosin is very low under physiological conditions in which $Mg^{2+}$ is present at several millimolar concentrations. It becomes very high only when F-actin is present. Muscle contraction is coupled with this high actomyosin-type ATPase reaction, the mechanism of which is described in the next chapter. The myosin ATPase reaction is very slow and its $K_m$ value is very low, so the intermediates are stable. The basic mechanism of actomyosin ATPase is the same as that of myosin ATPase, although some of the elementary steps are highly activated by F-actin. In this section, we describe the reaction mechanism of myosin ATPase, which is essential for understanding the reaction mechanism of actomyosin ATPase. Reviews of the mechanism of the myosin ATPase reaction have been published by Trentham, Eccleston & Bagshaw (1976) and Taylor (1979). See also Tonomura & Inoue (1974, 1975) and Inoue *et al.* (1979).

## *General properties of the myosin ATPase reaction*

When the myosin ATPase reaction is catalyzed in [$^{18}$O] water, the $^{18}$O atom is incorporated into phosphate. Therefore, in the myosin ATPase reaction the bond between oxygen and γ-phosphate is hydrolyzed.

The myosin-ATPase has a low substrate specificity. Myosin catalyzes not only the hydrolysis of various nucleotide triphosphates but the hydrolysis of ribose triphosphate and inorganic triphosphate as well. However, such ATP analogs do not induce muscle contraction, and hydrolysis of these ATP analogs by myosin is not accelerated by F-actin. The minimum requirement for a molecule to act as substrate for hydrolysis by myosin and for muscle contraction has been studied using various synthetic ATP analogs (see Tonomura, 1972). It has been shown that muscle contraction occurs if the substrate molecule (ATP) has a triphosphate chain, a hydrophobic ring (adenine) and a bridge (ribose) between them of adequate length. Modification of the triphosphate chain strongly affects the function of ATP. Furthermore, muscle contraction is not induced when a large group is incorporated into the adenine ring. However, changes in the backbone of the adenine ring do not alter the activity of ATP; ATP can induce tension when a large group is incorporated into ribose or even when ribose is replaced by a hydrocarbon chain. Therefore, the function of ribose must be to maintain the distance between the adenine ring and the triphosphate chain.

Divalent cations exert very strong effects on myosin ATPase. In the presence of millimolar concentrations of $Mg^{2+}$ (i.e., under physiological conditions), the activity of myosin ATPase is low, with its turnover rate being only several per minute, but it increases almost 100-fold when $Ca^{2+}$ replaces $Mg^{2+}$. Moreover, when divalent cations are removed by the addition of the chelating agent EDTA, the ATPase activity in the presence of $K^+$ or $NH_4^+$ increases almost 1000-fold over that in the presence of $Mg^{2+}$. In the presence of $Na^+$, however, the activity remains low even after the removal of divalent cations. Muscle contraction occurs in the presence of $Mg^{2+}$ or $Mn^{2+}$, but does not occur in the presence of $Ca^{2+}$ (in the absence of $Mg^{2+}$) or in the absence of divalent cations.

Ionic strength, temperature and pH also affect the activity of the myosin and actomyosin ATPase reactions. The ATPase activity of skeletal muscle myosin decreases with increasing KCl concentration. Therefore, high KCl concentration is used for the study of the myosin ATPase reaction. Furthermore, at high ionic strength the myosin filament is dissolved into monomers, and activation of myosin ATPase by F-actin does not take place.

## Reaction intermediates of myosin ATPase

In the presence of $Mg^{2+}$, the $V_{max}$ of the myosin ATPase reaction is low, and $K_m$ is lower than 1 μM. These results indicate that intermediates of the ATPase reaction are stable. The reaction intermediates of the $Mg^{2+}$-ATPase of myosin and the order of their formation have been studied; it was found that a complex of myosin with ADP and Pi (myosin–phosphate–ADP complex or E–P–ADP) is formed as a key intermediate in the myosin ATPase reaction. It has not been established whether both heads of myosin form E–P–ADP or not, and we will discuss this problem below. However, it was found that the actomyosin ATPase reaction, which is coupled with muscle contraction, occurs via E–P–ADP. Therefore, in this section we will describe the mechanism of myosin ATPase reaction via E–P–ADP.

*Formation of E–P–ADP* When the time course of the liberation of inorganic phosphate (Pi) in the myosin ATPase reaction is measured after the reaction has been stopped with trichloroacetic acid (TCA), it is found that 1 mol of Pi/mol of myosin is rapidly released during the initial phase of the ATPase reaction, as shown in Fig. 2.10 (Kanazawa & Tonomura, 1965). This initial rapid liberation of Pi is called the 'initial burst'. It occurs by the rapid formation of the complex of myosin with Pi and ADP and its denaturation by TCA. It was later shown from the measurements of the time courses of the release of free ADP and Pi from myosin during the initial phase of the ATPase reaction that, when myosin reacts with ATP, an intermediate in which ADP and phosphate are bound with myosin is rapidly formed, followed by the slow release of ADP and Pi from it. The rate of E–P–ADP formation determined from the time courses of the TCA–Pi burst increases as the concentration of ATP increases, with its $K_m$ ($K_f$) and $V_{max}$ ($V_f$) being about 50 μM and 55/s, respectively (Lymn & Taylor, 1970).

Another effective way to analyze the myosin ATPase reaction is to measure the change in the UV absorption spectrum caused by the reaction of myosin with ATP (Morita, 1967). ATP induces an increase in the UV absorption derived from the tryptophan (Trp) residues. As can be expected from this, ATP also enhances the Trp fluorescence of myosin. It was shown that reaction of myosin with ATP causes three Trp residues to be buried in the hydrophobic region of the molecule, and ADP causes one Trp residue to be buried (Werber, Szent-Györgyi & Fasman, 1972). Furthermore, it was found from the change in the absorbance of pH indicator dye when a stopped-flow apparatus was used, that a rapid liberation of $H^+$ occurs during the initial phase of the myosin ATPase reaction.

In Fig. 2.11 the rate of $H^+$ release and the rate of change in the UV spectrum were compared with the rate of TCA–Pi formation. It was found that the rate of $H^+$ release is equal to the rate of change in the UV absorption spectrum, and both rates are faster than the rate of the initial burst of TCA–Pi. These results suggest that the liberation of $H^+$ and the change in UV spectrum are caused by the formation of a complex of myosin and ATP ($E_b{}^*$ATP) and that this complex is later converted to E–P–ADP by a cleavage reaction. This is also supported by the fact that an unhydrolyzable ATP analog, AMPPNP, can also induce the release of $H^+$ and the increase in UV spectrum.

Fig. 2.12 shows the double reciprocal plot of the rate of change in the UV absorption against ATP concentration. The rate of change in UV absorption was not increased linearly with increase in ATP concentrations, and the $K_m$ and $V_{max}$ of $E_b{}^*$ATP formation are 200 µM and

Fig. 2.10 A schematic representation of the initial phase of the myosin ATPase reaction. An excess of ATP was added to myosin at a high concentration of KCl and several mM $MgCl_2$. After stopping the reaction, the amount of Pi liberated was measured. There is an initial rapid liberation of Pi due to the formation of the myosin–phosphate–ADP complex.

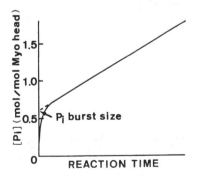

Fig. 2.11 Initial phase of the myosin ATPase reaction. Dashed line, Pi liberated that was measured after stopping the reaction with TCA; dotted line, H⁺ concentration; solid line, change in UV absorption of myosin.

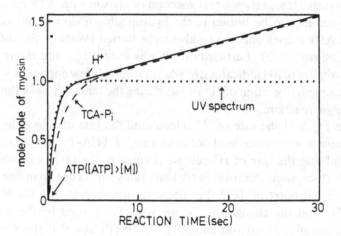

Fig. 2.12 Double reciprocal plot of the initial rate of formation of $E_b^*$ ATP deduced from change in UV absorption against ATP concentration. Conditions: 3 mg/ml HMM, 0.2 M KCl, 50 mM Tris-HCl, pH 7.8, 4 °C. The $MgCl_2$ concentration was 2.5 mM higher than that of ATP (Inoue & Tonomura, 1973).

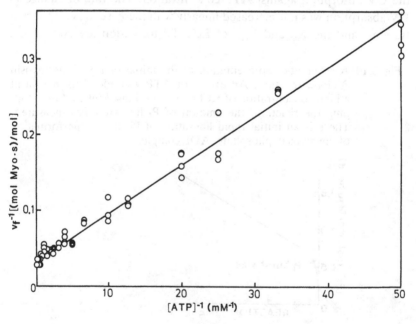

100/s, respectively. Therefore, $E_b$*ATP is formed through more than two steps. The rate of formation of slowly exchangeable bound ATP (tight ATP) was measured by initiating the reaction by adding [$\gamma$-$^{32}$P]ATP to myosin and stopping the reaction by chasing unlabeled ATP (Chock, Chock & Eisenberg, 1979). The rate of tight ATP formation was faster than the rates of $H^+$ release or Trp fluorescence change. Thus, two kinds of tight ATP complexes exist: one that does not cause the change in Trp fluorescence and UV absorption ($E_a$*ATP) and the other that does cause the change in Trp fluorescence and UV absorption ($E_b$*ATP). Since there is no lag phase in the formation of TCA–Pi and no change in the UV spectrum due to the step (E + ATP→ $E_a$*ATP), the tight ATP complex ($E_a$*ATP) is formed via loosely bound ATP (E–ATP) which is in rapid equilibrium with E + ATP. Fig. 2.13 shows the reaction mechanism of the myosin ATPase reaction. The myosin–phosphate–ADP complex (E–P–ADP) is formed via three kinds of myosin–ATP complexes: loose ATP complex (E–ATP), tight ATP complex without spectrum change ($E_a$*ATP), and tight ATP complex with spectrum change ($E_b$*ATP).

*Equilibrium between E*ATP and E–P–ADP*   In the myosin–phosphate–ADP complex, P is not covalently bound to the ATPase active site, since E–P–ADP readily releases Pi when acid is added, and the oxygen atom in water is incorporated during the formation of E–P–ADP but not during its decomposition. Nevertheless, E–P–ADP is in rapid equilibrium with the myosin–ATP complex (E*ATP). Thus, when excess S-1 is added to ATP, a small amount of E*ATP is formed with E–P–ADP especially when the concentration of KCl is low and at low pH (Bagshaw & Trentham, 1973). At 0.1 M KCl and pH 7.8, the ratio (E–P–ADP)/ (E*ATP) at equilibrium is 15–30. The equilibrium shifts strongly to the

Fig. 2.13 Reaction mechanism of myosin $Mg^{2+}$-ATPase via the myosin–phosphate–ADP complex (E–P–ADP). See text for details.

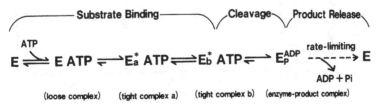

E–P–ADP side at high KCl concentrations. Therefore, when the KCl concentration of the reaction medium is rapidly decreased, the E*ATP complex is formed from E–P–ADP (see Fig. 2.14; Inoue, Arata & Tonomura, 1974).

*Product release from E–P–ADP*   When the $Mg^{2+}$-ATPase reaction is coupled with the pyruvate kinase system (as ADP + phosphoenolpyruvate → ATP + pyruvate) the rate of ADP release can be measured by the rate of release of pyruvate. It was concluded from this measurement that, in qualitative terms, ADP is released slowly from E–P–ADP (Imamura, Tada & Tonomura, 1966). Later, the rates of release of ADP and Pi from E–P–ADP was measured using various methods, such as rapid column chromatography on Sephadex G-25 and flow dialysis methods. It was shown from these measurements that Pi is slowly released from E–P–ADP at a rate almost identical to that of ADP.

Fig. 2.14   KCl-change-induced formation of ATP from E–ADP–P. The reaction was started by adding $3\,\mu M$ $[\gamma\text{-}^{32}P]ATP$ to $3\,mg/ml$ HMM in $1.5\,M$ KCl and $2\,mM\,MgCl_2$ at pH 7.8 and $0\,°C$. After 90 s the reaction mixture was poured into 30 vol of KCl-free solution at $0\,°C$ in the absence (open circles) and presence (filled circles) of $1\,mM$ unlabelled ATP. In the control series the reaction mixture was poured into 30 vol of a solution containing $1.5\,M$ KCl (crosses). (From Inoue, Arata & Tonomura, 1974.)

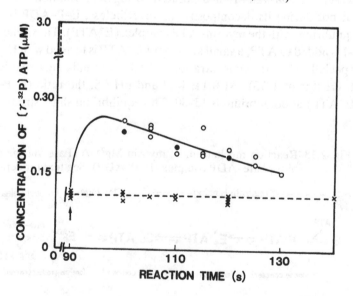

The products ADP and Pi can bind to the ATPase active site with dissociation constants of several μM and about 1 mM, respectively. They inhibit the myosin ATPase reaction. It is thought that E–P–ADP is decomposed via E–ADP and E–Pi complexes. However, the rates of decomposition of E–ADP and E–Pi are much higher than the rate of decomposition of E–P–ADP, and they are formed transiently during the decomposition of E–P–ADP.

Accompanying the decomposition of E–P–ADP, about 0.6 mol of $H^+$ is liberated. Because the rate of $H^+$ liberation is slower than the rate of E–P–ADP decomposition, we (Tonomura & Inoue, 1974) proposed that E–P–ADP is decomposed via E°, and that E° is slowly transformed to the original state, E. This model was supported by the finding that the $V_{max}$ of the steady-state rate of the ATPase reaction via E–P–ADP (with low $K_m$ value) is the same as the rate of $H^+$ liberation from E–P–ADP but lower than the rate of ADP liberation from E–P–ADP. However, later it was found that at high ATP concentrations, E–P–ADP is formed even at the steady-state, and E° exists only at very low concentrations of ATP.

The divalent $Mg^{2+}$ cations that form a complex with ATP bind tightly with myosin when myosin forms E–P–ADP and are liberated as a complex with ADP. Therefore, when EDTA is added to E–P–ADP the high EDTA($K^+$)-ATPase activity is observed after a lag phase due to the slow decomposition of E–P–ADP. When $Mg^{2+}$ is replaced with $Mn^{2+}$ or $Ca^{2+}$, the rate of decomposition of E–P–ADP becomes high, but the rate of Pi liberation from E–P–ADP also becomes high and the intermediate, E–Pi, is not observed. Thus, a conformational change of myosin may occur in the decomposition step of E–P–ADP.

*Reverse reaction of ATPase*  It is assumed that ATP is synthesized from ADP and Pi by the reverse reaction of the myosin ATPase reaction shown in Fig. 2.9. In fact, Mannherz, Schenck & Goody (1974) demonstrated that when ADP and Pi are added to S-1, a tight ATP complex is formed by the reaction $E + ADP + Pi \rightleftharpoons E·ADP·Pi \rightleftharpoons$ E–P–ADP $\rightleftharpoons E^*ATP$.

Furthermore, Arata, Inoue & Tonomura (1975a, b) analyzed the kinetics of release of ATP from $E^*ATP$ and E–P–ADP by allowing myosin to form E–P–ADP with added $[\gamma^{32}P]ATP$ and then adding an excess of unlabeled ATP. They thus showed that ATP is released from E–P–ADP $\rightleftharpoons E^*ATP$ as E–P–ADP $\rightleftharpoons E^*ATP \rightarrow$ E–ATP $\rightarrow$ E + ATP, and that the rate constant for the step E*ATP → E–ATP, which is the rate-limiting step of the reverse reaction, is very slow (0.002/s at 10 °C). It is interesting that the rate of this step is also enhanced by F-actin.

The reversibility of each elementary step of the ATPase reaction can easily be examined by measuring exchange reactions (see Sleep & Smith, 1981). The reaction E*ATP + $H_2O$ ⇌ E–P–ADP is in a rapid equilibrium. Therefore, when the ATPase reaction is carried out in $H_2{}^{18}O$, $^{18}O$ is incorporated into the product Pi at the ratio of 4 mol/mol of Pi (intermediate exchange). On the other hand, when ADP and Pi are added to myosin in the presence of $H_2{}^{18}O$, $^{18}O$ is slowly incorporated into Pi by the reversible reaction E–ADP + Pi + $H_2O$ ⇌ E–P–ADP + $H_2O$ ⇌ E*ATP (medium exchange). The exchange reaction between free Pi and free ATP through the ATPase reaction was also observed.

*Free energy change during the ATPase reaction*  When ATP is hydrolyzed into ADP + Pi, changes in free energy ($G$), enthalpy ($H$) and entropy ($S$) of −7.7 kcal/mol, −4.7 kcal/mol, and +9.9 e.u. occur. Since the rate constants of the forward and backward reactions ($\kappa_j$ and $\kappa_{-j}$, respectively) or the equilibrium constant, $K_{-j}$ ($\kappa_j/\kappa_{-j}$), have been determined for each step of the ATPase reaction, the free energy change $\Delta G_j$ at step $j$ can be calculated. In addition, $H_j$ and $S_j$ can also be obtained from the temperature dependence of $K_j$. Figure 2.15 summarizes the changes in $G$, $H$ and $S$ at each of the reaction steps obtained in this way (Arata, Inoue & Tonomura, 1975b). An important feature here is that $H$ increases by about 13 kcal/mol when E*ATP is formed from E + ATP and decreases by about 35 kcal/mol when E–P–ADP is converted to E–ATP. Contrary to $H$, $S$ increases at the step E + ATP → E*ATP and decreases at the step E–P–ADP → E–ADP. At physiological concentrations of ADP and $P_i$, $G$ decreases markedly at the step E–ATP → $E_6$*ATP, and at E–P–ADP → E + ADP + Pi.

$\Delta H$ has also been determined directly by measuring the heat released at each step of the ATPase reaction. It has been shown that the decomposition of E–P–ADP is accompanied by the production of a large amount of heat (Yamada, Shimizu & Suga, 1981).

## The stereochemistry of myosin ATPase

Studies of the stereochemistry of the hydrolysis of ATP by myosin and of the reaction of stereospecific ATP analogs with myosin have provided valuable information on the properties of the reaction intermediates of ATPase.

It has been shown that when the ATPase reaction was carried out in $H_2{}^{17}O$ using a synthetic stereospecific ATP analog in which $^{16}O$, $^{18}O$ and S are bound with the $\gamma$-P of ATP, and when $H_2O$ directly attacks ATP, the inversion-type phosphate is produced (Knowles, 1980; Cohn, 1982). From an analysis of the thiophosphate obtained by a myosin ATPase reaction, Webb & Trentham (1980) showed that, in the myosin ATPase reaction $H_2O$ directly attacks the $\gamma$-P of ATP. As shown in Chapters 6 and 7, in the $Na^+$, $K^+$-ATPase and $Ca^{2+}$-ATPase reactions, ATP is decomposed via EP in which P is covalently bound to the carboxyl group of the enzyme. The Pi produced by these ATPases is the retention type. However, in the case of the myosin ATPase reaction, there exists

Fig. 2.15 Reaction profile for free energy ($G$), enthalpy ($H$) and entropy ($S$) changes at various points in the myosin ATPase reaction. Conditions for standard state are 1 M reactants and products, 0.5 M KCl, pH 7.8 and 4 °C.

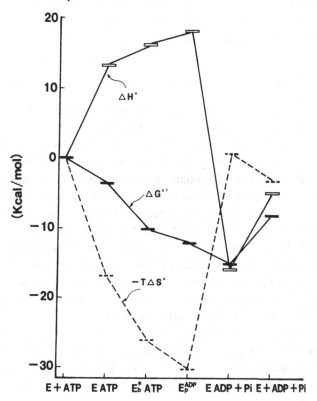

a rapid equilibrium between the enzyme–ATP complex, the enzyme-–P–ADP complex and ATP in which the oxygen atom of the phosphate group is substituted with sulfur. There is then no formation of an E–P–ADP complex and the contraction of muscle cannot be induced. It would be interesting to know what kind of reaction occurs in the reversible reaction of $E^*ATP + H_2O$ and E–P–ADP.

Eckstein & Goody (1976) showed that of the two stereoisomers of ATP$\beta$S (an ATP analog in which sulfur is substituted for oxygen in the $\beta$-phosphate of ATP), S-1 catalyzes the hydrolysis of only one form. Thus, the active site of S-1 has stereospecificity for the $\beta$-phosphate of ATP.

Rotatability of the glycosyl linkage between adenine and ribose rings strongly affects the reaction mechanism of ATPase. Takenaka, Ikehara & Tonomura (1978) showed that ATP analogs in which the linkage between adenine and ribose is fixed in the *cis* or *trans* position were also hydrolyzed by myosin, but that only those ATP analogs in which the linkage is freely rotatable can form E–P–ADP as a stable intermediate and induce the contraction of myofibrils. This result suggests that the position of the base relative to the ribose at the active site changes during the ATPase reaction.

## Nucleotide-induced conformational change in myosin

Conformational change of myosin induced by ATP has been detected by various methods. The reactivity of $SH_2$ is enhanced by nucleotide and the reactivity of 1 mol of reactive Lys residue (RLR) is suppressed by the binding of ADP or (pyrophosphate) PPi to myosin. Additional support for this comes from the observations that the EPR spectrum of the spin label bound with the $SH_1$ thiol residue and the absorption spectrum of the trinitrophenyl (TNP) group bound to RLR are affected by ATP, ADP and Pi. The UV-absorption and fluorescence spectra of the Trp residues of S-1 change during the ATPase action. However, most of these changes involve the local conformation of myosin, and a large change in the structure of myosin has not yet been detected. Furthermore, these changes are not ATP specific, and AMPPNP, ADP and PPi have a similar effect.

From the studies on the UV spectrum of myosin at various temperatures, Morita & Ishigami (1972) showed that the reaction intermediates of ATPase exist in two states and that the UV-spectrum change occurs only in one state. This suggestion was later supported for the S-1–ADP and S-1–AMPPNP complexes by analysis using $^{31}$P NMR (nuclear magnetic resonance) and by kinetic studies of the binding of S-1 with nucleotide. However, the physiological significance of these two states is not clear.

## Functional differences between the two heads of the myosin molecule

Whether the two myosin heads are identical or not is extremely important in considering the molecular mechanism of muscle contraction. We have so far proceeded with our discussion by tacitly assuming that the two heads of myosin are identical. However, workers in this field have not agreed on whether the two heads of myosin are identical, with both heads able to form E–P–ADP as a stable intermediate, or whether they are non-identical, with only one head forming E–P–ADP (see Taylor, 1977; Tonomura & Inoue, 1977). Our results suggest that myosin has two non-identical heads.

### Amount of reaction intermediate in the myosin ATPase reaction

The myosin–phosphate–ADP complex, E–P–ADP, is formed rapidly and decomposed slowly in the presence of $Mg^{2+}$. Therefore, the easiest way to determine the amount of E–P–ADP is to measure the size of the initial burst of phosphate (TCA–Pi) liberation after stopping the reaction with TCA. In skeletal muscle myosin, the burst size of TCA–Pi is slightly larger than 0.5 mol/mol myosin head. However, the myosin ATPase reaction has a pathway for the direct decomposition of E–ATP into E + ADP + Pi. Accordingly, when the concentration of $Mg^{2+}$ is lowered, the apparent size of the TCA–Pi burst increases. Fig. 2.16 shows the time course of Pi (TCA–Pi) release measured by stopping the reaction with TCA and that of free-Pi release measured using the double-membrane filtration method. The amount of Pi bound with myosin during the ATPase reaction, which was given as the difference between the amount of TCA–Pi and that of free Pi, gave a constant value of 0.5 mol/mol myosin head throughout the measurement (Furukawa, Inoue & Tonomura, 1981).

This small TCA–Pi burst is not due to the denaturation of myosin, since fresh preparations of myosin and myofibrils also show the same

amount of TCA–Pi burst. As described in the previous section,
E–P–ADP is in rapid equilibrium with E*ATP, but the equilibrium shifts
greatly to the E–P–ADP side under physiological conditions. Therefore,
the result described above shows that E–P–ADP is formed only in half
of the myosin head.

## Two-route mechanism of myosin ATPase reaction

In the simplest mechanism of myosin ATPase reaction

$$E + ATP \xrightarrow{k_f} E\text{-}P\text{-}ADP \xrightarrow{k_d} E + ADP + Pi$$

Fig. 2.16 Time courses of free-Pi and TCA–Pi liberation in the presence
of 0.1 mM free $Mg^{2+}$, 2.4 mg/ml myosin, 109 μM [γ-$^{32}$P]ATP,
0.5 M KCl, 0.1 mM free $Mg^{2+}$, 50 mM Tris HCl, pH 7.8, 20 °C.
TCA-Pi (open circles) was measured after stopping the reaction
with 5% TCA. Free-Pi liberation (filled circles) was measured
by the double-membrane filtration method.

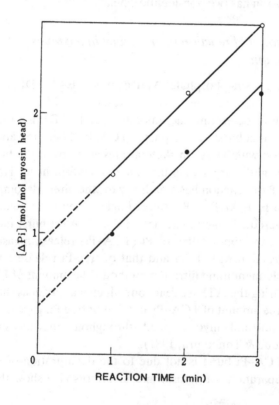

$K_m$ is given as the ATP concentration at $k_f = k_d$, and $V_{max} = k_d$. Since the rate of E–P–ADP formation, $k_f$, is very high, while $k_d$ is very small, $K_m$ is very low (less than 0.04 μM). However, the $K_m$ value of the ATPase reaction at the steady state was reported to be 0.5–2 μM, which is much higher than the calculated value. It was later found that the rates of ATPase reactions over a wide range of ATP concentrations were given as the sums of two ATPase reactions with different $K_m$ and $V_{max}$ values (Inoue, Shibata-Sekiya & Tonomura, 1972). These different values are attributed to the ATPase reaction taking place via E–P–ADP and also via a simple myosin–ATP complex (E–ATP).

The myosin molecule has two heads. Each head (S-1) can bind with F-actin and is dissociated from F-actin by 1 mol of ATP/mol of S-1. This result indicates that each head contains one high-affinity ATP-binding site. However, the amount of E–P–ADP formed by S-1 is 0.5 mol/mol of head. Therefore, the remaining 0.5 mol of head may dissociate from F-actin by formation of E–ATP. We denote the head that forms E–P–ADP as the burst head or head B and the other head, which forms E–ATP as a stable intermediate, as the non-burst head or head A. We proposed that the non-burst head may catalyze the ATPase reaction with high $K_m$ and $V_{max}$ values.

The existence of E–P–ADP and E–ATP in equimolar amounts has been confirmed by the measurement of the amount of ADP and ATP bound to myosin during the ATPase reaction (Inoue & Tonomura, 1973). Fig. 2.17 shows the amounts of ADP and ATP bound to myosin during

Fig. 2.17 Amount of nucleotide bound to HMM (heavy meromysin) during the $Mg^{2+}$-ATPase reaction as a function of ATP concentration. Conditions: 1.7 mg/ml (5 μM) HMM, 4 mg/ml PK, 2 mM PEP, 30 mM KCl, 10 mM $MgCl_2$, 20 mM Tris-HCl, pH 7.5, 20 °C.

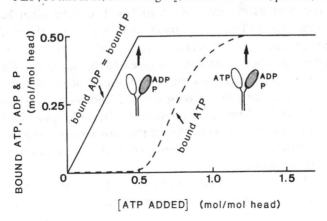

the myosin ATPase reaction. The amounts were measured as the functions of the amount of ATP added. The measurement was carried out by coupling ATPase with excess pyruvate kinase so that free ADP is immediately converted to ATP. When the amount of ATP added was increased, 0.5 mol of bound ADP/mol of myosin head was first observed. This ADP was attributed to the formation of E–P–ADP, since the amount of bound ADP equalled the amount of bound P. Bound ATP was observed when the amount of ATP added exceeded the saturated amount of bound ADP (0.5 mol/mol head) and the amount of bound ATP reached the maximum value of 0.5 mol/mol head.

## Separation of burst head (S-1B) and non-burst head (S-1A)

It was shown that transition between S-1B and S-1A does not take place as they have different chemical structures and can be separated by their different functions.

The S-1 preparation is an equimolar mixture of S-1B and S-1A, and both types of S-1 bind tightly with F-actin. Complete dissociation of S-1 from F-actin is induced by adding 1 mol of ATP/mol of myosin: one fraction (S-1B) forms E–P–ADP and the other (S-1A) forms E–ATP. However, the observed dissociation constant for formation of E–P–ADP is much lower than that for E–ATP. Therefore, when 0.5 mol of ATP is added to acto–S-1 in the presence of an ATP-regenerating system, S-1 can be separated by ultracentrifugation into two equimolar fractions: S-1 that is dissociated from F-actin (supernatant) and S-1 that is bound to F-actin (precipitate). The S-1 fraction that dissociates from F-actin consists mainly of S-1B (0.7 mol), whereas the S-1 fraction still bound to F-actin consists mainly of S-1A (0.7 mol). On the other hand, the rates of the $Mg^{2+}$-ATPase reaction in the steady state of the supernatant and the precipitate were almost the same as that of S-1 before separation. When the separation procedure was repeated the size of the initial burst of S-1 in the supernatant increased to 0.8 mol/mol of S-1, while that of S-1 in the precipitate decreased to 0.25 mol/mol of S-1 (Inoue & Tonomura, 1975). When S-1 was separated after adding various amounts of ATP to acto–S-1, the burst size of the supernatant was almost equal to that estimated from the dependence of the TCA–Pi burst size on the amount of ATP added.

The S-1 preparation was also separated into S-1B and S-1A by the use of column chromatography on immobilized F-actin. The S-1 preparation was applied to a column of a mixture of immobilized F-actin and immobilized pyruvate kinase and eluted with a low concentration of ATP in the presence of PEP. S-1B with formation of E–P–ADP was eluted ahead of S-1A with formation of M–ATP (Inoue & Tonomura, 1976).

S-1 can be separated into equimolar fractions of S-1B and S-1A by making use of the differences in their chemical structures. As described earlier in this chapter, when myosin is treated with PCMB and then with DTT, 1 mol of PCMB/mol of myosin binds tightly on head A, and the PCMB–S-1A thus formed binds strongly with F-actin and does not dissociate from F-actin even in the presence of ATP. Therefore, when PCMB–S-1 was mixed with F-actin and then centrifuged in the presence of ATP, only S-1B is dissociated from F-actin (Shibata-Sekiya & Tonomura, 1976). Another method is the modification of reactive Lys residues (RLR) with trinitrobenzene sulfonate (TNBS). In the presence of ADP, TNBS reacts selectively with a RLR on S-1B and forms TNP–S-1B. TNP–S-1B and S-1A can be separated by DEAE cellulose column chromatography.

## Differences in the chemical structures of burst and non-burst heads

The two heavy chains of myosin have very similar structures, but there is considerable microheterogeneity in their amino-acid sequences, though it is not clear whether this microheterogeneity is related to functional differences between the burst and non-burst heads of myosin. As described earlier in this chapter, only the RLR in head A is modified with TNBS in the presence of ADP or PPi. As shown in Fig. 2.18, it was found that the chemical structures around the modified RLR differ in the two heads. The fifth residue from TNP-Lys toward

Fig. 2.18 The difference in the chemical structures around the reactive Lys residue of the burst and non-burst heads.

**Burst Head**

TNP
Glu-Asp-Gln-Val-Phe-(Pro)-Met-Asn-Pro-Pro-Lys-Tyr-Asp-Lys-Ile-Glu-Asp-Met-Ala-Met-Met

TNP
(Ser)-Met-Asn-Pro-Pro-Lys-Tyr-Asp-Lys-Ile-Glu-Asp-Met-Ala-Met

**Nonburst Head**

the N-terminus is proline (Pro) in the burst head, while it is serine (Ser) in the non-burst head (Miyanishi *et al.*, 1980). Furthermore, it was shown that the SH$_r$ to which PCMB binds strongly and does not dissociate from even in the presence of DTT, is located only on S-1A. These results show clearly that the burst and non-burst heads have different chemical structures. Contrary to this, the difference in the contents of light chains LC$_1$ and LC$_3$ bears no relation to the burst and non-burst heads.

## Non-identical two-headed structure of myosin

Since the S-1 preparation was shown to be an equimolar mixture of the burst and non-burst heads, the next important question is whether the burst head (S-1B) and the non-burst head (S-1A) correspond to the two heads of one myosin molecule. It was strongly suggested from the study of ATP-induced dissociation of actomyosin that myosin has a heterodimer structure, A–B, with respect to the heads, not a homodimer structure, A–A and B–B. The evidence may be summarized (see Fig. 2.19): (1) Actomyosin dissociates completely into F-actin and

Fig. 2.19 Evidence for the heterodimer structure of skeletal muscle myosin, obtained by using myosin in which the non-burst head was modified with PCMB at a specific thiol group (SH$_r$). See text for details. (1) Control. (2) Myosin modified with PCMB is digested to S-1. Only PCMB-bound S-1A bound tightly to F-actin in the presence of ATP. (3) Myosin modified with PCMB.

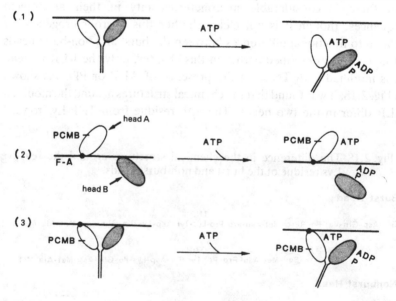

Table 2.1 *ATPase activities of the burst and the non-burst heads of myosin*

| ATPase | Burst head (B) | Non-burst head (A) |
|---|---|---|
| EDTA(K$^+$)-ATPase | + | − |
| Ca$^{2+}$-ATPase | + | − |
| Mg$^{2+}$-ATPase | + | + |
| AM-ATPase | + | − |

myosin on adding 1 mol of ATP/mol of myosin (0.5 mol/mol of myosin head), and 1 mol of E–P–ADP/mol of myosin is produced. The S-1 preparation can be separated into S-1A-rich and S-1B-rich fractions by ultracentrifugation in the presence of various amounts of ATP, whereas actomyosin can not be separated by this method. (2) The conclusive evidence is obtained from the chemical modification of myosin and heavy meromyosin (HMM) with PCMB (see Fig. 2.19). When myosin was modified with PCMB and then treated with DTT, 0.5 mol of PCMB bound tightly to 1 mol of myosin. This PCMB bound to the non-burst head (S-1A). Therefore, when PCMB-myosin was digested into S-1 only PCMB-bound S-1A bound tightly to F-actin even in the presence of ATP. (3) On the other hand, in the case of HMM or myosin, HMM with 0.5 mol of bound PCMB did not dissociate from F-actin even in the presence of ATP (Shibata-Sekiya & Tonomura, 1976). This result can be explained if PCMB-HMM has a heterodimer structure of S-1A–PCMB and S-1B.

## Functions of the burst and non-burst heads

The functions of the burst and non-burst heads were studied with separated heads; they are summarized in Table 2.1 (Inoue *et al.*, 1979). Both the burst and non-burst heads bind with F-actin and dissociate from it when ATP is added. In addition, both heads possess Mg$^{2+}$-ATPase activity. However, there are various functional differences between the two heads. Only head B forms E–P–ADP as a ternary complex of ATPase reaction. The Ca$^{2+}$- and EDTA(K$^+$)-ATPase activities are seen only on head B. On the other hand, change in reactivity of the RLR by ATP is observed only on the non-burst head (head A). Furthermore, only head A has a specific SH residue (SHa). In particular, the actomyosin-type ATPase reaction is observed only for the burst head.

This result indicates that the burst head functions in energy transduction in muscle. On the other hand, it was suggested that the non-burst head plays an important role in the regulation of muscle contraction by ATP and $Ca^{2+}$, as will be discussed in the next chapter.

In connection with this, it has been found that all of the energy-transducing ATPases discussed in this volume possess two or more subunits. Particularly, dynein ATPase (Chapter 5) has two chemically different heavy chains, both of which have ATPase activity. It is interesting that the energy-transducing ATPases have two or more subunits for their function, though the reason has yet to be understood.

# References

Arata, T., Inoue, A. & Tonomura, Y. (1975*a*). Standard free energy changes for formation of various intermediates in the reaction of H-mero-myosin ATPase. *J. Biochem.*, **77**, 895–900.

Arata, T., Inoue, A. & Tonomura, Y. (1975*b*). Thermodynamic and kinetic parameters of elementary steps in the reaction of H-meromyosin adenosine triphosphatase: remarkably large increase in the standard entropy for formation of the reactive H-meromyosin-phosphate-adenosine diphosphate complex. *J. Biochem.*, **78**, 277–86.

Bagshaw, C. R. (1977). On the location of the divalent metal binding sites and the light chain subunits of vertebrate myosin. *Biochemistry*, **16**, 59–67.

Bagshaw, C. R. & Trentham, D. R. (1973). The reversibility of adenosine triphosphate cleavage by myosin. *Biochem. J.*, **133**, 323–8.

Balint, M., Sreter, F. A., Wolf, I., Nagy, B. & Gergely, I. (1975). The structure of heavy meromyosin: the effect of $Ca^{2+}$ and $Mg^{2+}$ on the tryptic fragmentation of heavy meromyosin. *J. Biol. Chem.*, **250**, 6168–77.

Banga, I. & Szent-Györgyi, A. (1941). Preparation and properties of myosin A and B. *Stud. Inst. Med. Chem. Univ. Szeged.*, **1**, 5–15.

Bourne, G. H. (1973). *The Structure and Function of Muscle*, 2nd edn, vol. 3. New York & London: Academic Press.

Burke, M. & Sivaramakrishnan, M. (1981). Subunit interactions of skeletal muscle myosin and myosin subfragment. I. Formation and properties of thermal hybrids. *Biochemistry*, **20**, 5908–13.

Chock, S. P., Chock, P. B. & Eisenberg, E. (1979). The mechanism of the skeletal muscle myosin ATPase. II. Relationship between the fluorescence enhancement induced by ATP and the initial Pi burst. *J. Biol. Chem.*, **254**, 3236–43.

Cohn, M. (1982). $^{18}$O and $^{17}$O effects on $^{31}$PNMR as probes of enzymatic reactions of phosphate compounds. *Ann. Rev. Biophys. Bioenerg.*, **11**, 23–42.

*Cold Spring Harbor Symposium on Quantitative Biology.* (1973). Vol. 37. *The Mechanism of Muscle Contraction.* Cold Spring Harbor Laboratory, Cold Spring Harbor, N.Y.

Collins, J. H. (1976). Homology of myosin DTNB light chain with alkali light chains, troponin C and parvalbumin. *Nature.* **259,** 699–700.

Eckstein, F. & Goody, R. S. (1976). Synthesis and properties of diastereoisomers of adenosine 5'-(O-1-thiotriphosphate) and adenosine 5'-(O-2-thiotriphosphate. *Biochemistry*, **15,** 1685–91.

Elliott, A. & Offer, G. (1978). Shape and flexibility of the myosin molecule. *J. Mol. Biol.,* **123,** 505–19.

Elzinga, M. & Collins, J. H. (1977). Amino acid sequence of a myosin fragment that contains SH-1, SH-2, and $N_t$-methylhistidine. *Proc. Natl. Acad. Sci. (USA),* **74,** 4281–4.

Engelhardt, W. A. & Ljubimova, M. N. (1939). Myosin and adenosine triphosphatase. *Nature,* **144,** 668–9.

Frank, G. & Weeds, A. G. (1974). Amino acid of the alkali light chains of rabbit skeletal muscle myosin. *Eur. J. Biochem.,* **44,** 317–34.

Furukawa, K.-I. Inoue, A. & Tonomura, Y. (1981). Extra burst of Pi liberation and formation of the myosin-phosphate-ADP complex at various concentrations of $Mg^{2+}$ ions. *J. Biochem.,* **89,** 1283–92.

Gordon, A. M., Huxley, A. F. & Julian, F. S. (1966). The variation in isometric tension with sarcomere length in vertebrate muscle fibre. *J. Physiol.,* **184,** 170–92.

Huxley, A. F. (1957). Muscle structure and theories of contraction. *Prog. Biophys. Mol. Biol.,* 7, 255–313.

Huxley, A. F. & Niedergerke, R. (1954). Structural changes in muscle during contraction. *Nature,* **173,** 971–3.

Huxley, H. E. (1957). The double array of filaments in cross-striated muscle. *J. Biophys. Biochem. Cytol.,* 3, 631–48.

Huxley, H. E. (1960). Muscle cells. In *The Cell,* vol. 4, ed. J. Branchet & E. A. Mirsky, pp. 365–481. New York: Academic Press.

Huxley, H. E. (1963). Electron microscope studies on the structure of natural and synthetic protein filaments from striated muscle. *J. Mol. Biol.,* 7, 281–308.

Huxley, H. E. (1969). The mechanism of muscular contraction. *Science,* **164,** 1556–66.

Huxley, H. E. & Brown, W. (1967). The low-angle X-ray diagram of vertebrate striated muscle and its behaviour during contraction and rigor. *J. Mol. Biol.,* **30,** 383–434.

Huxley, H. E. & Hanson, J. (1954). Changes in the cross-sections of muscle during contraction and stretch and their structural interpretation. *Nature,* **173,** 973–6.

Imamura, K., Tada, M. & Tonomura, Y. (1966). The pre-steady state of the myosin-adenosine triphosphate system. IV. Liberation of ADP from the myosin-ATP system and effects of modifiers on the phosphorylation of myosin. *J. Biochem.,* **59,** 280–9.

Inoue, A., Arata, T. & Tonomura, Y. (1974). KCl jump induced formation of adenosine triphosphate from the reactive myosin-phosphate-ADP complex. *J. Biochem.*, **76**, 661–6.

Inoue, A., Shibata-Sekiya, K. & Tonomura, Y. (1972). The pre-steady state of the myosin-adenosine triphosphate system. XI. Formation and decomposition of the reactive myosin-phophate-ADP complex. *J. Biochem.*, **71**, 115–24.

Inoue, A., Takenaka, H., Arata, T. & Tonomura, Y. (1979). Functional implications of the two-headed structure of myosin. *Adv. Biophys.*, **13**, 1–194.

Inoue, A. & Tonomura, Y. (1973). Kinetic properties of the myosin-phosphate-ADP complex. *J. Biochem.*, **73**, 555–66.

Inoue, A. & Tonomura, Y. (1975). The amounts of adenosine di- and triphosphates bound to H-meromyosin and the adenosinetriphosphatase activity of the H-meromyosin-F-actin-relaxing protein system in the presence and absence of calcium ions. Physiological functions of the two routes of myosin adenosine triphosphatase in muscle contraction. *J. Biochem.*, **78**, 83–92.

Inoue, A. & Tonomura, Y. (1976). Separation of subfragment-1 of H-meromyosin into two equimolar fractions with and without formation of the reactive enzyme-phosphate-ADP complex, *J. Biochem.*, **79**, 419–34.

Kanazawa, T. & Tonomura, Y. (1965). The pre-steady state of the myosin-adenosine triphosphate system. I. Initial rapid liberation of inorganic phosphate. *J. Biochem.*, **57**, 604–15.

Karn, J., Brenner, S. & Barnett, L. (1983). Protein structural domains in the *Caenorhabditis elegans unc-54* myosin heavy chain gene are not separated by introns. *Proc. Natl. Acad. Sci. (USA)*, **80**, 4253–7.

Knowles, J. R. (1980). Enzyme-catalyzed phosphoryl transfer reactions. *Ann. Rev., Biochem.*, **49**, 877–919.

Kubo, S., Tokura, S. & Tonomura, Y. (1960). On the active site of myosin A-adenosine triphosphatase. I. Reaction of the enzyme with trinitro-benzenesulfonate. *J. Biol. Chem.*, **235**, 2835–9.

Kuhne, W. (1959). Untersuchungen uber Bewegungen und Veraderungen der contractilen Substanzen. *Arch. Anat. U. Physiol. Med.* **748**.

Lowey, S. (1971). Myosin: molecule and filament. In *Subunits in Biological Systems*, part A, ed. S. N. Timasheff & G. C. Fasman, pp. 201-59. New York: Academic Press.

Lymn, R. W. & Taylor, E. W. (1970). Transient state phosphate production in the hydrolysis of nucleotide triphosphate by myosin. *Biochemistry* **9**, 2975–83.

Mannherz, H. G., Schenck, H. & Goody, R. S. (1974). Synthesis of ATP from ADP and inorganic phosphate at the myosin-subfragment 1 active site. *Eur. J. Biochem.*, **48**, 287–95.

Margossian, S. S. & Lowey, S. (1982). Preparation of myosin and its subfragments from rabbit skeletal muscle. In *Methods in Enzymology*, vol. 85, part B, ed. D. W. Frederiksen & L. W. Cunningham, pp. 55–71. New York & London: Academic Press.

Matsuda, G., Maita, Y., Suzuyama, M., Setoguchi, M. & Umegane, T. (1977).

Amino acid sequence of the L-2 light chain of rabbit skeletal muscle myosin. *J. Biochem.*, **81**, 809–11.

Matsuda, G., Maita, Y. & Umegane, M. (1981). The primary structure of L-1 light chain of chicken fast skeletal muscle myosin and its genetic implication. *FEBS Lett.*, **126**, 111–13.

Mendelson, R. A., Putnam, S. & Morales, M. F. (1975). Time-dependent fluorescence depolarization and lifetime studies of myosin subfragment-one in the presence of nucleotide and actin. *J. Supramol. Struct.*, **3**, 162–8.

Miyanishi, T., Inoue, A. & Tonomura, Y. (1979). Differential modification of specific lysine residues in the two kinds of subfragment-1 of myosin with 2,4,6-trinitrobenzene sulfonate. *J. Biochem.*, **85**, 747–53.

Miyanishi, T., Maita, T., Matsuda, G. & Tonomura, Y. (1980). Differences in chemical structure around the reactive lysine residues in the burst and the nonburst heads of skeletal muscle myosin. *J. Biochem.*, **9**, 1845–53.

Morales, M. F., Borejdo, J., Botts, J., Cooke, R., Mendelson, R. A. & Takashi, R. (1983). Some physical studies of the contractile mechanism in muscle. *Ann. Rev. Phys. Chem.*, **33**, 319–51.

Morita, F. (1967). Interaction of heavy meromyosin with substrate. I. Difference in ultraviolet absorption spectrum between heavy meromyosin and its Michaelis–Menten complex. *J. Biol. Chem.*, **242**, 4501–6.

Morita, F. & Ishigami, F. (1972). Temperature dependence of the decay of the UV absorption difference spectrum of heavy meromyosin induced by adenosine triphosphate and inosine triphosphate. *J. Biochem.*, **81**, 305–12.

Morkin, E., Flink, I. L. & Banerjee, S. K. (1979). Phenylglyoxal modification of cardiac myosin S-1: evidence for essential arginine residues at the active site. *J. Biol. Chem.*, **254**, 12647–52.

Natori, R. (1954). The property and contraction process of isolated myofibrils. *Jikeikai Med. J.*, **1**, 119–26.

Needham, D. M. (1971). *Machina Carnis. The Biochemistry of Muscular Contraction in its Historical Development*. Cambridge University Press.

Perkins, J., Wells, J. A. & Yount, R. G. (1980). Fluorescence characterization of 1,$N^6$-etheno-ADP bound and trapped at the active site of myosin subfragment-1. *Fed. Proc.*, **39**, 1935.

Perry, S. V. (1967). The structure and interactions of myosin. *Progr. Biophys.*, **17**, 325–81.

Porter, K. R. & Palade, G. E. (1957). Studies on the endoplasmic reticulum. III. Its form and distribution in striated muscle cells. *J. Biophys. Biochem. Cytol.*, **3**, 269–300.

Sekine, T., Barnett, L. M. & Kielly, W. W. (1963). The active site of myosin adenosine triphosphatase. *J. Biol. Chem.*, **237**, 2769–72.

Sekine, T. & Yamaguchi, M. (1963). Effect of ATP on the binding of N-ethylmaleimide to SH groups in the active site of myosin ATPase. *J. Biochem.*, **54**, 196–8.

Shibata-Sekiya, K. & Tonomura, Y. (1975). Desensitization of substrate inhibition of acto-H-meromyosin ATPase by treatment of H-meromyosin with *p*-chloromercuribenzoate. Relation between the extent of desensitization and the amount of bound *p*-chloromercuribenzoate. *J. Biochem.*, **77**, 543–57.

Shibata-Sekiya, K. & Tonomura, Y. (1976). Structure and function of two heads of the myosin molecule, shown by the effect of modification of head A with *p*-chloromercuribenzoate in the interaction of head B with F-actin. *J. Biochem.*, **80**, 1371–80.

Sivaramakrishnan, M. & Burke, M. (1982). The free heavy chain of vertebrate skeletal myosin subfragment 1 shows full enzymatic activity. *J. Biol. Chem.*, **257**, 1102–5.

Slayter, H. & Lowey, S. (1967). Substructure of the myosin molecule as visualized by electron microscopy. *Proc. Natl. Acad. Sci. (USA)*, **58**, 1611–18.

Sleep, J. S. & Smith, S. J. (1981). Actomyosin ATPase and muscle contraction. In *Current Topics in Bioenergetics*, vol. 11, ed. D. R. Sanadi & L. P. Vernon, pp. 239–86. New York & London: Academic Press.

Squire, J. (1981). *The Structural Basis of Muscular Contraction*. New York: Plenum Press.

Starr, R. & Offer, G. (1973). Polarity of the myosin molecule. *J. Mol. Biol.*, **81**, 17–31.

Sutoh, K., Yamamoto, K. & Wakabayashi, T. (1984). Electron microscopic visualization of the $SH_1$ thiol of myosin by the use of an avidin–biotin system. *J. Mol. Biol.*, **178**, 323–39.

Szent-Györgyi, A. (1949). *Chemistry of Muscle Contraction*. New York: Academic Press.

Takahashi, K. (1978). Topography of the myosin molecule as visualized by an improved negative staining method. *J. Biochem.*, **83**, 905–8.

Takenaka, H., Ikehara, M. & Tonomura, Y. (1978). Interaction between actomyosin and 8-substituted ATP analogs. *Proc. Natl. Acad. Sci. (USA)*, **75**, 4229–33.

Tao, T. & Lamkin, M. (1981). Excitation energy transfer studies on the proximity between $SH_1$ and the adenosine triphosphatase site in myosin subfragment 1. *Biochemistry*, **20**, 5051–5.

Taylor, E. W. (1977). Myosin ATPase action. *Trends in Biochem. Sci.*, **2**, N32–5.

Taylor, E. W. (1979). Mechanism of actomyosin ATPase and the problem of muscle contraction. *CRC Crit. Rev. Biochem.*, **6**, 103–64.

Thomas, D. D. (1978). Large-scale rotational motions of proteins detected by electron paramagnetic resonance and fluorescence. *Biophys. J.*, **24**, 439–62.

Tong, S. W. & Elzinga, M. (1983). The sequence of the $NH_2$-terminal 204-residue fragment of the heavy chain of rabbit muscle myosin. *J. Biol. Chem.*, **258**, 13100–10.

Tonomura, Y. (1972). *Muscle Proteins, Muscle Contraction and Cation Transport*. Tokyo & Baltimore: Japan Scientific Society Press & University Park Press.

Tonomura, Y. & Inoue, A. (1974). The substructure of myosin and the reaction mechanism of its adenosine triphosphatase. *Mol. Cell. Biochem.*, **5**, 127–43.

Tonomura, Y. & Inoue, A. (1975). Energy transducing mechanisms in muscle. In *MTP International Review of Science: Biochemistry Series One, vol. 3: Energy Transducing Mechanisms*, ed. E. Racker, pp. 121–61. London: Butterworths.

Tonomura, Y. & Inoue, A. (1977). Myosin ATPase action. *Trends in Biochem. Sci.*, **2**, N32–5.

Trentham, D. R., Eccleston, J. F. & Bagshaw, C. R. (1976). Kinetic analysis of ATPase mechanism. *Q. Rev. Biophys.*, **9**, 217–81.

Trinick, J. & Elliott, A. (1979). Electron microscope studies of thick filaments from vertebrate skeletal muscle. *J. Mol. Biol.*, **131**, 133–6.

Ueno, H. & Harrington, W. F. (1981). Conformational transition in the myosin hinge upon activation of muscle. *Proc. Natl. Acad. Sci. (USA)*, **78**, 6101–5.

Wagner, P. D. & Giniger, E. (1981). Hydrolysis of ATP and reversible binding of F-actin by myosin heavy chains free of all light chains. *Nature*, **292**, 560–2.

Webb, M. R. & Trentham, D. R. (1980). The stereochemical course of phosphoric residue transfer during the myosin ATPase reaction. *J. Biol. Chem.*, **255**, 8627–32.

Wells, J. A. & Yount, R. G. (1979). Active site trapping of nucleotide by cross-linking two sulfhydryls in myosin subfragment 1. *Proc. Natl. Acad. Sci. (USA)*, **76**, 4966–70.

Werber, M. M., Szent-Györgyi, A. G. & Fasman, G. (1972). Fluorescence studies on heavy meromyosin: substrate interaction. *Biochemistry*, **11**, 2872–83.

Yagi, K. (1975). Myosin-subfragment-1. *Adv. Biophys.*, **8**, 1–34.

Yamada, T., Shimizu, H. & Suga, H. (1981). Calorimetic studies of adenosine 5′-triphosphate hydrolysis by heavy meromyosin. *Biochemistry*, **20**, 4484–8.

# 3   Actomyosin ATPase reaction and the mechanism of muscle contraction

The very low activity of myosin ATPase is dramatically accelerated by F-actin, and the contraction of myofibrils occurs only under conditions in which a high activity of this actomyosin-type ATPase is expressed. The study of the reaction mechanism of actomyosin ATPase has been based on the mechanism of myosin ATPase which was described in Chapter 2. In particular, the steps of the myosin ATPase reaction which are accelerated by F-actin and those of the ATPase reaction in which binding and dissociation occur between F-actin and the myosin head have been studied. Based on these studies, the coupling mechanism of ATPase with contraction has been considered. In this chapter, we will first discuss the structures of F-actin and the complex of F-actin with the myosin head. Then we will discuss the mechanism of the actomyosin ATPase reaction and that of the control of actomyosin ATPase by $Ca^{2+}$. Finally, we will discuss the molecular mechanism of muscle contraction on the basis of the reaction mechanism of actomyosin ATPase.

## F-actin

Actin, another contractile protein, was first discovered by Straub in 1942. It accounts for 20–25% of myofibrillar proteins. In myofibrils actin constitutes the thin filament together with two regulatory proteins of muscle contraction, tropomyosin (TM) and troponin (TN).

Straub (1942) found that actin extracted from acetone-dried powder of muscle exists as a monomer (G-actin), but it polymerizes to form a filament (F-actin) in the physiological ion environment (see Fig. 3.1). He also showed that ATP binds tightly to G-actin, and is hydrolyzed to ADP + Pi on polymerization of actin. Thus, in F-actin each actin monomer contains one molecule of tightly bound ADP. However, ATP hydrolysis is not essential for the polymerization of actin. G-actin, free from bound nucleotide, can also form a physiologically functional F-actin. The function of nucleotide bound to actin may be described as follows: The nucleotide stabilizes the structure of actin, since G-actin without bound nucleotide denatures very rapidly. It regulates the polymerization of actin, since G-actin with bound ATP can polymerize more rapidly than that with bound ADP.

G-actin is a globular protein composed of a single polypeptide chain. Its primary structure (Fig. 3.2) has been determined by Elizinga & Collins

Fig. 3.1 The model structure of G- and F-actin.

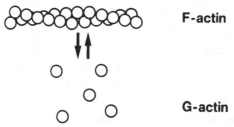

Fig. 3.2 Amino acid sequence of rabbit skeletal muscle actin (Elzinga & Collins, 1973). (1), 3-methyl histidine; (2), cysteine residue that is specifically modified with an affinity-labeling ATP analog (see text); (3), reactive thiol group.

①                                                                    ㊿

Ac-DETEDTALVCDDGSGLVLAGFAGDDAPRAVFPSIVGRPRHQGVMVGMGQK

(1)

DSTVGDEAQSKRGILTLKYPIGXTGIITNDDMEKIWHHTFYDELRVAPEE

HPTLLTEAPLNPKANREKMTQIMFETFNVPAMTVAIQAVLSLTASGRTTG

IVLDSGDGVTHNVPIYEGYALPHAIMRLDLAERDLTDYLMKILTERGYSF

(2)

VTTAEREIVRDIKQKLCYVALDFENEMATAASSSLEKSYELPDGEVITIG

NERFRCPETLFQPSFIGMESAGIHETTYDSIMKCDIDIRKDLYANNVMSG

GTTMYPGTADRMQKEITALAPSTMKIKIIAPPERKYSVWIGGSILASLST

(3)

FQQMWITKQEYDEAGPSIVHRKCF
374

(1973). Actin is made up of 374 amino acid residues, and its molecular weight was determined to be 42 000. Actin contains five Cys residues. Among them, the Cys-373 residue which is the one next to the C-terminus (see (3) of Fig. 3.2) reacts rapidly with SH reagents. Since the functions of actin are unaffected by the chemical modification of this residue, actin preparations in which this Cys residue is modified with fluorescent dye or spin label are often used in the study of the molecular dynamics of F-actin. The Cys residue at position 217 (see 2 of Fig. 3.2) is thought to form the ATP-binding site, since an ATP analog, 5'-dinitrophenylmer-captopurine ribose-5'-triphosphate, binds covalently to this residue and is hydrolyzed to diphosphate and Pi upon polymerization of actin.

G-actin binds strongly with DNase-I of pancreas at the molar ratio of 1:1. Since this complex readily crystallizes, the tertiary structure of G-actin was studied by X-ray diffraction analysis (Suck, Kabsch & Mannheriz, 1981). Figure 3.3 shows the three-dimensional structure of actin monomer deduced from the structure of G-actin–DNase-I. G-actin has dimensions $6.7 \times 4.0 \times 3.7$ nm and consists of two large domains. The ATP-binding site is assumed to be the cleft region between these two domains (shown by arrow in Fig. 3.3).

G-actin polymerizes to form F-actin when KCl or $Mg^{2+}$ is added. Oosawa & Asakura (1975) studied the mechanism of polymerization of actin. They showed that for formation of F-actin from G-actin a critical

Fig. 3.3 Balsa wood model of the actin monomer at 6 Å resolution. Courtesy of D. Suck and W. Kabsch.

actin concentration which depends on the medium is required before the polymerization occurs; above the critical concentration, excess G-actin is converted into the F-form. Such a condensation type of polymerization is seen in several systems, such as tubulin and flagellin. Oosawa & Asakura explained this type of polymerization by postulating that F-actin is a linear helical polymer.

The structure of F-actin has been studied mainly by electron microscopy. It was found that F-actin is made up of actin monomers with a diameter of 5 nm polymerized into a double-helical form (see Fig. 3.1). The half pitch of the F-actin helix is about 35 nm and it contains 13 actin monomers. It has been shown from X-ray diffraction analysis of living muscle that actin monomers are arranged with a spacing of 5.4 nm. It was shown by electron microscopy after negative staining that F-actin decorated with S-1 or HMM has a unidirectionally pointed arrowhead structure (Fig. 3.4), suggesting that F-actin has polarity. Furthermore, electron microscopy of the muscle fibers decorated with HMM reveals that the direction of the arrowhead differs on both sides of the Z-line, and the arrowhead tips point away from the Z-line toward the A-band. This suggests that the interaction between F-actin and the myosin head occurs with polarity as expected from the sliding filament theory.

F-actin has a very rigid structure. The rigidity of F-actin is reported to be about one-tenth that of steel rod. Therefore, ADP bound to actin

Fig. 3.4 Electron micrograph of acto–S-1 complex. Scale bar, 100 nm.

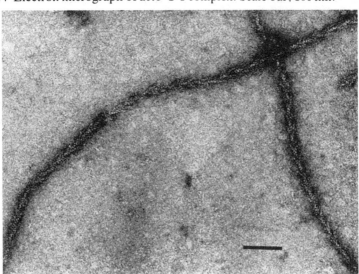

and dye or spin label attached to Cys-373 of actin in muscle fiber demonstrate a fixed angle against the direction of the fiber. F-actin is also rigid on the direction of fiber, and the length of the thin filament does not change during contraction of rapid extension of the fiber.

## Formation of actomyosin

In the absence of nucleotide, myosin and its proteolytic subfragments, HMM and S-1, bind strongly to F-actin at a molar ratio of myosin head:actin monomer of 1:1. This complex corresponds to the crossbridge in muscle fiber in the rigor state.

Since the binding constant of myosin head (S-1) with F-actin is very high, its accurate value is not easy to measure directly. Recently, Yasui, Arata & Inoue (1984) measured the rate constant of binding of S-1 with F-actin $(k_1)$ and that of the dissociation of acto–S-1 $(k_{-1})$. They estimate the binding constant as $K = k_1/k_{-1}$. In 0.1 M KCl and at 20 °C, S-1 binds with F-actin at the rate of $2.6 \times 10^7/\text{M/s}$ and dissociates very slowly (0.5/min). The binding constant was calculated to be $3.2 \times 10^9/\text{M}$. The value of $\Delta G°$ was estimated from this value to be $-11.3 \, \text{cal/mol}$. From the temperature dependence of the binding constant of acto–S-1, the values of $\Delta H°$ and $\Delta S°$ accompanying the binding were calculated to be $+2.3 \, \text{kcal/mol}$ and $+46 \, \text{e.u.}$, respectively. Thus, the motive force of binding is the increase in entropy, which may be due to the conformational change in the myosin head.

The tertiary structure of the complex of F-actin with the myosin head has been studied by three methods. The first method chemically crosslinks actin and S-1. Yamamoto & Sekine (1979) showed that when trypsin-treated S-1, in which the m.w. 100 000 heavy chain was digested into m.w. 51 000, 25 000 and 21 000 segments, was crosslinked with F-actin by a bifunctional SH reagent, dimethyl suberimidate, only the m.w. 51 000 segment was crosslinked with F-actin. Later, Mornet *et al.* (1981) showed that when acto–S-1 is treated with zero-length crosslinker, 1-ethyl-3 (3-dimethylaminopropyl) carbodiimide, both the 51 000 and the 21 000 segments are crosslinked to actin. From this result they concluded that one S-1 molecule binds to a pair of actin monomers on F-actin with m.w. 51 000 and 21 000 segments, respectively. However, Sutoh (1983) later showed that only one of the m.w. 51 000 or 21 000 segments binds to one actin monomer. It should be noted that the digestion of the m.w. 51 000–21 000 junction by trypsin or the modification of $SH_1$ on the m.w. 21 000 segment was inhibited by F-actin.

The second method involves electron microscopy. Fairly detailed information has been obtained on the tertiary structure of the complex of actin and myosin head from a three-dimensional reconstitution of the electron microscopic image of acto–S-1 obtained by negative staining methods (Moore, Huxley & De Rossier, 1970; Craig & Knight, 1983). It has been shown from these studies that S-1 binds with actin with a tilting angle of about 75°.

The third method examines the binding between actin and S-1 by energy transfer. For example, Takashi (1980) has shown that the distance between the SH residue at position 373 of actin and the $SH_1$ thiol residue of S-1 is 5.8 nm.

## Dissociation of actomyosin induced by nucleotide

During muscle contraction the myosin head repeatedly dissociates from the thin filament and recombines to it; this action is coupled with ATP hydrolysis. The mechanism of dissociation of the myosin head from F-actin is, however, not easy to study directly using muscle fibers or myofibrils. Even in the case of actomyosin, both myosin and actin form a filament at low ionic strength, and actomyosin shows superprecipitation on addition of ATP.

However, when the ionic strength is higher than 0.3 M, the myosin filament is dissolved into monomer, and the activation of myosin ATPase by F-actin does not take place. The dissociation of actomyosin is also induced by unhydrolyzable ATP analogs, such as PPi, ADP and AMPPNP. Since the mechanism of dissociation of actomyosin by these ATP analogs was simple, the mechanism of dissociation of actomyosin was first studied using such unhydrolyzable ATP analogs. In this section, we will first describe the mechanism of dissociation of actomyosin by unhydrolyzable ATP analogs. We will then describe the mechanism of dissociation of actomyosin by ATP at high ionic strength. Finally, we will discuss the relationship between the ATPase reaction and the dissociation and recombination of F-actin with myosin head studied using HMM or S-1 which are soluble even at low ionic strengths.

## Dissociation of actomyosin by unhydrolyzable ATP analogs

It has been shown from the studies on the dissociation of actomyosin by PPi at high ionic strength that actomyosin dissociates via a ternary complex of actomyosin with PPi (Morita & Tonomura, 1960):

$$AE + PPi \rightleftharpoons AE\text{-}PPi \rightleftharpoons A + E\text{-}PPi$$

Studies on the dissociation of actomyosin or acto–S-1 by ADP or AMPPNP have also shown that actomyosin forms a ternary complex with these ATP analogs. It was concluded from these studies that the F-actin-binding site and the ATP-binding site on the myosin molecule are separated from each other.

When AMPPNP is added to acto–S-1 (AE), there exists an equilibrium, as shown in Fig. 3.5 (Inoue & Tonomura, 1980; M. Yasui, T. Arata & A. Inoue, personal communication). S-1 binds to actin with a dissociation constant $(K_{a'})$ of $3.2 \times 10^{-9}$ and to AMPPNP with $K_{s'}$ of $2.6 \times 10^{-7}$. On the other hand, F-actin binds to the S-1–AMPPNP complex with a dissociation constant $(K_a)$ of $2.0 \times 10^{-5}$, and AMPPNP binds to acto–S-1 with a dissociation constant $(K_s)$ of $1.1 \times 10^{-3}$. Thus, $K_a/K_{a'} = K_s/K_{s'}$ is about $1.6 \times 10^4$. The extent of dissociation of S-1 from F-actin (which is equivalent to (free S-1)/(added S-1)) is given by $1/[1 + \{(FA)/K_a\} \{1 + K_s/(AMPPNP)\}]$. Therefore, the values of $K_a$ and $K_s$ can be calculated from the dependence of $\alpha$ on the concentration of AMPPNP and F-actin. It was actually demonstrated that in the pre-

Fig. 3.5 Mechanism of dissociation of acto–S-1 by AMPPNP. Values show the dissociation constants (M. Yasui, T. Arata & A. Inoue, personal communication).

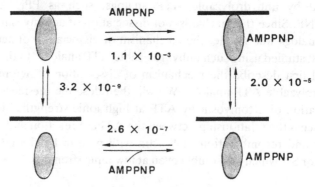

sence of $Mn^{2+}$, S-1 binds to both F-actin and AMPPNP when high concentrations of F-actin and AMPPNP are present.

## Dissociation of actomyosin by ATP

The time course of dissociation of actomyosin after the addition of ATP has been measured by Tonomura & Watanabe (1954) from the changes in the light-scattering intensity. As shown in Fig. 3.6, when ATP is added to actomyosin at high ionic strength, the light-scattering intensity decreases rapidly (phase 1), remains at the decreased level for a certain period (phase 2), and then returns slowly to the original intensity (phase 3). Phase 1 is attributed to the dissociation of actomyosin. The time in phase 2 is related to the concentration of ATP added, and is given by the amount of ATP added/rate of ATPase reaction. Phase 3 is the recombination of myosin with F-actin after the hydrolysis of ATP.

The rate of dissociation of actomyosin was determined from the decrease in light-scattering intensity (phase 1). At a low concentration of ATP, the rate of dissociation was almost the same as that of the change in the UV spectrum ($E_b$*ATP). However, it is much higher than the rate of $E_b$*ATP formation (about 200/s) and is of the order of 1000/s (Lymn & Taylor, 1971). Therefore, the dissociation of actomyosin is considered to be caused by the formation of the $E_a$*ATP complex:

$$AE + ATP \rightleftharpoons AE\text{–}ATP \rightarrow AE_a\text{*}ATP$$
$$\downarrow$$
$$A + E_b\text{*}ATP \rightarrow A + E_b\text{*}ATP \rightarrow A + E\text{–}P\text{–}ADP$$

As described in Chapter 2, myosin has two non-identical heads. It

Fig. 3.6 Time course of the change in the light-scattering intensity after the addition of ATP to actomyosin at high ionic strength. See text for details.

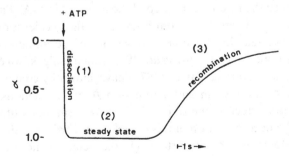

was found that at high ionic strength, actomyosin or acto–HMM is dissociated by 1 mol of ATP/mol of myosin or HMM, when myosin or HMM
forms E–P–ADP with one head (burst head). These results suggested
that there is an interaction between the two heads of the myosin molecule
when myosin binds with F-actin.

Acto–S-1 is dissociated by 1 mol of ATP/mol S-1, with one head forming E–P–ADP and the other head forming E-ATP. There is not a large
difference between the burst head and non-burst head (see Chapter 2)
in the rate of dissociation of acto–S-1 by ATP and its analogs. This
suggests that both heads dissociate by formation of the $E_a$*ATP complex,
although the rate-limiting steps of the ATPase reaction are the decomposition of E–P–ADP in the burst head and formation of E–P–ADP in
the non-burst head.

# Reaction mechanism of actomyosin ATPase

When ATP is added to actomyosin at low ionic strength,
actomyosin aggregates and precipitates into a large mass. This phenomenon is known as superprecipitation, and is one of the models of muscle
contraction. The rate of the actomyosin ATPase reaction varies during
the course of superprecipitation: it is low before the superprecipitation,
very high during it, and becomes low after the superprecipitation. In
addition, it is difficult to analyze the binding of myosin heads with F-actin
during the superprecipitation. Therefore, the analysis of ATPase is carried out mainly with the proteolytic fragments, HMM or S-1, which
are soluble even at low ionic strength.

## *Acceleration of decomposition of E–P–ADP by F-actin*

Since the rate-limiting step of the myosin $Mg^{2+}$-ATPase reaction is the release of products from E–P–ADP, the acceleration of the
myosin ATPase reaction by F-actin relies on the acceleration of this
step. Furthermore, we (Imamura *et al.*, 1965; Inoue, Shigekawa & Tonomura, 1973), and Lymn & Taylor (1971) have shown by comparing the
initial phases of the myosin and actomyosin ATPase reactions that F-
actin exerts little effect on the step of E–P–ADP formation but strongly
accelerates the decomposition of this complex. Accordingly, the following two problems became important: (1) the detail of the mechanism

of acceleration of E–P–ADP decomposition, and (2) the relation between the binding of F-actin with the myosin head and the elementary steps of the ATPase reaction.

## *Mechanism of actomyosin ATPase reaction*

The study of the mechanism of the actomyosin ATPase reaction was based on the mechanism of myosin ATPase which was described in Chapter 2 and the mechanism of dissociation of actomyosin which was described in the previous section. It was found that the mechanism of the actomyosin ATPase reaction depends largely on the experimental conditions, as the shortening rate and tension development vary depending on the conditions. For example, the shortening of muscle is fast but the tension developed is low when a relatively high concentration of KCl is present. On the other hand, when the concentration of KCl is low, the shortening of muscle is slow but the tension developed is high. Corresponding to these differences in the physiological properties of muscle contraction, it has been shown that the main route of actomyosin ATPase reaction also differs in the above conditions. Therefore, we discuss the mechanism of actomyosin ATPase reaction for each of these two conditions.

*Actomyosin ATPase reaction at relatively high ionic strength and/or low temperature* At relatively high concentrations (0.05–0.2 M) of KCl and/or low temperatures most of the myosin heads are dissociated from F-actin during the actomyosin-type ATPase reaction. In the presence of high concentrations of the ATP the rate of dissociation of acto–S-1 or acto–HMM is much higher than the rate of ATPase reaction, and the rate of the overall reaction of actomyosin-type ATPase in steady state ($\Delta V_0$) is limited by the rate of rebinding of HMM–P–ADP or S-1–P–ADP with F-actin ($v_{recomb}$). Lymn & Taylor (1971) proposed the following mechanism for the main route of the actomyosin (AE) ATPase reaction based on the mechanism of myosin ATPase and that of dissociation of actomyosin by ATP:

$$AE + ATP \rightleftharpoons AE\text{–}ATP \qquad AE\text{–}ADP\text{–}P \rightarrow AE + ADP + Pi$$
$$\downarrow \qquad \qquad \uparrow$$
$$A + E\text{–}ATP \rightarrow A + E\text{–}P\text{–}ADP$$

According to this mechanism, the binding between F-actin and

E–P–ADP is the rate-limiting step in the reaction. Therefore, the rate of ATPase reaction should be proportional to the concentration of F-actin. However, Chock, Chock & Eisenberg (1976) observed that even under such conditions, both the rate of actomyosin-type ATPase reaction ($\Delta v_0$) and the rate of recombination of E–P–ADP with F-actin ($v_{recomb}$) approach a maximal level as the concentration of F-actin increases. They explained this result by postulating that E–P–ADP in a refractory state that cannot react with F-actin is formed by the reaction of actomyosin with ATP, and that the conversion of E–P–ADP in the refractory state (R) to that in the non-refractory (N) state is the rate-limiting step of the ATPase reaction at high concentrations of F-actin:

$$AE + ATP \rightarrow E\text{--}P\text{--}ADP(R) \rightarrow A + E\text{--}P\text{--}ADP(N) \rightarrow AE + ADP + Pi$$
$$\text{(refractory state)} \qquad\qquad \text{(non-refractory state)}$$

In the following discussion it must be noted that although not explicitly mentioned, the presence of the refractory state is tacitly assumed in the form of the rate of binding of F-actin with E–P–ADP, as a function of F-actin concentration.

*Actomyosin ATPase reaction at very low ionic strength and room temperature* The above mechanism is valid under the conditions in which almost all the crossbridges are dissociated in the steady state. However, this mechanism cannot explain the kinetic properties of the actomyosin ATPase reaction at low KCl concentrations and room temperatures where the myosin head is bound to F-actin even during the ATPase reaction in the steady state (Stein *et al.*, 1979; Inoue, Ikebe & Tonomura, 1980). Under these conditions, the rate of ATPase reaction ($v$) is proportional to the amount of the complex of F-actin with HMM or S-1, AE–P–ADP, as shown in Fig. 3.7. In addition, the rate of ATPase reaction ($v$) is much faster than the rate of binding of S-1–P–ADP or HMM–P–ADP with F-actin ($v_r$) (Inoue, Ikebe & Tonomura, 1980). Thus, under these conditions the main route of the actomyosin ATPase reaction is the hydrolysis of ATP without the accompanying dissociation of actomyosin as given by the following mechanism:

$$AE + ATP \rightleftharpoons AE\text{--}ATP \rightleftharpoons AE\text{--}P\text{--}ADP \rightarrow AE + ADP + Pi.$$

As described in the previous chapter, the rate of dissociation of AE–ATP into A + E–ATP is very large (more than 1000/s), and AE–ATP is in rapid equilibrium with A + E–ATP. This E–ATP is considered to correspond to the $E_a$*ATP in the previous chapter. The observed dissociation constant for the binding of HMM or S-1 with F-actin in the presence of ATP is about 50 μM under these conditions. Since the rate of direct decomposition of AE–P–ADP is about ten times that of recombination of E–P–ADP with F-actin, the dissociation constant of the step, AE–ATP → A + E–ATP, was estimated to be 5 μM. Therefore, the rate of binding of E-ATP with F-actin must be very large.

*Mechanism of actomyosin ATPase under general conditions*    Thus far, we have described the mechanism of actomyosin ATPase reaction for two relatively extreme conditions. Under more general conditions the ATPase reactions via the above two routes co-exist. In other words, ATP is hydrolyzed via two routes, shown schematically in Fig. 3.8, one without dissociation of actomyosin and the other with the dissociation and recombination of actomyosin. The rate-limiting step of the former route is the direct decomposition of the ternary complex, AE–P–ADP, to AE + ADP + Pi. The rate-limiting step of the latter route is the rebinding of E–P–ADP with F-actin. According to this mechanism, the rate of actomyosin ATPase reaction ($\Delta v_0$) is given by the equation

Fig. 3.7 Comparison of the rate of acto-HMM ATPase reaction at the steady state, $\Delta v_0$ and the extent of binding of HMM with F-actin, $\alpha$, with the rate of recombination of HMM–P–ADP with F-actin, $v_{recomb}$ as a function of F-actin concentration in the absence of KCl and at 20 °C, 1 mM MgCl$_2$, pH 7.8. Triangles, $\Delta v_0$; crosses, $1-\alpha$; open circles, $v_{recomb}$ (Inoue, Ikebe & Tonomura, 1980).

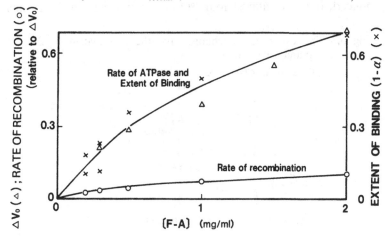

$$\Delta v_0 = (1 - \alpha) \Delta V_0 + \alpha v_{\text{recomb}}$$

where $\alpha$ is the extent of dissociation (free HMM or S-1/added HMM or S-1), $v_{\text{recomb}}$ is the rate of the rebinding of E–P–ADP with F-actin, and $\Delta V_0$ is the rate constant of the step, AE–P–ADP $\to$ AE + ADP + Pi, which is equal to $\Delta v_0$ at an infinite concentration of F-actin. The presence of AE–P–ADP in the outer route is ignored in Fig. 3.8, since the binding of E–P–ADP with F-actin is the rate-limiting step of this route, and AE–P–ADP formed in this route is immediately decomposed. It must be noted that a large part of the E–P–ADP in the outer cycle is in the refractory state which cannot bind with F-actin, and A + E–P–ADP of the outer cycle is not in equilibrium with the AE–P–ADP of the inner cycle.

Figure 3.9 shows that the above 'two-route' mechanism of actomyosin ATPase reaction occurs over a wide range of conditions. In the presence of 10 mM K-Pi and 1 mM $Mg^{2+}$ at 12 °C and pH 7 and at low concentrations of F-actin, the ATPase reaction occurs mainly via the outer cycle and $\Delta v_0$ is almost the same as $v_{\text{recomb}}$. As the concentration of F-actin increases, the hydrolysis of ATP through AE–P–ADP becomes dominant, and $(1 - \alpha)\Delta V_0$ becomes the major component of $\Delta v_0$. Thus, $\Delta v_0$ is equal to $(1 - \alpha) \Delta V_0 + \alpha v_{\text{recomb}}$ over all F-actin concentrations.

The two-route mechanism of the actomyosin ATPase reaction was supported by the measurement of [18]O exchange reaction during the ATPase reaction (Midelfort, 1981). When the myosin ATPase reaction was catalyzed in $H_2{}^{18}O$, almost four [18]O atoms are incorporated into product Pi by the equilibrium step of $E^*ATP + H_2O \to E-P-ADP$. It was shown that in the actomyosin ATPase reaction two kinds of Pi were produced, one with almost four [18]O atoms and the other with nearly

Fig. 3.8 The two-route mechanism for the actomyosin ATPase. For explanation see text.

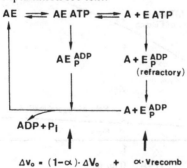

one $^{18}$O atom. These two kinds of Pi are attributed to the Pi produced by the ATPase reaction through direct decomposition of AE–P–ADP and the Pi produced through dissociation and recombination of actomyosin, respectively.

## Ca$^{2+}$ regulation of actomyosin ATPase

It is now established that muscle contraction is switched on and off by a 0.1–1 μM range of Ca$^{2+}$. In muscle cells Ca$^{2+}$ is actively transported into the sarcoplasmic reticulum, which is distributed along

Fig. 3.9 The relationship between the rate of acto–HMM ATPase reaction, $\Delta v_0$, the extent of dissociation of HMM from F-actin, $\alpha$, the rate of recombination of HMM–P–ADP with F-actin, $v_{recomb}$, and $\Delta v_0$, at infinite F-actin concentration, $\Delta V_0$, at various F-actin concentrations in the absence of KCl and at 12 °C. Open circles, $\alpha \cdot v_{recomb}$; filled circles, $(1-\alpha)\Delta V_0$; triangles, $\Delta v_0$. Conditions: 2 mM K-PEP, 1 mM MgCl$_2$, pH 7.8, 12 °C (Inoue, Ikebe & Tonomura, 1980).

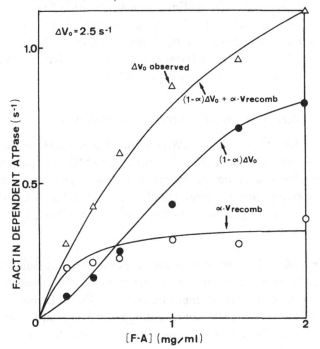

myofibrils. $Ca^{2+}$ is also pumped out through the plasma membrane by the $Ca^{2+}$-ATPase or the $Na^+$–$Ca^{2+}$ exchange reaction. Therefore, the concentration of intracellular $Ca^{2+}$ is very low, and $Ca^{2+}$ is stored in the sarcoplasmic reticulum. Excitation of the cell membrane of skeletal muscle fibers is transmitted into the sarcoplasmic reticulum through the transverse (T)-system, and $Ca^{2+}$ release from the sarcoplasmic reticulum is induced. The mechanism of release of $Ca^{2+}$ has not yet been clarified.

In the case of skeletal muscle, muscle contracts when $Ca^{2+}$ binds to troponin on the thin filament and relaxes when the $Ca^{2+}$ that was released into the fibers is transported back into the sarcoplasmic reticulum. However, the $Ca^{2+}$-receptor protein varies by muscle type. In scallop adductor muscle, $Ca^{2+}$ binds directly to myosin, which forms the thick filament; in smooth muscle, $Ca^{2+}$ binds to calmodulin, which binds with myosin light-chain kinase, and muscle contraction is induced by the $Ca^{2+}$-dependent phosphorylation of myosin light chain. Furthermore, a variation of the $Ca^{2+}$ regulatory system has been found in nonmuscle movement (see Chapter 4).

Excellent reviews have been published on the regulation of muscle contraction. The book edited by Ebashi, Maruyama & Endo (1980) contains a variety of works in this field. For skeletal muscle see the review articles by Ebashi & Endo (1968), Ebashi, Endo & Ohtsuki (1969), Weber & Murray (1973) and Ebashi (1980). For regulation of smooth muscle see books edited by Casteels, Godfraind & Ruegg (1977), Stephens (1977) and Crass & Barnes (1982), and for myosin-linked regulation see the review by Szent-Györgyi (1975).

## Structures of tropomyosin and troponin

Ebashi & Ebashi (1964) found that the $Ca^{2+}$ sensitivity of the actomyosin ATPase reaction is restored by adding a protein complex obtained from an acetone-dried powder of muscle. This protein complex (native tropomyosin) was later found to be a complex of tropomyosin and a $Ca^{2+}$-receptor protein, troponin (Ebashi & Kodama, 1965).

*Tropomyosin*　Tropomyosin, which was discovered by Bailey in 1948, binds tightly with F-actin but dissociates from F-actin at very low ionic strength or at high ionic strength (about $0.5\,M$ KCl). Tropomyosin has a molecular weight of 68 000 and is a dimer of two subunits of molecular weight 34 000. The molecule is nearly 100% $\alpha$-helical, and its length is 40 nm. Since an α-helix with molecular weight of 34 000 has a length of 40 nm, it was considered that the molecule is a side-by-side complex of two helices. Since SH groups at identical positions form disulfide links, two 100% α-helical subunits are aligned in parallel to form one molecule.

Skeletal muscle tropomyosin is composed of two isozymic forms, α and β. These two isoforms are functionally identical, and in a tropomyosin preparation α–α and α–β complexes coexist. Primary structures of both isoforms have already been determined (Mak, Smillie & Stewart, 1980) and are shown in Fig. 3.10. The primary structure of tropomyosin also suggests that tropomyosin subunits form a so-called super-coil in which two α-helical chains bind side by side and form a coiled structure. There are repeating structures every seven residues and hydrophobic residues at positions 2 and 6. As shown in Fig. 3.11, when the peptides form an α-helix, positions 2 and 6 are at one side of the helix. Therefore, two subunits can join end-to-end by an external overlap of the nonpolar zone. Furthermore, a repeat of about every 40 residues occurs in the amino acids along the chain. This is believed to correspond to the actin-binding sites.

Tropomyosin binds to F-actin at the ratio of 1 mol/7 mol of actin monomers. The length of tropomyosin (40 nm) is equal to seven times

Fig. 3.10 Amino acid sequences and subunits of tropomyosin from rabbit skeletal muscle (Mak, Smillie & Stewart, 1980). A regular pattern of nonpolar amino acid residues occurs in two series, I (squared) and II (circled). The residues in each series repeat at every seventh residue in the sequence.

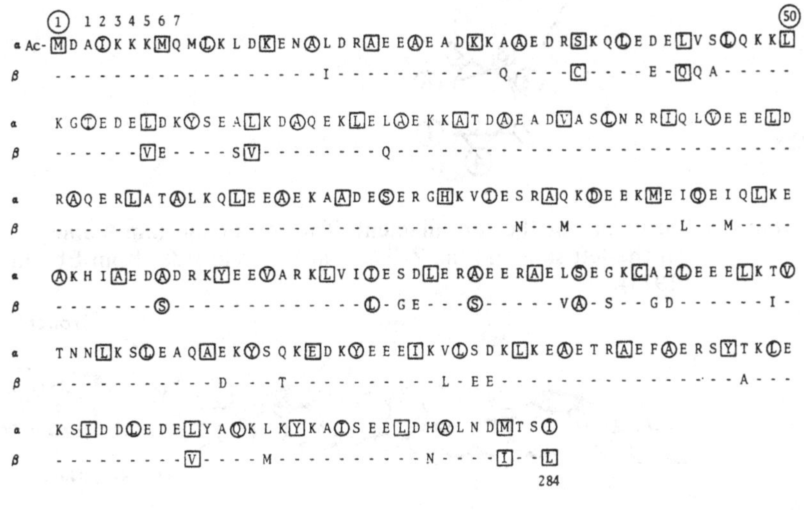

284

the diameter of the actin monomer. Tropomyosin was assumed to bind to the groove of the helix of F-actin as shown in Fig. 3.12 (Ebashi, 1974). This model was later supported by X-ray analysis and by electron microscopy of the thin filament. However, the accurate location of tropomyosin on the F-actin filament still remains unknown.

*Troponin*   Troponin, the $Ca^{2+}$ receptor of myofibrils, was discovered by Ebashi & Kodama (1965). It can restore the $Ca^{2+}$-sensitivity of the actomyosin ATPase reaction. Troponin binds with tropomyosin at a 1:1 molar ratio. Therefore, thin filaments contain 1 mol of troponin/ 7 mol of actin monomers. It has also been suggested that the binding portion of troponin (on tropomyosin) is one third of the way from the C-terminal end of tropomyosin.

Troponin is composed of three subunits (Hartshorne & Mueller, 1969;

Fig. 3.11 Double-stranded super-coil of tropomyosin. Lower figure shows the end-on-view of the α-helices in the super-coil. Nonpolar amino acid residues occur at positions 2 and 6.

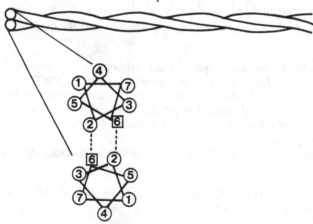

Fig. 3.12 Structure of the thin filament. The top of the thin filament is at the left side and the Z-line is at the right side (from Ebashi, 1974).

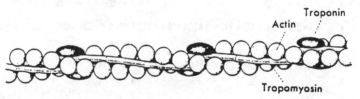

Greaser & Gergely, 1971): the $Ca^{2+}$-binding component (troponin-C), the inhibitory component (troponin-I) and the tropomyosin-binding component (troponin-T). Troponin can be reconstituted from these three subunits.

Troponin-C (TN-C) has a molecular weight of 18 000, and one molecule contains four $Ca^{2+}$-binding sites. Two sites have higher affinity for $Ca^{2+}$ and $Mg^{2+}$ than the other two sites. The sites of lower affinity for $Ca^{2+}$, however, bind specifically with $Ca^{2+}$. Since the dissociation of $Ca^{2+}$ from the former sites occurs very slowly, the binding of $Ca^{2+}$ to the latter sites is considered to regulate the contraction and relaxation of muscle (Johnson, Charlton & Potter, 1979). The amino acid sequence of TN-C has been determined by Collins *et al.* (1977) (Fig. 3.13). The positions of the $Ca^{2+}$-binding sites on the first-order structure of the molecule have been estimated from a comparison with parvalbumin, a $Ca^{2+}$-binding protein with known tertiary structure (see Krestsinger & Barry, 1975). It was suggested that $Ca^{2+}$ binds to site (I)–(IV). From studies on proteolytic segments of TN-C, it was suggested that sites (I) and (II) are the $Ca^{2+}$-, $Mg^{2+}$-binding sites, and (III) and (IV) are the $Ca^{2+}$-specific sites (Leaves *et al.*, 1978).

Troponin-I (TN-I) has a molecular weight of 21 000 and its amino acid sequence has been determined as shown in Fig. 3.13 (Wilkinson & Grand, 1975). In the presence of tropomyosin, TN-I inhibits the $Mg^{2+}$-ATPase of actomyosin. This is the most important function of troponin, and this inhibitory action is controlled by $Ca^{2+}$. Troponin-T (TN-T) has a molecular weight of 35 000 and the amino acid sequence of TN-T has been determined. TN-T contains 259 amino acid residues (Pearlstone *et al.*, 1977). This component plays a role in the binding of troponin subunits to tropomyosin.

The interaction among these three subunits of troponin in the presence and absence of $Ca^{2+}$ has been studied by an ultracentrifugal method (Potter & Gergely, 1974) and by measuring the effects of the subunits on the actomyosin ATPase reaction (Shigekawa & Tonomura, 1973). The results obtained by these two methods can be well explained by the model shown in Fig. 3.14. When $Ca^{2+}$ binds to TN-C, the bond between TN-I and TN-C becomes tight, TN-I dissociates from F-actin, and the inhibition of actomyosin ATPase by TN-I disappears. In this arrangement TN-C is bound to F-actin through TN-T and tropomyosin (TM). When $Ca^{2+}$ is removed, the bond between TN-C and TN-I becomes loose, TN-I binds tightly to TM and F-actin, and the actomyosin ATPase reaction is inhibited.

## Mechanism of $Ca^{2+}$ control by the tropomyosin–troponin system

The activity of actomyosin ATPase in the presence of tropomyosin and troponin decreases to that of myosin alone when $Ca^{2+}$ is removed in the presence of a high concentration of $Mg^{2+}$-ATP. The muscle relaxes under this condition. During relaxation stiffness of muscle fiber is very low, and the X-ray diffraction pattern of muscle fiber also suggests that during relaxation myosin heads dissociate from the thin filament.

It has also been shown in HMM or S-1, F-actin–tropomyosin–troponin

Fig. 3.13  Amino acid sequences of the troponin subunits of rabbit skeletal muscle. I–IV, $Ca^{2+}$-binding regions of TN-C. (TN-C, Collins *et al.*, 1977; TN-I, Wilkinson & Grand, 1975; TN-T, Pearlstone *et al.*, 1977).

①                                                                    ㊿

TN-C    DTQQAEARSTLSEEHIAEFKAAFDMFDADGGGDISVKELGTVMRMLGQTP
                                              (I)

TKEELDAIIEEVDEDGSGTIDFEEFLVMMVRQMKEDAKGKSEEELAECFR
                    (II)

IFDRNADGYIDAEELAEIFRASGEHVTDEEIESLMKDGDKNNDGRIDFDE
(III)                                                    (IV)

FLKMMEGVQ
159

TN-I    Ac-GDEEKRDRAITARRQHLKSVMLQIAATELEKEEGRREAEKQNYLAEHCPP

LSLPGSMAEVQELCLQLHAKIDAAEEEKYDMQIKVQKSSKELEDMNQKLF

DLRGKFKRPPLRRRVRMSADAMLKALLGSKHKVCMDLRANLKQVKKEDTE

KERDVGDWRKNIEEKSGMEGRKKMFESES
                              179

TN-T    Ac-SAEEVEHVEEEAEEEAPSPAEVHEPAPEHVVPEEVHEEEKPRKLTAPKIP

EGEKVDFDDPQKKRQNKDLMELQALIDSHFEARKKEEEELVALKEKIEKR

RAERAEQQRIRAEKERERQNRLAEEKARREEEDAKRRAEEDLKKKKALSS

MGANTSSTLAKADQKRGKKNTAREMKKKILAERRKPLNIDHLSDEKLRDK

AKELWDTLYQLETDKFEFDEKLKRQKYDIMNVRARVEMLAKFSKKAGTTA

KGKVGGRTK
259

complex (FA–TM–TN) and ATP systems in solution that the dissociation of myosin head from FA–TM–TN is induced by removal of Ca$^{2+}$ at physiological ionic strength (see Fig. 3.15) (Inoue & Tonomura, 1982). Furthermore, as shown in Fig. 3.16, it has been shown that the rate of recombination of the myosin head in the form of E–P–ADP with FA–TM–TN was reduced by removal of Ca$^{2+}$. On the other hand, at very low ionic strength and at room temperature, HMM or S-1 does not dissociate from FA–TM–TN even in the absence of Ca$^{2+}$, although the ATPase activity is very low. Under this condition the direct decomposition of the actomyosin–P–ADP complex into actomyosin + Pi + ADP (see Fig. 3.9) is also inhibited by removal of Ca$^{2+}$. Under this condition, muscle develops high tension and has a high degree of stiffness although the shortening rate is very low (Yanagida, Kuranaga & Inoue, 1983).

At present there are two hypotheses for the structural change in the thin filament caused by the binding and dissociation between troponin-C and Ca$^{2+}$. The first is that the binding of Ca$^{2+}$ to troponin induces a structural change in F-actin through structural changes in troponin and tropomyosin. However, what kind of change actually occurs in the structure of F-actin is not yet known.

The second hypothesis is that the position of tropomyosin (TM) on the thin filament changes upon binding of Ca$^{2+}$ to troponin, thereby causing steric inhibition of the binding of myosin head to F-actin. Tropomyosin is present in the groove of the double-stranded helical polymer of actin. The change in the position of tropomyosin depending on the

Fig. 3.14 Interactions among actin, tropomyosin and the three subunits of troponin in the presence and absence of Ca$^{2+}$. Wide and dotted lines indicate strong and weak interactions, respectively.

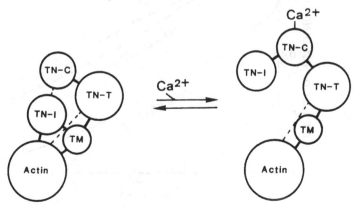

presence or absence of $Ca^{2+}$ was first speculated upon from the result of X-ray analysis of living muscle. Later, Wakabayashi *et al.* (1975) compared the three-dimensional reconstituted images of the F-actin–TM complex, that show the contraction reaction and the F-actin–TM–TN-T–TN-I complex that shows relaxation reaction, from electron microscope images. They showed that TM is in close contact with actin in the inactive filament, whereas its position moves 1 nm in the active filament. On the basis of this and the three-dimensional image of acto–S-1 reconstituted from electron microscope image, they proposed a model in which TM sterically blocks the binding between S-1 and F-actin. However, several problems arose from their work. The first is that the polarity of the thin filament was mistakenly reversed in this model. So the location of tropomyosin on the actin filament is quite different from where it is in their model. The second problem is that the identification of tropomyosin in the reconstituted image is still unclear, and the three-dimensional reconstituted image of acto–S-1 has since been extensively modified. Therefore, progress in the structural studies of the thin fila-

Fig. 3.15 The dependence on KCl concentration of the extent of HMM binding with F-actin–TM–TN complex in the presence of ATP. Conditions: 5 μM HMM, 6 mg/ml FA–TM–TN, 1 mM ATP, 3 mM $MgCl_2$, pH 6.6, 20 °C. Open circles, 0.1 mM $CaCl_2$; filled circles, 2 mM EGTA; crosses, ratio of extent of binding in 2 mM EGTA to 0.1 mM $CaCl_2$ (Inoue & Tonomura, 1982).

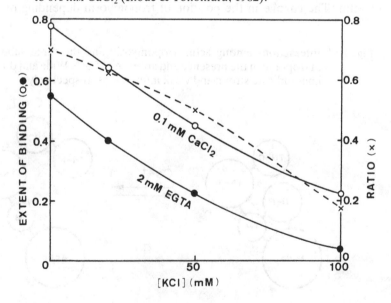

ments are required for clarification of the mechanism of $Ca^{2+}$ regulation of muscle contraction.

### Substrate inhibition of the actomyosin ATPase reaction

One of the charcteristics of the $Ca^{2+}$ regulation of the skeletal muscle actomyosin ATPase reaction by the tropomyosin–troponin (TM–TN) system is that it requires high concentrations of $Mg^{2+}$-ATP. Figure 3.17 shows the dependence of the acto–HMM $Mg^{2+}$-ATPase reaction on the concentration of ATP in the presence of TM–TN, when ATPase activity was measured using the ATP-regenerating system (Inoue & Tonomura, 1975). When $Ca^{2+}$ is removed, the ATPase activity at low ATP concentrations remains nearly intact, but the ATPase activity at high ATP concentrations is strongly inhibited. This phenomenon is called 'substrate inhibition' of the actomyosin ATPase reaction, and $Ca^{2+}$ actually controls the level of substrate inhibition of the actomyosin ATPase reaction.

Since one tropomyosin–troponin complex binds to seven actin

Fig. 3.16  Effect of $Ca^{2+}$ on the apparent first-order rate constant of recombination of S-1 with the FA–TM–TN complex. Conditions: 5 μM ATP, 4 μM S-1, 50 μM KCl, 2 mM MgCl₂, pH 7.8, 20 °C. Open circles, 0.1 mM CaCl₂; filled circles, 2 mM EGTA. Note that at low concentration of FA–TM–TN there is no $Ca^{2+}$-control. See text for detail (Inoue & Tonomura, 1982).

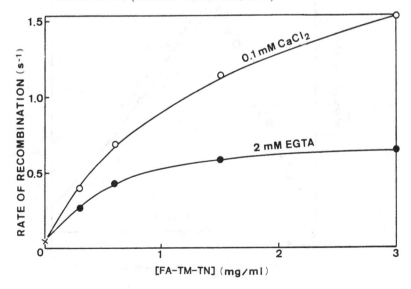

monomers and the interaction of all seven actin monomers with the myosin heads is controlled by the binding of $Ca^{2+}$ to one troponin molecule, Bremel, Murray & Weber (1973) and Weber & Murray (1973) assumed that the seven actin monomers that are bound to one tropomyosin molecule can be in only one of two states, etiher all are on or all are off. They explained the substrate inhibition as follows: In the presence of $Ca^{2+}$, the seven actin monomers that are bound to one tropomyosin are all in the on-state, and in the absence of $Ca^{2+}$, the seven actin monomers are all in the off-state. However, when ATP concentration is low and any of the seven actin monomers binds to the myosin head, all of the seven monomers turn to the on-state and exhibit high

Fig. 3.17  Dependence on ATP concentration of the rate of the acto–HMM ATPase reaction in the presence of the tropomyosin–troponin complex. Conditions: 3.4 mg/ml HMM, 3 mg/ml F-actin–TM–TN complex, 4 mg/ml pyruvate kinase, 1 mM PEP, 2 mM $MgCl_2$, 50 mM KCl, pH 7.8, 20 °C. Filled circles, 0.05 mM $CaCl_2$; open circles, 3 mM EGTA; crosses, ATPase activity of HMM alone (Inoue & Tonomura, 1975).

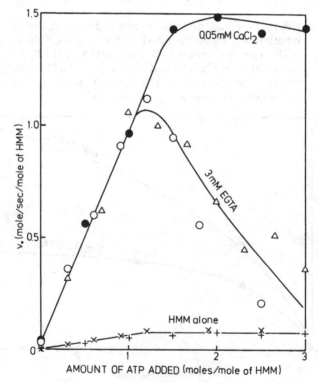

activity. When ATP concentration is high and the bound head no longer exists, all seven actin monomers are then in the off-state.

In fact, it has been shown that when ADP or AMPPNP are present and the concentration of S-1 is small, the amount of S-1 bound to the F-actin–tropomyosin–troponin complex (FA–TM–TN) is small and is controlled by the presence or absence of $Ca^{2+}$. However, when the concentration of free S-1 exceeds a certain level, the amount of S-1 bound to FA–TM–TN increases greatly and does not depend on the presence or absence of $Ca^{2+}$.

This substrate inhibition is closely related to the non-identical nature of the two myosin heads. As has been described in Chapter 2, skeletal muscle myosin has two non-identical heads. One head (burst head or head B) forms E–P–ADP as a stable intermediate and the other head (non-burst head or head A) forms E–ATP.

The amount of intermediate formed on the burst and non-burst heads was estimated from the measurements of amounts of ADP and ATP bound to myosin (HMM or S-1) during the ATPase reaction. When the ATP concentration is lower than 1 mol/mol of HMM, the ATPase activity of the HMM–FA–TM–TN system increases with an increase in ATP concentration and is independent of the presence or absence of $Ca^{2+}$ (see Fig. 3.17). Under this condition, only bound ADP is observed regardless of the presence or absence of $Ca^{2+}$. When ATP concentration is higher than 1 mol/mol of HMM, the ATPase activity in the presence of $Ca^{2+}$ remains at a high level, but the ATPase activity in the absence of $Ca^{2+}$ decreases with an increase in ATP concentration (see Fig. 3.17). Under this condition, up to 1 mol of ATP/mol of HMM binds in a manner proportional to the substrate inhibition of ATPase. Therefore, it was concluded that when the non-burst head binds to F-actin, a considerable number of actin monomers co-operatively assume the on-state and accelerate the decomposition of E–P–ADP on the burst head and when ATP concentration is increased and ATP also binds to the non-burst head, the non-burst head dissociates from F-actin and actin monomers enter into the off-state (see Fig. 3.18).

## Variety of $Ca^{2+}$-regulatory systems of muscle contraction

The contraction of every muscle cell so far studied is controlled by $Ca^{2+}$, and, as described in the next chapter, almost every type of cell motility seen in animals and plants is controlled by $Ca^{2+}$, although some of these are reversely inhibited by $Ca^{2+}$. In skeletal muscle, the contraction of muscle is controlled by tropomyosin and troponin which are linked to F-actin, as described above. However, the mechanism of $Ca^{2+}$ control varies widely among different types of muscle.

*Cardiac muscle*   Heart muscle contracts rhythmically and works as a pump. Contraction of cardiac muscle is also switched on or off by the binding of $Ca^{2+}$ to the tropomyosin–troponin system, as in skeletal muscle. However, contractility of cardiac muscle is regulated by hormones and other factors. In heart cells, $Ca^{2+}$ flows into the cell during the excitation of cell membrane and flows out mainly by a $Na^+$–$Ca^{2+}$-exchange reaction. Thus, the amount of $Ca^{2+}$ in sarcoplasmic reticulum (SR) vesicles varies by the frequency of contraction. Furthermore, the $Ca^{2+}$-pump activity itself is regulated by a cAMP-dependent protein kinase. Therefore, the amount of $Ca^{2+}$ released to the cell may be controlled by hormones. On the other hand, the affinity of troponin-C for $Ca^{2+}$ is reduced when troponin-I is phosphorylated by cAMP-dependent protein kinase.

*Molluscan muscle*   Kendrick-Jones, Lehman & Szent-Györgyi (1970) found that the acceleration of the $Mg^{2+}$-ATPase activity of scallop adduc-

Fig. 3.18  A schematic diagram showing the control of the state of seven actin monomers bound to one tropomyosin–troponin complex by $Ca^{2+}$ and ATP. The actomyosin ATPase reaction is catalyzed by the burst head (head B). Formation of an enzyme–ATP complex on the non-burst head (head A) of myosin induces the conversion of actin monomers from an on-state to an off-state.

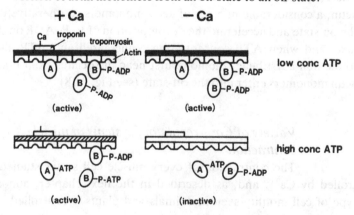

tor muscle myosin by purified F-actin is controlled by trace amounts of $Ca^{2+}$. This result suggests that $Ca^{2+}$ control is myosin linked. This $Ca^{2+}$ regulation does not require the tropomyosin–troponin system, and molluscan muscles do not contain troponin. Lehman & Szent-Györgyi (1975) showed that such myosin-linked regulation is seen widely in the animal kingdom.

In molluscan muscle, $Ca^{2+}$ binds directly to myosin, and the light chains play an important role in the binding of $Ca^{2+}$ to myosin (Szent-Györgyi, Szentkiralyi & Kendrick-Jones, 1973; Szent-Györgyi, 1975, 1980). Scallop adductor muscle myosin contains 4 mol of light chains (LC)/mol of myosin, each with a molecular weight of about 18 000. Of these, 2 mol are dissociated by EDTA at high temperature and thus are called EDTA-LC. When EDTA-LC is removed from myosin, the actomyosin ATPase shows high activity even in the absence of $Ca^{2+}$. When EDTA-LC is added to this myosin, the original $Ca^{2+}$ sensitivity is restored. It was shown that this $Ca^{2+}$ regulation is restored by the foreign EDTA-LC of molluscan muscle and also by LC of smooth muscle myosin. If its myosin has one mole of EDTA-LC, skeletal muscle $LC_2$ can also recover the $Ca^{2+}$ regulation.

EDTA-LC has one $Ca^{2+}$-binding site. However, this site can also bind $Mg^{2+}$ and the binding and dissociation of $Ca^{2+}$ occurs slowly. Thus, this site does not play an essential role in the $Ca^{2+}$ regulation, but the removal of $Ca^{2+}$ and $Mg^{2+}$ from this site by EDTA induces the dissociation of EDTA-LC from the myosin heavy chain. It was thought that the $Ca^{2+}$-binding site is located on the heavy chain of myosin and that the affinity for $Ca^{2+}$ of the myosin heavy chain is strengthened by the binding of EDTA-LC.

It was shown that EDTA-LC binds to the S-1–S-2 joint (Flicker, Walliman & Vibert, 1983) and that S-1 loses $Ca^{2+}$ sensitivity even when it contains EDTA-LC. However, it is still uncertain how binding of $Ca^{2+}$ to myosin controls the interaction between myosin and F-actin. It is quite interesting that the actomyosin ATPase reaction of scallop adductor muscle is also $Ca^{2+}$ insensitive at low concentrations of ATP and that the inhibition of ATPase by removal of $Ca^{2+}$ requires high concentrations of ATP, both characteristics of skeletal muscle, although the $Ca^{2+}$-regulatory mechanisms are quite different. Therefore, there may exist some interaction between the burst head, which catalyzes actomyosin ATPase, and the non-burst head.

*Smooth muscle* The structure of smooth muscle is quite different from that of skeletal muscle. Smooth muscle cells have single nuclei and do not have the striations seen in skeletal muscle (see book edited by Stephens, 1977). The rate of contraction of smooth muscle is low. Furthermore, there is a delay between the stimulation and development of tension (Fay, 1977). This is because the mechanism of $Ca^{2+}$ control of smooth muscle is different from that of skeletal muscle.

Two types of models have been proposed for the $Ca^{2+}$ control of smooth muscle. Mikawa, Nonomura & Ebashi (1977) and Nonomura (1980) proposed that contraction of smooth muscle is controlled by the binding of $Ca^{2+}$ to leiotonin, which binds with thin filaments. Sobieszek (1977), Bremel, Sobieszek & Small (1977) and several other groups (Aksoy *et al.*, 1976; Chacko, Conti & Adelstein, 1977; Ikebe, Onishi & Watanabe, 1977) proposed that the contraction and relaxation of smooth muscle are controlled by the phosphorylation and dephosphorylation of a myosin light chain (LC). In this case, $Ca^{2+}$ binds to a $Ca^{2+}$-binding protein, calmodulin, which gives $Ca^{2+}$ sensitivity to various protein systems, and the $Ca^{2+}$–calmodulin complex activates myosin light-chain kinase. Muscle contracts when the 20 000-dalton light-chain of myosin is phosphorylated by myosin light-chain kinase and relaxes when the phosphorylated light chain is dephosphorylated by myosin light-chain phosphatase (see Fig. 3.19).

An interesting observation has been made by Suzuki *et al.* (1982). They showed that in the presence of ATP, phosphorylation and dephosphorylation of the myosin light chain lead to the formation and disassem-

Fig. 3.19 Control of state of smooth muscle myosin by $Ca^{2+}$-dependent phosphorylation and $Ca^{2+}$-independent dephosphorylation of myosin light chain.

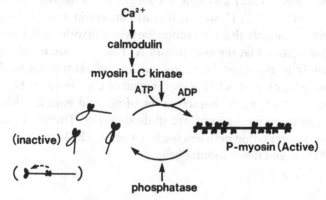

bly, respectively, of the thick filaments. The physicochemical properties of the myosin dissolved by the dephosphorylation of LC in the presence of ATP are different from those of myosin dissolved by high ionic strength. Onishi & Wakabayashi (1982) observed myosins dissolved either by dephosphorylation of LC in the presence of ATP or by high ionic strength and found that the structure of myosin dissolved at high ionic strength is the same as that of skeletal muscle myosin. However, when myosin was dissolved by dephosphorylation in the presence of ATP, the myosin molecule is folded by the binding of a certain portion of the tail with head–tail junction. However, it is still uncertain whether the thick filaments of the smooth muscle dissolve *in vivo* during relaxation. Thus, how the above morphological change in the myosin molecule induced by the phosphorylation and dephosphorylation of myosin light chain relates to the contraction and relaxation of smooth muscle still remains unknown. On the other hand, as will be described in Chapter 4, it is thought that in some nonmuscle cells the movement of cells is controlled by the formation and dissolution of thick filaments induced by the phosphorylation and dephosphorylation of myosin.

## Molecular mechanism of muscle contraction

We have so far discussed the structures and functions of two contractile proteins, myosin and F-actin, which constitute the thick and thin filaments, respectively, and tropomyosin and troponin, the regulatory proteins of skeletal muscle. In the myofibrils, the two filaments are arranged in a specific order, and the reaction between myosin and actin brings about muscle contraction. In this section we will first discuss the physiological properties of muscle contraction and then describe how the actomyosin ATPase reaction is coupled with muscle contraction.

### *Physiological properties of muscle contraction*

Muscle contracts with developing tension, and the energy derived from ATP hydrolysis is transduced into mechanical work which is given as tension ($P$) × shortening distance ($X$). Energy transduction in muscle is very efficient and is reported to be about half that of the total energy change.

To understand the function of the crossbridges during muscle contraction, it is important to know the physiological properties of muscle contraction. The mechanical properties of muscle contraction, especially the relationships among the shortening rate, tension development and heat production, have been studied in detail by Hill (1938, 1970) and the basic properties of muscle contraction have been clarified:

(1) The following relationship exists between the shortening rate ($v$) and the tension developed ($P$):

$$(P + a)\,(v + b) = b(P_0 + a) = \text{constant}$$

where a and b are constants and $P_0$ is the maximum tension (isometric tension), which is developed at the isometric contraction. Increase in shortening rate ($v$) and decrease in tension ($P$) occur in parallel and shortening rate becomes maximal at $P = 0$.

(2) Heat production ($Q$) is dependent on the shortening distance ($x$) and can be given by

$$Q = A + \alpha x,$$

where $A$ denotes the amount of heat produced in the isometric state and $\alpha$ is a constant. This equation, however, is modified as described later.

(3) The rate of energy liberation ($E$) is proportional to the reduction in the load ($P$) and can be given by

$$E = \beta(P_0 - P) + A$$

where A denotes the rate of heat production in the isometric state, and $\beta$ is a constant.

Since ATP hydrolysis is the source of energy for muscle contraction the equation in (3) indicates that ATP hydrolysis increases as the tension decreases and the shortening rate increases. It approaches a maximum level when the shortening rate approaches the maximum level and tension becomes zero. This phenomenon is called the Fenn effect. However, these relationships were later modified: it was shown that $\alpha$ is not constant but is given by $\alpha = 0.16\,P_0 + 0.18\,P$. Therefore, energy liberation, $E$, decreases as the load approaches zero.

It has actually been shown that the ATPase activity of myofibrils during shortening is higher than that of the myofibrils after shortening or that of skinned muscle in the isometric state (Arata, Mukohata & Tonomura, 1977).

In living muscle, the ADP produced is rapidly regenerated into ATP by the creatine kinase system, and the amount of creatine liberated is shown to be proportional to the sum of mechanical work done and heat liberated (Carlson, Haedy & Wilky, 1963). By blocking creatine kinase with 1-fluoro-2,4-dinitrobenzene (FDNB), Cain, Infante & Davies (1962) showed that the amount of ATP that is actually decomposed is proportional to the mechanical work and heat that is produced.

The mechanical properties of muscle contraction were later studied by analyzing the transient changes in the tension or length of muscle fiber in the contraction state, and more direct information on the mechanical properties of crossbridges has been obtained. Podolsky *et al.* (1969) studied the change in the length of muscle fiber when the tension is suddenly altered, and Huxley & Simmons (1971) later studied the transient changes in the tension when the length of muscle is suddenly changed. Huxley & Simmons (1971, 1973) showed that when the length of muscle is suddenly changed, tension returns to the original level through the following four steps: (1) The level of tension decreases rapidly from the original level, $P_0$, to $P_1$. Since this change is proportional to the change in muscle length (L), and the levels of both $P_0$ and $P_1$ are proportional to the overlap between thick and thin filaments, this step is believed to be caused by the elastic element that exists in the crossbridges. When tension is suddenly altered, the length of the fiber is rapidly changed by about 1%. (2) The tension increases rapidly from $P_1$ to $P_2$. The level of $P_2$ is also proportional to the overlap between the two filaments. Since $P_2$ decreases to zero when $\Delta L$ is about 15 nm/half the length of the sarcomere it is considered that crossbridges move about 15 nm. $P_2$ is not related linearly to $\Delta L$. Therefore, it was suggested that there exist more than two states of crossbridges during contraction. (3) The recovery of tension stops for 5–20 ms, and (4) the tension returns slowly to the steady-state level $P_0$.

A rapid change in length also causes changes in the rate of ATP hydrolysis. Arata, Mukohata & Tonomura (1977) showed that when glycerinated muscle is stretched only slightly (2–3%) for 0.3 s, then suddenly released to return to the original length, and then allowed to stand for a certain period ($\alpha$ s), the ATPase activity increases with increase in $\alpha$. The most pronounced activation of ATPase occurs when $\alpha$ is approximately the same as the turnover rate of ATPase (see Fig. 3.20). This result also suggests that the state of crossbridges is altered when the length of muscle is rapidly changed.

## Molecular model of muscle contraction

Since the structures of the two filaments of muscle have been clarified to some extent, several molecular mechanisms of muscle contraction have been proposed based on the physiological properties of muscle contraction.

It was shown by A. F. Huxley in 1957 that the mechanical properties of muscle contraction studied by Hill can be explained by a simple model of muscle contraction. In his model, the sites (M) on the thick filament can bind to the sites (A) on the thin filament. M can move about a fixed distance on the thick filament by thermal motion. The rate of binding of M to A, $f$, the rate of dissociation of MA, $g$, and the tension developed by the M–A complex, $P$, are all given as a function of the distance ($x$) between the location of M and the fixed point, O, on the thick filament. Thus, when muscle contracts at rate, $v$, the fraction of MA at each point $x$ and total tension can be calculated from $f(x)$, $g(x)$ and $P(x)$. Furthermore, the amount of energy liberated ($E$) is also calculated by assuming that 1 mol of ATP is hydrolyzed during a cycle of

Fig. 3.20 ATPase activity of oscillated muscle fibers as a function of duration of the isometric phase following sudden release. The ATPase activity was measured by coupling it with a creatine kinase system. Conditions: 50 mM KCl, pH 7.8, 0 °C. Open and closed circles represent two different series of experiments with different fiber preparations (Arata, Mukohata & Tonomura, 1979).

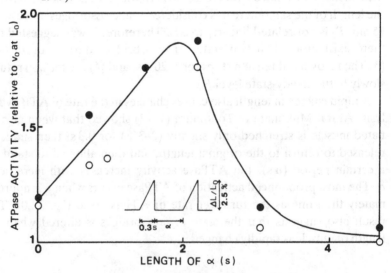

binding and dissociation of myosin. He showed that the mechanical properties of muscle contraction reported by A. V. Hill (1938) can be explained by assigning proper functions for $f(x)$, $g(x)$ and $P(x)$. This model became the basis of many molecular models of muscle contraction proposed later.

However, it is difficult to explain several results of physiological work using this model. One problem is that the rate of total energy liberation ($E$) decreases with an increase in the shortening rate ($v$) when the shortening rate approaches its maximal. This model also cannot explain the results of the transient change in length or tension after causing rapid change in tension or length. To explain these results, Huxley & Simmons (1971) and Huxley (1974) proposed a new model in which there are two or more states of the complex of myosin head with F-actin and the levels of tension differ in these states. This new model argues that the driving force for tension development is the transition between the two states and the rapid tension recovery seen in the mechanical transient state (state 2) is due to this transition. This model also assumes that ATP is not hydrolyzed in dissociation and rebinding in one of the states. Therefore, when $v$ approaches its maximum, ATP hydrolysis ($E$) decreases with increase in $v$. An improvement on this model was later attempted by T. L. Hill (1974, 1975) and Nishiyama *et al.* (1977). They showed that the mechanical properties of muscle contraction can be explained by this improved model.

Thus we have a framework for the molecular mechanism of muscle contraction and an explanation of the mechanical properties of muscle contraction as a simple mechanism. However, large variations still remain, and it is not easy to measure the rate constants of dissociation and recombination of crossbridges and tension development as functions of relative locations of myosin head to F-actin.

## Movement of crossbridges during muscle contraction

One of the central problems in this field is to understand what type of crossbridge movement occurs during muscle contraction, and this has been studied by various methods including X-ray diffraction analysis, measurement of the polarization or fluctuation of fluorescent dye bound to myosin head, or ESR spectrum of spin label bound to myosin head (see Morales *et al.*, 1982; Highsmith & Cooke, 1983; Huxley & Farugui, 1983). However, no concordant conclusion has yet been drawn.

Therefore, in the following discussion on the coupling mechanism of ATPase reaction with muscle contraction we will make an assumption about the movement of crossbridges during contraction. It was proposed by H. E. Huxley (1969) that muscle contraction is induced by the change in the tilting angle between myosin head and thin filament ($\alpha$ and $\beta$), and muscle develops tension through transition from $\alpha$ to $\beta$. We adopt this model to explain the coupling mechanism of ATPase with muscle contraction. It must be noted, however, this model is not the only model applicable to the following discussion. Actually, various models have been proposed in which tension development is induced by a conformational change of myosin head bound to actin, by a rotational movement perpendicular to the fiber axis of the head, or by a structural change in myosin head.

Mechanical studies on muscle contraction showed that the elasticity of muscle resides in the crossbridges. It has been shown from structural studies of the myosin molecule that the S-2–LMM junction is flexible over a length of about 30 nm. This part is susceptible to proteolysis, and its $\alpha$-helix content is low. Furthermore, the susceptibility of the S-2–LMM junction to proteolysis increases when glycerinated muscle develops tension, as described in Chapter 2. Therefore, we assume that the elastic component of crossbridges resides in the S-2–LMM junction.

## Coupling of the actomyosin ATPase reaction with muscle contraction

Based on the model of movement of crossbridges, various molecular mechanisms of muscle contraction have been proposed differing in their mechanism of actomyosin ATPase reaction (Lymn & Taylor, 1971; Tonomura, 1972; Inoue *et al.*, 1979; Eisenberg & Greene, 1980).

The simplest mechanism among them is that of Lymn & Taylor (1971). They proposed a mechanism based on their reaction mechanism of actomyosin ATPase. It includes the following four steps: (1) dissociation of myosin head from F-actin by formation of E–ATP, (2) transition of E–ATP into E–P–ADP, (3) recombination of E–P–ADP with F-actin, and (4) change in the angle of the crossbridge accompanying the decomposition of AE–P–ADP into AE + ADP + Pi, and then the cycle is repeated.

However, as described earlier in this chapter, it was shown that ATP is hydrolyzed via two routes in the actomyosin ATPase reaction:

$$AE + ATP \rightleftharpoons AE-ATP \rightleftharpoons A + E-ATP$$

$$ADP + Pi \quad\quad AE-P-ADP \quad A + E-P-ADP$$

The ratio of ATPases through the inner and outer cycle occurring in the *in vitro* system varies depending on the reaction conditions. At low ionic strength and at room temperature, when the tension development is large but the shortening rate is low, the acto–HMM ATPase reaction occurs mainly via the inner cycle. At relatively high ionic strength, when the tension development is small but the shortening rate is fast, on the other hand, the outer cycle becomes the main route.

We (Tonomura, 1972; Inoue *et al.*, 1979) proposed a mechanism of muscle contraction based on the above mechanism of actomyosin ATPase by assuming that the ATP hydrolysis through the outer cycle corresponds to the cycle of dissociation–reassociation of crossbridges, while the inner cycle is involved in tension development. When muscle shortens rapidly and performs mechanical work, both routes must occur at the proper ratio, and the most effective muscle contraction is achieved when the two routes cycle alternately. Therefore, in this model, muscle contraction occurs by the following steps: (1) the structure of crossbridges in $\alpha$-form is transformed into $\beta$-form by formation of AE–P–ADP by the inner cycle of ATPase reaction, and tension is developed. (2) AE–P–ADP decomposes when the tension of crossbridges is decreased by the shortening of the fiber. (3) ATP induces the dissociation of crossbridges in $\beta$-form by formation of E–ATP, which is then transformed into E–P–ADP and (4) E–P–ADP recombines with F-actin in the $\alpha$-form and the cycle is repeated. Thus, the structure of the crossbridges, $\alpha$ or $\beta$, determines the route of the ATPase reaction, and 2 mol of ATP are required for one cycle of crossbridge action.

However, this model cannot explain the following result: the rate-limiting step of the outer cycle of ATPase reaction is the transition of E–P–ADP in a refractory state which cannot bind with F-actin, to a non-refractory state, which can bind with F-actin. The rate of this step is found to be 1–2/s, which is much lower than the rate of ATPase in glycerol-treated muscle fiber during contraction. Therefore, muscle contraction may be coupled with the inner cycle of the ATPase reaction. This agrees well with the finding that higher tension is developed under conditions in which there is a high ratio of inner-cycle ATP hydrolysis to outer-cycle ATP hydrolysis.

If we take the model that muscle contraction is coupled with the inner cycle of the ATPase reaction, we need to make clear two problems. One problem is at the steps in which dissociation of myosin heads from F-actin and their recombination take place. As described earlier in this chapter, AE–ATP is in rapid equilibrium with A + E–ATP and the rate of forward reaction is in the order of 1000/s, which is much faster than the rate of transition of E–ATP to E–P–ADP, 100–200/s. Therefore, we assume that dissociation of crossbridges takes place by formation of E–ATP and that E–ATP rebinds to F-actin very rapidly.

Another important problem is at the step in the ATPase reaction in which the change in the tilting angle of crossbridges occurs. Arata & Shimizu (1981) studied the formation of the reaction intermediate associated with tension development using change in the mobility of spin-label bound to myosin in glycerinated muscle fibers. On the other hand, Goldman *et al.* (1982) challenged the rapid-mixing experiment inside the muscle by using caged ATP, 3P-1-(2-nitro)phenylethyl ATP, from which ATP is rapidly released by photolysis. However, clear results have not yet been obtained from these studies.

On the other hand, there has been an interesting observation on the acto–S-1–nucleotide system. Arata (1984) studied the structure of AE, AE–AMPPNP, AE–P–ADP, and AE–ADP, using chemical crosslinking. As shown in Fig. 3.21, when AE or AE–ADP was crosslinked with 1-ethyl-3,3-dimethylaminopropyl-carbodiimide (EDC), acto–S-1 showed high ATPase activity. However, when acto–S-1 in AE–AMPPNP or AE–P–ADP was crosslinked, the product acto–S-1 showed lower ATPase activity. This result suggested that AE and AE–ADP have different structures from AE–AMPPNP or AE–P–ADP.

Another approach has been to study the change in free energy, $\Delta G$, in the elementary steps of the actomyosin ATPase reaction. Recently, M. Yasui, T. Arata & A. Inoue (personal communication) measured the binding of S-1 with F-actin in the presence of ATP, AMPPNP, and ADP. Then, using the value of $\Delta G$ in the elementary steps of the myosin ATPase reaction (Chapter 2), they calculated the $\Delta G$ in the elementary steps of the actomyosin ATPase reaction. As shown in Fig. 3.22, it was found that more than 80% of effective free energy is liberated at the step of conversion of AE–P–ADP to AE–ADP or AE. Therefore, we assume here that the change in the tilting angle of crossbridges occurs at the step AE–P–ADP→ AE–ADP or AE.

A simple molecular model built on this assumption is shown in Fig. 3.23. (1) When AE–P–ADP decomposes to AE–ADP or AE (AE-none), the state of myosin is transduced from the $\alpha$- to the $\beta$-state. At the same time, the crossbridge rotates, the elastic element is stretched, and tension is developed. Sliding of the two filaments past each other takes place and the elastic element returns to its original position. (2) When ATP reacts with the crosssbridge in the $\beta$-state, the crossbridge is dissociated by the formation of AE–ATP, and is recombined with F-actin in the $\alpha$-state, and then the cycle is repeated.

When muscle is activated at a fixed length (isometric state), a large degree of tension is developed, but the ATPase activity is very low. The state of crossbridges of this muscle is considered to be fixed in some state. Therefore, the step of the ATPase reaction that is coupled with movement of crossbridges can be determined by studying the state

Fig. 3.21 Dependence of the ATPase activities of covalent acto–S-1 complexes formed in the presence of nucleotide on the concentration of KCl in the crosslinking medium. Crosslinking was performed at 0 °C and pH 7.0 for 60–90 nm by 2 mM 1-ethyl-3,3-dimethyl-aminopropyl-carbodiimide (EDC). ATPase activity of the covalent complex was measured in 50 mM KCl, 5 mM Mg at 20 °C. Crosses, no nucleotide; open circles, +ADP; triangles, +AMPPNP; filled circles, +ATP (Arata, 1984).

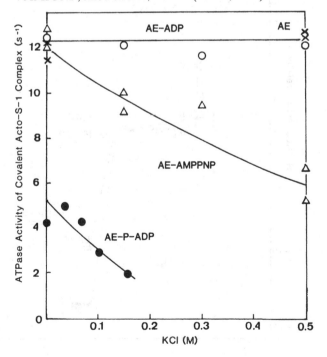

of the crossbridges in isometric contraction. According to the above mechanism, crossbridges of muscle during isometric contraction exist at AE, which is the same state as crossbridges in the rigor state. However, there remains the possibility that crossbridges are fixed before the transformation of crossbridge angles. Therefore, according to this model, crossbridges during isometric contraction exist at AE–ATP and AE–P–ADP, which are formed by a reaction of actin with E–ATP or E–P–ADP.

In skeletal muscle, one of the two myosin heads (the non-burst head) is not involved in energy transduction. Instead, it forms E–ATP and participates in the regulation of contraction by $Ca^{2+}$ and ATP, as previously discussed in this chapter. In addition, the non-burst head is in the following equilibria with actin and ATP:

Fig. 3.22 Change in the effective free energy, $\Delta G$, in the elementary steps of the actomyosin ATPase reaction. The value of $\Delta G'$ is given for a downward reaction. 5 mM ATP, 0.1 mM ADP, and 1 mM Pi were used for calculation (M. Yasui, T. Arata & A. Inoue, personal communication).

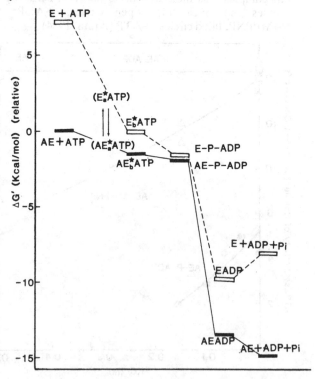

$$AE + ATP \rightleftharpoons AE\text{–}ATP \rightleftharpoons A + E\text{–}ATP$$

Acto–non-burst head dissociates after forming the ternary complex AE–ATP by reacting with ATP. However, in the case of actomyosin both heads dissociate when only the burst head forms the complex E–P–ADP. Thus, the existence of some kind of interaction between the two heads is suggested.

Therefore, we assume here that when the burst head is in the $\alpha$-state there is a steric hindrance between the two heads and the non-burst head dissociates from actin. When the burst head is in the $\beta$-state, the steric hindrance is relieved and therefore the non-burst head can bind to actin. Figure 3.24 shows a molecular model built on this assumption that contraction is induced by the two non-identical heads. When the burst head B enters the $\beta$-state by forming AE–ADP or AE, the non-burst head A becomes capable of binding to F-actin and plays the role of preventing the slipping of crossbridges. When the burst head in the $\beta$-state reacts with ATP and dissociates from the thin filament, the non-burst head also dissociates from the thin filament and thus escapes from again becoming a resistance to movement between the two filaments.

Thus, studies on the structure of myosin have elucidated the action of crossbridges during muscle contraction, and studies on the reaction mechanism of actomyosin ATPase have given us information about the coupling mechanism of ATPase with muscle contraction. It is hoped

Fig. 3.23 A simple mechanism of muscle contraction induced by the burst head (head B). See text for details.

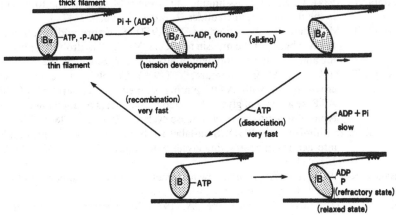

Fig. 3.24 Molecular mechanism of muscle contraction induced by the two non-identical heads (non-burst head A and burst head B). See text for details.

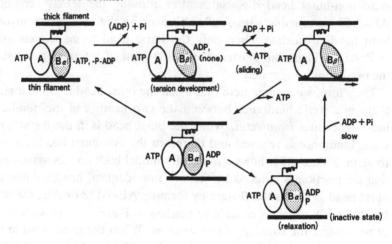

that the molecular mechanism of muscle contraction will soon be clarified by further progress in this field.

# References

Aksoy, M. O., Williams, D., Sharkey, E. M. & Hartshorne, D. J. (1976). A relationship between $Ca^{2+}$ sensitivity and phosphorylation of gizzard actomyosin. *Biochem. Biophys. Res. Commun.*, **69**, 35–41.

Arata, T. (1984). Chemical cross-linking of myosin subfragment-one to F-actin in the presence of nucleotides. *J. Biochem.*, **96**, 337–47.

Arata, T., Mukohata, Y. & Tonomura, Y. (1977). Structure and function of the two heads of the myosin molecule. VI. ATP hydrolysis, shortening and tension development of myofibrils. *J. Biochem.*, **82**, 801–12.

Arata, T., Mukohata, Y. & Tonomura, Y. (1979). Coupling of movement of cross-bridges with ATP splitting studied by acceleration of the ATPase activity of glycerol-treated muscle fibers by applying various types of repetitive stretch-release cycles. *J. Biochem.*, **86**, 525–42.

Arata, T. & Shimizu, H. (1981). Spin-label study of actin-myosin-nucleotide interactions in contracting glycerinated muscle fibers. *J. Mol. Biol.*, **151**, 411–37.

Bailey, K. (1948). Tropomyosin: a new asymmetrical protein component of the muscle fibril. *Biochem. J.*, **43**, 271–9.

Bremel, R. D., Murray, J. M. & Weber, A. (1973). Manifestation of cooperative behaviour in the regulated actin filament during actin-activated ATP hydrolysis in the presence of calcium. *Cold Spring Harbor Symp. Quant. Biol.*, **37**, 267–75.

Bremel, R. D., Sobieszek, A. & Small, J. V. (1977). Regulation of actin-myosin interaction in vertebrate smooth muscle. In *The Biochemistry of Smooth Muscle*, ed. N. L. Stephens, pp. 533–48. Baltimore: University Park Press.

Cain, D. F., Infante, A. A. & Davies, R. E. (1962). Adenosine triphosphate and phosphoryl creatine as energy supplier for single contractions of working muscle. *Nature*, **196**, 214–17.

Carlson, F. D., Naedy, D. J. & Wilkie, D. R. (1963). Total energy production and phosphocreatine hydrolysis in the isotonic twitch. *J. Gen. Physiol.*, **46**, 851–82.

Casteels, R., Godfraind, T. & Ruegg, J. E. (1977). *Excitation-contraction Coupling in Smooth Muscle*, New York & Amsterdam: Elsevier North-Holland.

Chacko, S., Conti, M. & Adelstein, R. S. (1977). Effect of phosphorylation of smooth muscle myosin on actin activation and $Ca^{2+}$ regulation. *Proc. Natl. Acad. Sci. USA*, **74**, 129–33.

Chock, S. P., Chock, P. B. & Eisenberg, E. (1976). Pre-steady state kinetic evidence for a cyclic interaction of myosin subfragment one with actin during the hydrolysis of adenosine 5'-triphosphate. *Biochemistry*, **15**, 3244–53.

Collins, J. H., Greaser, M. L., Potter, J. M. & Horn, M. J. (1977). Determination of the amino acid sequence of troponin C from rabbit skeletal muscle. *J. Biol. Chem.*, **252**, 6356–62.

Craig, R. & Knight, P. (1983). Myosin molecules, thick filaments and the actin-myosin complex. In *Electron Microscopy of Proteins. Macromolecular Structure and Function*, vol. 4, ed. J. R. Harris, pp. 97–203. London: Academic Press.

Crass, M. F. & Barnes, C. D. (1982). *Vascular Smooth Muscle: Metabolic, Ionic, and Contractile Mechanisms*. New York & London: Academic Press.

Ebashi, S. (1974). Regulatory mechanism of muscle with special reference to the $Ca^{2+}$-troponin-tropomyosin system. In *Essays in Biochemistry*, vol. 10, ed. P. N. Champbell & F. Dickens, pp. 1–36. London: Academic Press.

Ebashi, S. (1980). The Croonian Lecture, 1979. Regulation of muscle contraction. *Proc. R. Soc. Lond. B.*, **207**, 259–86.

Ebashi, S. & Ebashi, F. (1964). A new protein component participating in the superprecipitation of myosin B. *J. Biochem.*, **55**, 604–13.

Ebashi, S. & Endo, M. (1968). Calcium ion and muscle contraction. *Prog. Biophys. Mol. Biol.*, **18**, 123–83.

Ebashi, S., Endo, M. & Ohtsuki, I. (1969). Control of muscle contraction. *Q. Rev. Biophys.*, **2**, 351–84.

Ebashi, S. & Kodama, A. (1965). A new protein factor promoting aggregation of tropomyosin. *J. Biochem.*, **58**, 107–8.

Ebashi, S., Maruyama, K. & Endo, M. (1980). *Muscle Contraction: Its Regulatory Mechanisms*. Tokyo & Berlin: Japan Scientific Society Press & Springer-Verlag.

Eisenberg, E. & Greene, L. E. (1980). The relation of muscle biochemistry to muscle physiology. *Ann. Rev. Physiol.*, **42**, 293–309.

Elzinga, M. & Collins, J. H. (1973). The amino acid sequence of rabbit skeletal muscle actin. *Cold Spring Harbor Symp. Quant. Biol.*, **37**, 1–7.

Fay, F. S. (1977). Isometric contractile properties of single isolated smooth muscle cells. *Nature*, **265**, 553–6.

Flicker, P. F., Walliman, T. & Vibert, P. (1983). Electron microscopy of scallop myosin location of regulatory light chains. *J. Mol. Biol.*, **169**, 723–41.

Goldman, Y. E., Hibberd, M. G., MacCray, J. A. & Trentham, D. R. (1982). Relaxation of muscle fibres by photolysis of caged ATP. *Nature*, **300**, 701–5.

Greaser, M. L. & Gergely, J. (1971). Reconstitution of troponin activity from three protein components. *J. Biol. Chem.*, **246**, 4226–33.

Hartshorne, D. J. & Mueller, A. (1969). The preparation of tropomyosin and troponin from natural actomyosin. *Biochim. Biophys. Acta*, **175**, 301–19.

Highsmith, S. & Cooke, R. (1983). Evidence for actomyosin conformational changes involved in tension generation. In *Cell and Muscle Motility*, vol. 4, ed. R. Dowben & J. Shay, pp. 207–37. New York: Plenum.

Hill, A. V. (1938). The heat of shortening and the dynamic constants of muscle. *Proc. R. Soc. B*, **6**, 136–95.

Hill, A. V. (1970). *First and Last Experiments in Muscle Mechanics*. London: Cambridge University Press.

Hill, T. L. (1974). Theoretical formation for the sliding filament model of contraction of striated muscle, Part 1. *Prog. Biophys. Mol. Biol.*, **28**, 267–340.

Hill, T. L. (1975). Theoretical formation for the sliding filament model of contraction of striated muscle, Part 2. *Prog. Biophys. Mol. Biol.*, **29**, 105–59.

Huxley, A. F. (1957). Muscle structure and theories of contraction. *Prog. Biophys. Chem.*, **7**, 255–318.

Huxley, A. F. (1974). Muscular contraction. *J. Physiol.*, **243**, 1–43.

Huxley, A. F. & Simmons, R. M. (1971). Proposed mechanism of force generation in striated muscle. *Nature*, **233**, 533–8.

Huxley, A. F. & Simmons, R. M. (1973). Mechanical transients and the origin of muscular force. *Cold Spring Harbor Symp. Quant. Biol.*, **37**, 669–80.

Huxley, H. E. (1969). The mechanism of muscle contraction. *Science*, **164**, 1356–66.

Huxley, H. E. & Faruqui, A. R. (1983). Time-resolved X-ray diffraction studies on vertebrate striated muscle. *Ann. Rev. Biophys. Bioenerg.*, **12**, 381–412.

Ikebe, M., Onishi, H. & Watanabe, S. (1977). Phosphorylation and dephosphorylation of a light chain of the chicken gizzard myosin molecule. *J. Biochem.*, **82**, 299–302.

Imamura, K., Kanazawa, T., Tada, M. & Tonomura, Y. (1965). The pre-steady state of the myosin-adenosine triphosphate system. III. Properties of the intermediate. *J. Biochem.*, **57**, 627–36.

Inoue, A., Ikebe, M. & Tonomura, Y. (1980). Mechanism of the $Mg^{2+}$- and $Mn^{2+}$-ATPase reactions of acto-H-meromyosin and acto-subfragment-1 in the absence of KCl at room temperature: direct decomposition of the complex of myosin-P-ADP with F-actin. *J. Biochem.*, **88**, 1663–77.

Inoue, A., Shigekawa, M. & Tonomura, Y. (1973). Direct evidence for the two route mechanism of the acto-H-meromyosin-ATPase reaction. *J. Biochem.*, **74**, 923–34.

Inoue, A., Takeneka, H., Arata, T. & Tonomura, Y. (1979). Functional implication of the two beaded structure of the myosin molecule. *Adv. Biophys.*, **13**, 1–194.

Inoue, A. & Tonomura, Y. (1975). The amounts of adenosine di- and triphosphates bound to H-meromyosin-F-actin-relaxing protein system in the presence and absence of calcium ions. The physiological functions of the two routes of myosin adenosine triphosphatase in muscle contraction. *J. Biochem.*, **78**, 83–92.

Inoue, A. & Tonomura, Y. (1980). Dissociation of acto-H-meromyosin and that of acto-subfragment-1 induced by adenyl-5′-yl-imidodiphosphate: evidence for a ternary complex of F-actin, myosin head, and substrate. *J. Biochem.*, **88**, 1653–62.

Inoue, A. & Tonomura, Y. (1982). Regulation of binding of myosin subfragments with regulated actin by calcium ions in the presence of magnesium ATP. *J. Biochem.*, **91**, 1231–9.

Johnson, I. D., Charlton, S. C. & Potter, J. D. (1979). A fluorescence stopped flow analysis of $Ca^{2+}$ exchange with troponin C. *J. Biol. Chem.*, **254**, 3497–502.

Kendrick-Jones, J., Lehman, W. & Szent-Györgyi, A. G. (1970). Regulation in molluscan muscles. *J. Mol. Biol.*, **54**, 313–26.

Kretsinger, R. H. & Barry, C. D. (1975). The predicted structure of the calcium-binding component of troponin. *Biochem. Biophys. Acta*, **405**, 40–52.

Leaves, P. C., Rosenfeld, S. S., Gergeley, J., Grabarek, Z. & Drabikowski, W. (1978). Proteolytic fragments of troponin C. Localization of high and low affinity of $Ca^{2+}$ binding sites and interactions with troponin I and troponin T. *J. Biol. Chem.*, **253**, 5452–9.

Lehman, W. & Szent-Györgyi, A. G. (1975). Regulation of muscular contraction: distribution of actin control and myosin control in the animal kingdom. *J. Gen. Physiol.*, **66**, 1–30.

Lymn, R. W. & Taylor, E. W. (1971). Mechanism of adenosine triphosphate hydrolysis by actomyosin. *Biochemistry*, **10**, 4617–24.

Mak, A., Smillie, L. B. & Stewart, G. R. (1980). A comparison of the amino acid sequences of rabbit skeletal muscle α- and β-tropomyosins. *J. Biol. Chem.*, **255**, 3647–55.

Midelfort, C. F. (1981). On the mechanism of actomyosin ATPase from fast muscle. *Proc. Natl. Acad. Sci. USA*, **78**, 2067–71.

Mikawa, T., Nonomura, Y. & Ebashi, S. (1977). Does phosphorylation of myosin light chain have a direct relation to regulation in smooth muscle? *J. Biochem.*, **82**, 1789–91.

Moore, P. B., Huxley, H. E. & De Rosier, D. J. (1970). Three-dimensional reconstruction of F-actin, thin filaments and decorated thin filaments. *J. Mol. Biol.*, **50**, 279–95.

Morales, M. F., Borejdo, J., Botts, J., Cooke, R., Mendelson, R. & Takashi, R. (1982). Some physical studies of the contractile mechanism in muscle. *Ann. Rev. Phys. Chem.*, **33**, 319–51.

Morita, F. & Tonomura, Y. (1960). Kinetic analysis of the pyrophosphate-myosin B system by the use of the light-scattering method. *J. Am. Chem. Soc.*, **82**, 5172–7.

Mornet, D., Bertrand, R., Panel, P., Audemard, E. & Kassab, R. (1981). Structure of the actin-myosin interface. *Nature*, **292**, 301–6.

Nishiyama, K., Shimizu, H., Kometani, K. & Chaen, S. (1977). The three state model for the elementary process of energy conversion in muscle. *Biochim. Biophys. Acta*, **460**, 523–36.

Nonomura, Y. (1980). Leiotonin, regulatory protein of smooth muscle. In *Muscle Contraction: Its Regulatory Mechanisms*, ed. S. Ebashi, K. Maruyama & M. Endo, pp. 329–45. Tokyo & Berlin: Japan Scientific Society Press & Springer-Verlag.

Onishi, H. & Wakabayashi, T. (1982). Electron microscopic studies of myosin molecules from chicken gizzard muscle. I. The formation of the intramolecular loop in the myosin tail. *J. Biochem.*, **92**, 871–9.

Oosawa, F. & Asakura, S. (1975). *Thermodynamics of the Polymerization of Protein*. New York & London: Academic Press.

Pearlstone, J. R., Johnson, P., Carpenter, M. R. & Smillie, L. B. (1977). Primary structure of rabbit skeletal muscle troponin-T: sequence determination of the $NH_2$-terminal fragment CB 3 and the complete sequence of troponin-T. *J. Biol. Chem.*, **252**, 983–9.

Podolsky, R. J., Nolan, A. C. & Zaveler, S. A. (1969). Cross-bridge properties derived from muscle isotonic velocity transient. *Proc. Natl. Acad. Sci. USA*, **64**, 504–11.

Potter, J. D. & Gergely, J. (1974). Troponin, tropomyosin and actin interactions in the $Ca^{2+}$ regulation of muscle. *Biochemistry*, **13**, 2697–703.

Shigekawa, M. & Tonomura, Y. (1973). Interrelation among the three components troponin and tropomyosin studied by their kinetic effects on the ATPase reaction of actomyosin. *J. Biochem.*, **73**, 1135–48.

Sobieszek, A. (1977). Ca-linked phosphorylation of a light chain of vertebrate smooth-muscle myosin. *Eur. J. Biochem.*, **73**, 477–83.

Stein, L. A., Schwartz, R. P., Chock, P. B. & Eisenberg, E. (1979). Mechanism of actomyosin adenosine triphosphatase. Evidence that adenosine

5'-triphosphate hydrolysis can occur without dissociation of the actomyosin complex. *Biochemistvry*, **18**, 3895–909.

Stephens, N. L. (1977). *The Biochemistry of Smooth Muscle*. Baltimore: University Park Press.

Stone, D. & Smillie, L. B. (1978). The amino acid sequence of rabbit skeletal α-tropomyosin. The $NH_2$-terminal half and complete sequence. *J. Biol. Chem.*, **253**, 1137–48.

Straub, F. B. (1942). Actin. *Stud. Inst. Med. Chem. Univ. Szeged.*, **2**, 3–15.

Suck, D., Kabsch, W. & Mannherz, H. G. (1981). Three-dimensional structure of the complex of skeletal muscle actin and bovine pancreatic DNase I at 6 Å resolution. *Proc. Natl. Acad. Sci. (USA)*, **78**, 4319–23.

Sutoh, K. (1983). Mapping of actin-binding sites on the heavy chain of myosin subfragment 1. *Biochemistry*, **22**, 1579–85.

Suzuki, H., Kamata, T., Onishi, H. & Watanabe, S. (1982). Adenosine triphosphate-induced reversible change in the conformation of chicken gizzard myosin and heavy meromyosin. *J. Biochem.*, **91**, 1699–705.

Szent-Györgyi, A. G. (1975). Calcium regulation of muscle contraction. *Biophys. J.*, **15**, 707–23.

Szent-Györgyi, A. G. (1980). Role of regulatory light chains in myosin-linked regulation. In *Muscle Contraction: Its Regulatory Mechanisms*, ed. S. Ebashi, K. Maruyama & M. Endo, pp. 375–89. Tokyo & Berlin: Japan Scientific Society Press & Springer-Verlag.

Szent-Györgyi, A. G., Szentkiralyi, E. M. & Kendrick-Jones, J. (1973). The light chains of scallop myosin as regulatory subunits. *J. Mol. Biol.*, **74**, 179–203.

Takashi, R. (1980). Fluorescence energy transfer between subfragment-1. *Biochemistry*, **18**, 5164–9.

Tonomura, Y. (1972). *Muscle Proteins, Muscle Contraction and Cation Transport*. Tokyo & Baltimore: Japan Scientific Society Press & University Park Press.

Tonomura, Y. & Watanabe, S. (1954). Effect of adenosine triphosphate on the light-scattering of actomyosin solution. *Nature*, **169**, 112–13.

Wakabayashi, T., Huxley, H. E., Amos, L. A. & Klug, A. (1975). Three-dimensional image reconstruction of actin-tropomyosin complex and actin-tropomyosin-troponin T-troponin I complex. *J. Mol. Biol.*, **93**, 477–97.

Weber, A. & Murray, J. M. (1973). Molecular control mechanisms in muscle contraction. *Physiol. Rev.*, **53**, 612–73.

Wilkinson, J. M. & Grand, R. J. A. (1975). The amino acid sequence of troponin I from rabbit skeletal muscle. *Biochem. J.*, **149**, 493–6.

Yamamoto, K. & Sekine, T. (1979). Interaction of myosin subfragment-one with actin. II. Location of the actin binding site in a fragment of subfragment-one heavy chain. *J. Biochem.*, **86**, 1863–8.

Yanagida, T., Kuranaga, I. & Inoue, A. (1983). Interaction of myosin with thin filaments during contraction and relaxation: Effect of ionic strength. *J. Biochem.*, **92**, 407–12.

*Actomyosin in ATPase reaction and the mechanism of muscle contraction*

Yasui, M., Arata, T. & Inoue, A. (1984). Kinetic properties of binding of myosin subfragment-one with F-actin in the absence of nucleotide. *J. Biochem.*, **96**, 1673–80.

# 4

# Actin and myosin
# in nonmuscle cells

In the preceding chapters, the actin–myosin system has been shown to be the fundamental system in muscle contraction. This system, however, is not exclusive to muscle cells: it exists in all the nonmuscle cells in the animal kingdom. Moreover, actin and myosin have been isolated from plants, protozoa and fungi and have also been reported to exist in bacteria. Furthermore, it has been revealed that the actin–myosin system is involved not only in cell motility but also in the determination of cell shape by playing a role in the cytoskeleton. These findings, along with the development of molecular-level technology, have made the actin–myosin system one of the most fascinating subjects in cell biology.

The cytoskeletal aspects of this system will first be described. We will discuss mainly the distributions of contractile and regulatory proteins in the cell and give a brief description of the dynamic changes in the morphology of actin *in vivo* and the actin-binding proteins that are believed to play major roles in these changes. For more general characteristics of the actin–myosin system and its control system in nonmuscle cells, the reader is referred to excellent reviews and monographs by Pollard & Weihing (1974), Pollard (1976), Clarke & Spudich (1977), Hitchcock (1977), Perry, Mergreth & Adelstein (1977), Allen & Allen (1978), Korn (1978), Goldman *et al.* (1979), Taylor & Condeelis (1979), Adelstein & Eisenberg (1980), Kamiya (1981), Kendrick-Jones & Scholey (1981) and Sheterline (1983).

## Actin and myosin in nonmuscle cells

Examples of movement in nonmuscle cells can be found in the following phenomena: cytoplasmic streaming, changes in cell shape, amoeboid movement, cytokinesis, mitosis, saltatory particle movement,

redistribution of cell surface components, endocytosis, exocytosis, orga-
nelle movement and movement of tissue cells. These seemingly unrelated
phenomena are currently being shown, one after another, to possess
actin and myosin as the molecular basis of motility. Their actin and
myosin are nearly identical to that of skeletal muscle.

## Characteristics of isolated contractile and regulatory proteins

Nonmuscle myosin was first isolated and purified from *Physarum polycephalum* (Hatano & Tazawa, 1968). The current method
of isolating nonmuscle myosin was developed by assuming that nonmuscle and muscle myosins are similar to each other in their physicochemical characteristics. The method generally involves extraction of
actomyosin from cells using 0.6 M KCl and precipitation at low ionic
strength.

Electron micrographs of rotary-shadowed myosin molecules isolated
from the slime mold are shown in Fig. 4.1a. This myosin is composed
of two heads (19.7 nm in longitudinal axis) and a tail (173 nm in length).
The structure of its heads is similar to that of skeletal muscle myosin
(Chapter 2), but the tail is about 20 nm longer, presumably corresponding
to the slightly heavier molecular weight of its heavy chain.

With the exception of only a few cases, nonmuscle myosin consists
of two heavy chains with a molecular weight of 200 000–220 000 each
and light chains with molecular weights of around 20 000. Similar to
skeletal muscle myosin, nonmuscle myosins exhibit actin-binding ability
and the activities of EDTA(K$^+$)-ATPase, Ca$^{2+}$-ATPase and Mg$^{2+}$-
ATPase. These ATPases, however, show some fluctuations in their specific activities depending on the cell source. Among the properties of myosin, one of the features that is important for cell motility is that its
Mg$^{2+}$-ATPase is activated by F-actin at low ionic strength.

The nonmuscle cell that actin was first isolated from and purified is
also *Physarum polycephalum* (Fig. 4.1b; Hatano & Oosawa, 1966). With
the development of a heavy meromyosin (HMM)-binding technique
(Ishikawa, Bischoff & Holtzer, 1969) and a fluorescent-HMM or fluorescent antibody technique (Sanger, 1975), the presence of actin has now
been confirmed in more than 100 types of nonumuscle cells. Actin is
a highly ubiquitous protein, and F-actins isolated from various types
of nonmuscle cells are all morphologically indistinguishable from skeletal
muscle F-actin; all are in the form of a 7-nm-wide, double-stranded

helix with a pitch of 35 nm. The chemical structure of nonmuscle F-actin is also similar to that of skeletal muscle F-actin: the molecular weight estimated by sodium dodecyl sulfate-polyacrylamide gel electrophoresis (SDS-PAGE) is 42 000, and 91–94% of the amino acid residues are identical to muscle actin. ADP is bound to nonmuscle F-actin at 1 mol/mol of monomer actin.

Although nonmuscle actins and muscle actin appear to have the same molecular weight when analyzed by SDS-PAGE, they can be separated into several different types by an isoelectric focusing technique. Mammalian actin can be separated into at least three isoactins, in order of decreasing acidity, $\alpha$ $\beta$ and $\gamma$. The $\alpha$-actin is the skeletal muscle actin.

Fig. 4.1 (*a*) Electron micrographs of *Physarum* myosin molecules at high ionic strength. Top and second rows are phosphorylated myosin, and third and fourth rows are dephosphorylated myosin (Takahashi *et al.*, 1983). (*b*) Electron micrograph of *Physarum* F-actin. The sample was negatively stained. The half-pitch of the helix is about 35 nm (Hatano & Tazawa, 1968).

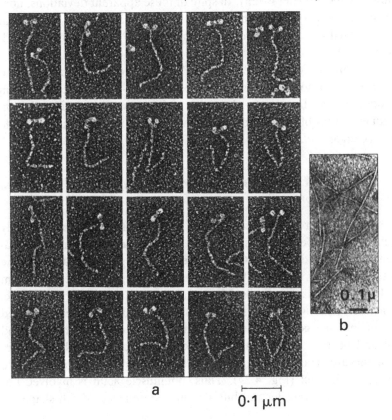

a

0·1 μm

b

The $\beta$- and $\gamma$-actins are found in nonmuscle cells, and the $\gamma$-actin comigrates with chicken gizzard smooth muscle actin. Peptide maps of these three isoactins are similar but distinguishable. The actin from *Acanthamoeba* shows an isoelectric point more basic than that of $\gamma$-actin, and thus is called the $\delta$-actin.

The isolation of actin and myosin from a prokaryotic cell source, *Escherichia coli* (Nakamura & Watanabe, 1978) has been reported. However, the presence of contractile proteins in bacteria has been argued against by many investigators.

Tropomyosin, which plays an important role in the regulation of skeletal muscle contraction, is also present in nonmuscle cells. However, all tropomyosins so far purified are of mammal source. The molecular weight of nonmuscle tropomyosin ranges from 26 000 to 30 000, which is clearly smaller than that of muscle tropomyosin (35 000). In addition, the axial periodicity seen in the paracrystals of nonmuscle tropomyosin is 34 nm, which is shorter than the paracrystal periodicity of skeletal muscle tropomyosin (40 nm). In spite of these apparent deviations, nonmuscle tropomyosin and muscle tropomyosin can form a hybrid that is functional for $Ca^{2+}$ control of actomyosin ATPase (Scholoss & Goldman, 1980).

$\alpha$-Actinin, which is present in the Z-line of skeletal muscle, has also been found in nonmuscle cells. Antibodies to $\alpha$-actinin prepared from smooth muscle have been shown to stain stress fibers and adhesion plaques in fibroblast cells. $\alpha$-Actinin was later isolated from nonmuscle cells (Geiger *et al.*, 1980).

### Intracellular distribution of actin, myosin and tropomyosin

Electron microscopy and a fluorescence-labeled antibody technique have been very useful in determining the intracellular distributions of actin and myosin. Cultured cell lines, e.g., fibroblasts, are the best suited to this technique because they adhere tightly to the substratum and form a thin layer of cells. Most of the actin filaments form networks in actively moving fibroblasts, and anti-actin staining appears diffused throughout the cell. On the other hand, in the cells that have spread out and become sessile on the surface of the culture dish, the actin filaments are arranged in straight bundles, some of which span the entire length of the cell (Fig. 4.2a). Thus, nonmuscle actin is involved not only in energy transduction but also in maintenance of cell shape, as

Fig. 4.2 Immunofluorescent localization of actin and myosin on stress fibers in cultured cells. Contractile network proteins, actin (*a*) (Lazarides, 1975) and myosin (*b*) (courtesy of K. Fujiwara) were visualized by immunofluorescent techniques. Note the periodic distribution of myosin over the entire length of the stress fibers.

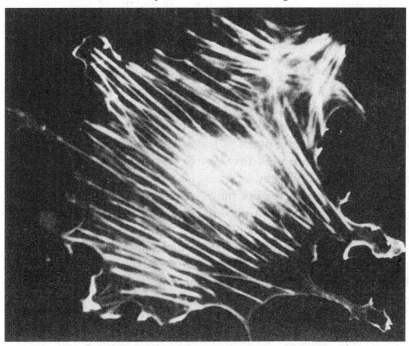

a

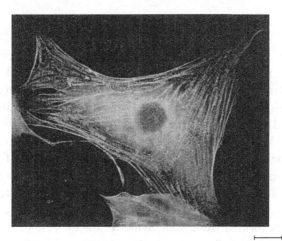

b

20 μm

part of the cytoskeleton. A typical example of actin in the cytoskeleton of cells is the stereo cilia of the inner ear cochlear hair where bundles of more than 3000 actin filaments form brush borders and move passively in response to the vibration of sound (Tilney, DeRosier & Mulroy, 1980).

The pattern of anti-myosin staining is similar to that of anti-actin staining; stress fibers are stained. However, the anti-myosin staining pattern reveals discontinuous lines with certain periodicity (Fig. 4.2b). Actively moving cells show a diffuse stain of myosin.

When fibroblasts are stained with anti-tropomyosin, the fluorescence appears uniform along the actin filament bundles. When the cells begin to migrate, the actively moving cell cortex and speudopodia emit very little tropomyosin fluorescence. Thus, it is believed that tropomyosin is not bound to the networks of actin filaments. For the present, the most plausible function of tropomyosin in nonmuscle cells is to bind to actin filaments, thereby increasing its structural stability. $\alpha$-Actinin is present at sites on the cell inner surface where the cell membrane attaches to the substratum to form adhesion plaques. At present, $\alpha$-actinin is assumed not to bind directly to the membrane but to mediate the anchorage of stress fibers on the cell membrane, through vinculin, a protein with a molecular weight of 130 000 (Geiger *et al.*, 1980). Also, like myosin and tropomyosin, $\alpha$-actinin stain exhibits a periodicity on the stress fibers.

## Changes in the actin organization *in vivo* and actin-binding proteins

### *Changes in the actin organization* in vivo

The content of actin in nonmuscle cell is very high. The molar ratio of actin to myosin has been estimated at about 4 in skeletal muscle, about 30 in smooth muscle, and about 100 in nonmuscle cells in general. Actin exists in cells not only as a single filament but also as various organized structures. In floating cultured cells actin forms a diffuse network. When the cells cease to move, it forms a highly ordered three-dimensional structure called a geodome. In completely flattened cells, straight bundles of actin called stress fibers are formed (Lazarides, 1976).

Various types of actin bundles have been reported in nonmuscle cells; in the acrosomal reaction of sperm a straight actin bundle with uniform polarity is formed from nonpolymerized actin (Tilney *et al.*, 1973) or a coiled bundle (Tilney, 1975). A major change in the organization of actin filaments upon fertilization also occurs in the egg; it has been reported that in sea urchins numerous bundles of actin filaments, called microvilli, are formed on the surface of the egg upon fertilization.

## Actin-binding proteins

Only a few examples of dynamic change in the morphology of actin *in vivo* have been discussed above. However, the morphology is highly variable among various types of nonmuscle actins. Variation ranges, for example, from bundles, in which actin filaments are arranged in a highly ordered manner, to networks such as those seen in the ruffling membranes of cultured cells, in which actin filaments are distributed in a completely disordered manner. In addition, it is generally believed that in nonmuscle cells about half of the actin molecules exist as monomers under physiological conditions.

Recently, a number of proteins have been found that might account for the polymorphic forms of actin, i.e., bundles, networks and monomers, and also for the transformation among these forms (Fig. 4.3). Most of these proteins have been purified to homogeneity, and

Fig. 4.3 Actin-binding proteins.

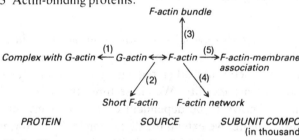

| | PROTEIN | SOURCE | SUBUNIT COMPOSITION (in thousands) |
|---|---|---|---|
| (1) | Profilin | Mammalian tissues | 1×13 |
| | DNase I | Bovine pancreas | 1×31 |
| (2) | Gelsolin* | Macrophage | 1×91 |
| | Fragmin* | *Physarum* | 1×43 |
| | Villin* | Intestinal epithelium | 1×95 |
| | Capping protein | *Acanthamoeba* | 28+31 |
| (3) | Fascin (55K protein) | Sea urchin eggs | 1×55 |
| | Villin* | Intestinal epithelium | 1×95 |
| | 95K protein* | *Dictyostelium* | 2×95 |
| (4) | Actinogelin* | Ehrlich tumor cells | 2×115 |
| | 90K protein | *Acanthamoeba* | 2×90 |
| | ABP | Macrophage | 2×270 |
| | Filamin | Fibroblasts | 2×250 |
| | 120K protein | *Dictyostelium* | 2×120 |
| | Gelactin I–IV | *Acanthamoeba* | 1×23–38 |
| | Actinotanglin (36K protein) | *Physarum* | 2×36 |
| | Fimbrin | Intestinal epithelium | 1×68 |
| (5) | α-actinin* | HeLa cells | 2×100 |
| | Spectrin | Erythrocyte | 1×240 |
| | | | 1×220 |
| | Vinculin | Fibroblast | 1×130 |

*Calcium-sensitive proteins

their relationships to actin have been analyzed in detail. Some of the actin-binding proteins bind actin in a $Ca^{2+}$-sensitive manner and, thus, may account for some physiological process where actin plays a main role. In fact, some of these proteins form F-actin, the morphology of which closely resembles that of microfilaments *in situ*. Unfortunately, however, very little is known about the physiological functions of these proteins, with the exception of only a few cases. Moreover, only a small number of these proteins are expected to be directly involved in coupled cell motility with ATP hydrolysis. Thus, in this chapter, description of actin-binding proteins is limited to the brief summary in Fig. 4.3 of those that are likely to be involved in a variety of changes in the morphology of actin. For more detailed descriptions of the actin-binding proteins, the reader should consult several recently published reviews (Craig & Pollard, 1982; Korn, 1982; Weeds, 1982). The proteins that are likely to be involved in the control of cell motility are discussed in some detail below.

## Mechanism of cytoplasmic streaming

Various types of cytoplasmic streaming are seen in a wide variety of cells (Kamiya, 1959; Allen, 1981). In this section, we will discuss how myosin, actin and other regulatory proteins interact with each other and generate motility. We chose three typical cell motility systems which are known to be dependent on the actin–myosin–ATP interaction. Furthermore, we extend the discussion in the following section on the evolutionary aspects of actomyosin and its regulatory system, which we have come to conceive by examining the properties of actomyosin in various systems ranging from nonmuscle to highly organized muscle.

### Contractile and regulatory proteins in amoebae

Because giant amoebae, *Amoeba proteus* and *Chaos carolinensis*, can be manipulated easily, the physiological aspects of the mechanism of amoeboid movement have been studied by many investigators. A detailed classification of amoeboid movement was reported by Allen (1972). Yet, as described below, no definitive conception of the mechanism has so far emerged. On the other hand, studies on the contractile and regulatory proteins isolated from amoebae have rapidly advanced in recent years, and they have revealed many interesting facts.

Acanthamoeba *myosin isoenzymes*    Myosins from *Physarum poly-cephalum, Dictyostelium discoideum* and macrophages are similar to skeletal muscle myosin in molecular weight and subunit composition. In contrast, myosin from *Acanthamoeba castelanii* has some unusual characteristics (Pollard & Korn, 1973). Three types of myosin isoenzymes, IA, IB and II, have been isolated from *Acanthamoeba castelanii.* In contrast to the double-headed myosins isolated from other cell sources, myosins IA and IB are believed to be single headed. These two kinds of myosin have no ability to form bipolar thick filaments. The molecular weights of the polypeptides that correspond to the so-called heavy chains are 130 000 and 125 000 for myosin IA and myosin 1B, respectively, whereas the molecular weights of the light chains are 17 000 and 14 000 for myosin IA, and 27 000 and 14 000 for myosin IB. Definition of these proteins as myosin comes from their characteristic properties such as $EDTA(K^+)$-, $Ca^{2+}$- and $Mg^{2+}$-ATPase activities, binding to F-actin and activation of their $Mg^{2+}$-ATPase activities by F-actin.

*Acanthamoeba* myosin II, on the other hand, is a double-headed myosin. It forms bipolar thick filaments and has two heavy chains with a molecular weight of 170 000 each and two kinds of light chains with molecular weights of 17 500 and 17 000. Interestingly, it is believed that these unusual, low molecular weight heavy chains of *Acanthamoeba* myosin are not the products of proteolytic digestion of larger molecules but are the products of different genes, not only because they are different from each other in molecular weight, subunit composition and enzymatic properties, but also because of the results of peptide mapping and immunological studies. Although very little is known about the individual functions of these myosin isoenzymes in a single cell, there has been evidence for their difference from their localization in the cell: myosin 1 is localized near the plasma membrane, whereas myosin 2 is not concentrated near the plasma membrane and, thus, possibly spans the entire cytoplasm (Pollard & Korn, 1973).

*Regulation of the ATPase activity by phosphorylation–dephosphorylation of myosin*    With advances in the study of the regulatory mechanism of smooth muscle contraction, the regulation of thick filament formation and that of actin-activated ATPase activity by phosphorylation and dephosphorylation of myosin has recently become known in various types of nonmuscle cells. In *Acanthamoeba* myosin I (Maruta & Korn, 1977) and *Physarum* myosin (Ogihara *et al.*, 1983), the myosin ATPases are strongly activated by F-actin when their heavy chains are phosphorylated by heavy chain kinases. Phosphorylated, but not dephosphorylated, *Physarum* myosin forms bipolar thick filaments. Contrary to these myosins, *Acanthamoeba* myosin II and *Dictyostelium* myosin are active in the dephosphorylated form. Platelet myosin and thymus myosin are regulated by mechanisms different from the above myosins; they are regulated by phosphorylation of their m.w. 20 000 light chains by the $Ca^{2+}$-calmodulin-dependent light chain kinase, the same mechanism as that seen in smooth muscle myosin. The above nonmuscle heavy chain kinases are all $Ca^{2+}$ insensitive, but recently Maruta *et al.* (1983) was the first to report the isolation of a $Ca^{2+}$-sensitive heavy chain kinase from aggregation *Dictyostelium discordeum* amoeba.

Whether the highly dramatic changes in the actomyosin-ATPase activity caused by the phosphorylation and dephosphorylation of myosin are due only to the assembly and disassembly of thick filaments or also to some kind of conformational change in the active site of ATPase is not known. However, it is possible that, in nonmuscle cells where the myosin content is low as compared with that in muscle cells, the 'effective' concentration of myosin relative to F-actin is lowered by the disassembly of thick filaments, and as a result, the actin-activated ATPase activity is lowered. Korn's group (Collins & Korn, 1981) reported that the actin-activated ATPase activity of dephosphorylated *Acanthamoeba* myosin II has a sharp sigmoidal dependence on the myosin concentration. Furthermore, the phosphorylation sites are not located near the catalytic site for ATPase on the heavy chain but in the terminal region of the tail. Phosphorylation–dephosphorylation of the tail portion of myosin may affect the actomyosin-ATPase activity through assembly and disassembly of thick filaments.

## The gelation factor and other actin-binding proteins in amoebae

When amoebae are homogeinized under appropriate conditions and then subjected to ultracentrifugation, the supernatant forms, in a temperature-dependent manner, a solid gel which is composed mainly of actin. This has been observed not only with amoebae but with numerous different types of cells as well. Morphologically, the gel exists in the form of either networks or bundles of actin filaments. This gelation reaction has been shown to be caused by the crosslinking of F-actin, and under appropriate conditions myosin causes the gel to contract. Efforts have been made to isolate crosslinking proteins from these motile extracts and more than 20 crosslinking proteins have now been isolated and purified from various types of cells (cf. Fig. 4.3).

Although amoeboid movement is a very complicated phenomenon, it seems to include at least five cycling steps if one pays attention to their locomoting nature, which is in contrast to the highly stable skeletal muscle fibers: (1) destruction of an already made cytoskeleton by severing long actin filaments into short filaments or even monomeric forms, (2) formation of new actin-nucleation sites on the plasma membrane, (3) polymerization of actin from those sites, (4) construction of gel-like cytoplasm by the crosslinking actin filaments, and (5) association of myosin with the gel to cause contraction. The last step may include activation of myosin via phosphorylation or dephosphorylation to make an active force-generating form of myosin, i.e., thick filaments as mentioned above.

The crosslinking proteins shown in Fig. 4.3, such as actin-binding protein (ABP), the m.w. 120 000 protein and gelactin I–IV, seem to explain well step (4). The proteins that regulate actin filament length such as capping protein seem to be participating in steps (1) and (3). The m.w. 120 000 protein is thought to work in step (3) as well, because of its ability to lower the critical concentration for actin polymerization. Little is known about the other steps. The $Ca^{2+}$ sensitivity of the gelation process and of the movement of several cell models has been explained in terms of these known actin modulators, but there is little agreement on the mechanism of cell movement.

One approach to the elucidation of the mechanism of amoeboid movement is to reconstitute the movement by combining these proteins with the actin–myosin system. Among the proteins shown in Fig. 4.3, actin-binding protein (also called filamin), the m.w. 120 000 protein and gelactins I–IV all inhibit the acto—HMM ATPase activity. Then, for the myosin filaments to work in the actin matrix and cause contraction, these crosslinkers would be inappropriate factors. To explain this contradiction, a mechanism has been proposed that the actin network made of actin filaments and the m.w. 120 000 protein is not itself contractile; contraction requires an eventual detachment or inactivation of the m.w. 120 000 protein from the actin filaments (Hellewell & Taylor, 1979). Later, Condeelis *et al.* (1984) showed that in submicromolar concentrations of $Ca^{2+}$, the m.w. 95 000 protein that is a $Ca^{2+}$-sensitive actin-gelling protein aligns actin filaments into bundles. The myosin thick filaments interact effectively with these bundled actin filaments to show a higher ATPase activity than randomly oriented actomyosin. On addition of $Ca^{2+}$, the m.w. 95 000 protein detaches actin bundles, leaving a highly arrayed actin filaments that are more free to slide with myosin, resulting in a higher ATPase activity than when m.w. 95 000 protein is bound.

*Hypotheses on the mechanism of amoeboid movement*   The simplest model built for the purpose of discussing the mechanism of nonmuscle motility is the brush border of the intestine (see Fig. 4.4; Mooseker & Tilney, 1975). Indeed, since the brush border exhibits fairly specialized structure and function, this simple model is not expected to explain amoeboid movement and other motility seen in nonmuscle cells. The reason for presenting this model here is to point out three important

Fig. 4.4 (*a*) Electron micrograph of an isolated brush border of intestinal epithelium. Each microvillus contains a core of actin filaments (Mooseker & Tilney, 1975). (*b*) The polarity of the actin filaments in the microvillus and in skeletal muscle as represented by arrowheads. Diagram on the left shows the functional organization of actin and myosin in the brush border (Mooseker, 1983).

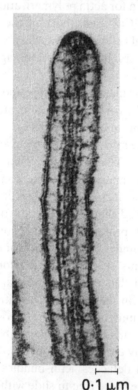

a

0·1 μm

features seen in nonmuscle cells. The first is the association between microfilaments and the plasma membrane. The second is the polarity of microfilaments that elongate from the membrane. The polarity of actin filaments revealed by HMM or myosin head (S-1) decoration is an essential feature for tension generation. The third feature is the presence of the bipolar myosin filaments. The sliding between myosin filaments and actin filaments in this configuration will result in movement and/or changes in cell shape via the plasma membrane.

A variety of hypotheses on the mechanism of amoeboid movement have been proposed. On the basis of current knowledge, these hypotheses

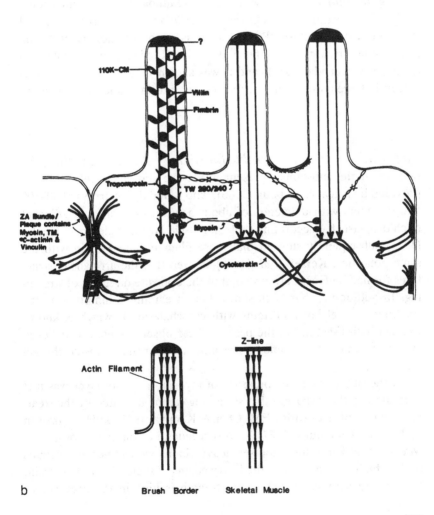

can grossly be classified into either 'the tail or ectoplasmic contraction model' or 'the frontal (fountain zone) contraction model' (Allen, 1961). However, many aspects of the mechanism still remain obscure, and thus no decisive hypothesis is yet available. The mechanism of amoeboid movement has already been reviewed (Komnick, Stockem & Wohlfarth-Bottermann, 1973).

## Sliding mechanism of cytoplasmic streaming in characean cells

*Sliding mechanism* The cytoplasmic streaming seen in the internodal cells of fresh water algae characeae (*Nitella* and *Chara*) is a rotational streaming called cyclosis. Precise analysis of the physiology of *Nitella* cytoplasmic streaming has been done by Kamiya and his collegues (Kamiya & Kuroda, 1956), and it was shown that the sliding force is tangentially generated only at the sol–gel interface and not inside the sol.

*Contractile proteins responsible for sliding* An answer to the question about the substances present inside the stationary gel layer was provided by Nagai & Rebhun (1966), who found dense arrays of 50–100 fine filaments with diameters of about 5 nm in the gel layer adjacent to chloroplast. By decorating skeletal muscle HMM, Palevitz & Hepler (1975) subsequently showed that these filaments are F-actin with the same polarities. Kersey *et al.* (1976) showed that the arrowhead structures formed by HMM always point in the opposite direction of streaming. In addition, it has been shown that streaming is inhibited by the treatment of gel layer proteins with cytochalasin B, which is known to react with F-actin. On the basis of these observations, it is believed at present that actin filaments are unipolarly arranged inside the gel layer.

On the other hand, the presence of myosin in the sol layer was first suggested by the discovery that streaming can be inhibited by the treatment of the sol layer with NEM (Chen & Kamiya, 1975) and later proven by Kato & Tonomura (1977), who succeeded in isolating myosin from *Nitella*. *Nitella* myosin possesses heavy chains with molecular weights slightly higher than those of the heavy chains of skeletal muscle myosin. Its ATPase activity at high ionic strength is high in the presence of

EDTA or Ca$^{2+}$, and low in the presence of Mg$^{2+}$. Accordingly, it becomes very important to know the *in vivo* organization of this *Nitella* myosin.

Using a special optical system called the Nomarski method, Kamitsubo (1972) reported that small particles in the vicinity of the subcortinal fibrils could flow at the streaming speed along these fibrils upon touching these fibrils, whereas particles not touching the fibrils showed only Brownian motion. Williamson (1975) reported the presence in the sol of small particles which bind to and dissociate from the cytoplasmic fibrils in a Mg$^{2+}$-pyrophosphate-sensitive manner.

Figure 4.5 shows an electron micrograph of the thin-sectioned particles bound to the cytoplasmic fibrils in the Mg$^{2+}$-ATP-deficient cell. These

Fig. 4.5 Subcortical fibers in a *Chara* cell. Fibrils are decorated with many endoplasmic organelles in the absence of ATP (*a*). ATP makes fibrils naked (*b*). In an electron micrograph of the fibrils (*c*), balloon-shaped organelles are linked to actin cables through electron-dense structures with some periodicity in ATP-deficient cells (Nagai & Hayama, 1979).

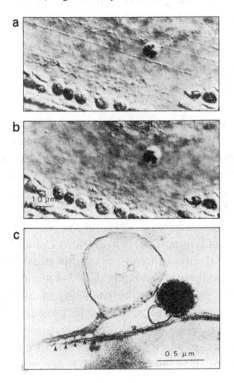

particles can also be dissociated from the fibrils by $Mg^{2+}$-ATP. They are all membrane-limited and can be separated into a balloon-shaped portion and a rod-shaped protuberance portion. In addition, they contain small, globular bodies with diameters of 20–30 nm, on the surface of the protuberance. It can also be seen that these globular bodies are in association with actin filaments on the gel side. The protuberances always point in the direction of streaming. From these observations, Nagai & Hayama (1979) have proposed that the myosin molecules arranged on the surface of the protuberance actively slide against the actin molecules and move on the cable of actin, dragging with them a bulk of cytoplasm (sol). Recently, Ogihara & Condeelis (1983) have found that anti-myosin antibody was localized on cytoplasmic vesicular structures in *Dictyostelium discoideum* amoeba cells and that the myosin containing vesicles touched the cytoplasmic actin filament.

It is known that the streaming in characean cells is inhibited by $Ca^{2+}$ influx into the cytoplasm and that it can continue in the presence of 50 mM EDTA. Thus this streaming is markedly different from the other, more generally seen cell motility systems that are activated by micromolar concentrations of $Ca^{2+}$. No protein that is involved in this reversed $Ca^{2+}$-control system has yet been found.

## Pressure-flow mechanism of cytoplasmic streaming in Physarum *plasmodia*

*Cytoplasmic streaming of Physarum* plasmodia   The plasmodium of a species of myxomycete, *Physarum polycephalum*, is one of the best suited materials for the study of cell motility. It is a mass of naked cytoplasm without cell walls and shows very active cytoplasmic streaming, thereby allowing an easy manipulation for physiological studies. It can also be cultured in a large scale for biochemical studies. The cytoplasmic streaming can easily be observed in the strand part of the plasmodium, where the sol (endoplasm) actively flows inside the tubular gel (ectoplasm). The most interesting aspect of this streaming is that it reverses the direction rhythmically at intervals of 2–4 minutes. Kamiya & Kuroda (1958) clearly showed that the streaming of the sol in the plasmodium is a passive movement caused by the pressure difference inside the sol.

The movement of cytoplasm in *Physarum* is activated by a micromolar range of $Ca^{2+}$ in a caffeine drop model (Hatano, 1970) and in a Triton model (Ogihara, 1982) and is inhibited by a millimolar range of $Ca^{2+}$ in the Triton model and in a saponin model (Yoshimoto, Matsumura & Kamiya, 1981). In the saponin model, Yoshimoto *et al.* showed that the concentration of $Ca^{2+}$ in the plasmodium oscillates with a periodicity identical to, but with a phase shifted by 180° from that of, the tension rhythm. The biochemical mechanism of activation by a micromolar range of $Ca^{2+}$ remains obscure. However, a factor which inhibits the superprecipitation of *Physarum* myosin B in a millimolar range of $Ca^{2+}$ has been found as will be described later.

*Contractile elements in the gel layer*    In 1962, Wohlfarth-Bottermann reported the presence of microfilaments with diameters of about 6 nm in the plasmodia of *Physarum*, which were subsequently identified as actin filaments by the HMM-binding technique. Later, Kamiya (1973) found that birefringent fibrils located at the anterior end of the plasmodium repeatedly appear and disappear during streaming (Fig. 4.6). We (Ogihara & Kuroda, 1979) showed that these fibrils are mainly composed of actin. The birefringent actin fibrils correspond to the actin bundles (Fig. 4.7*a*); these bundles appear when the strand part contracts and disappear when the gel layer relaxes. During relaxation the microfilaments form an intertwining tangled network configuration as shown in Fig. 4.7*b*. Furthermore, Nagai, Yoshimoto & Kamiya (1978) claimed

Fig. 4.6 Polarization micrographs of the advancing frontal region of the living *Physarum* plasmodium. Birefringent fibrils appear (*a*) and disappear (*b*) cyclically in accordance with the shuttle streaming. Photographs were taken at 2-minute intervals. Scale bar, 100 μm (Ogihara & Kuroda, 1979).

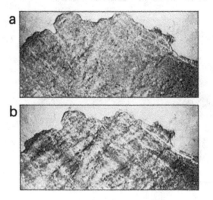

*Actin and myosin in nonmuscle cells*

Fig. 4.7 Electron micrographs of longitudinal sections of the plasmodial strand in a contracting phase (*a*) and at the onset of the relaxing phase (*b*). Microfilaments are straight and bundled in the contracting phase. Among them, thicker filaments are scattered sporadically. In the relaxing phase, bundles of straight filaments have been converted into a network of randomly oriented microfilaments (Nagai, Yoshimoto & Kamiya, 1978). (*c*) Electron micrograph of a negatively stained complex of actin filaments and the m.w. 36 000 actin-binding protein purified from *Physarum* plasmodia. The actin filaments run parallel to adjacent filaments (arrows), and they are also heavily curled and intertwined with each other to form densely packed regions (arrowheads) (Ogihara & Tonomura, 1982).

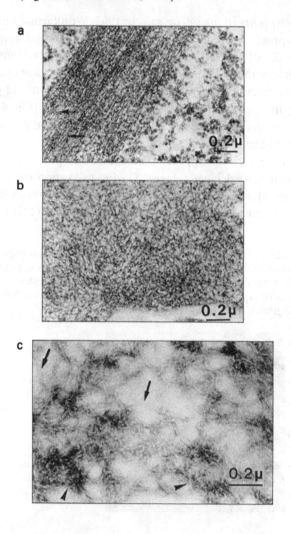

that filaments thicker than the microfilaments are present in the microfilament bundles during the contraction phase but are absent in the relaxation phase. They proposed that these thicker filaments may be composed of myosin and that in plasmodium the myosin filaments undergo assembly and disassembly during one round of the contraction–relaxation cycle.

## Contractile and regulatory proteins

Actin was first isolated from *Physarum* by Hatano & Oosawa (1966). The amino acid sequence of *Physarum* actin has already been determined and only 33 (9%) out of the 375 amino acids in skeletal muscle actin are substituted in *Physarum* actin.

Myosin has been purified to homogeneity by several groups. Nevertheless, variations have been large among reports concerning some of its characteristics, particularly the degree of activation of its $Mg^{2+}$-ATPase by F-actin and the degree of formation of the thick filaments. We (Ogihara *et al.*, 1983) have recently shown that phosphorylation of myosin heavy chains is required for thick filament formation, ATP-induced superprecipitation and the activation of ATPase by F-actin. These results, together with the electron microscopic observations by Nagai, Yoshimoto & Kamiya (1978) described above, strongly suggest that in plasmodium, myosin undergoes phosphorylation and dephosphorylation during one round of the contraction–relaxation cycle, leading to the assembly and disassembly of thick filaments and thus to the periodic generation of motive force.

Various proteins have been reported for the regulatory proteins of the actin–myosin system in *Physarum*. The first was fragmin, reported by Hasegawa *et al.* (1980). Fragmin has a molecular weight similar to that of actin and is capable of fragmenting F-actin in the presence of 1 μM or more of $Ca^{2+}$. When fragmin is present, G-actin cannot undergo a complete polymerization and thus forms F-actin with very short lengths. Hasegawa *et al.* assumed that fragmin is involved in the sol–gel transformation in *Physarum* cytoplasm.

The second protein, actinotanglin (a m.w. 36 000 protein), was purified by us (Ogihara & Tonomura, 1982) from a fraction that was rich in tangled microfilaments (Fig. 4.7*b*). F-actin bound to actinotanglin forms a large aggregate (Fig. 4.7*c*). When analyzed by high-magnification electron microscopy, each F-actin is curled and, occasionally, two actin filaments run in parallel, crosslinking each other. This organization closely resembles the *in situ* organization of microfilaments during relaxation as shown in Fig. 4.7*b*. The important point here is that the F-actin that forms an aggregate by binding to actinotanglin is no longer capable of activating the $Mg^{2+}$-ATPase activity of myosin nor of inducing super-precipitation. Thus, it is quite conceivable that in *Physarum*, actinotanglin functions not only as a regulator of the cytoskeleton but also as a regulator of the contractile events.

The cytoplasmic streaming of *Physarum* is inhibited by a millimolar range of $Ca^{2+}$, as described previously. A factor that inhibits the ATP-induced superprecipitation in the presence of 0.1 mM $Ca^{2+}$ has been prepared by Kohama (1981) from *Physarum* myosin B. This factor becomes functional only after the heavy chains of myosin are phosphorylated, thus, the $Ca^{2+}$ inhibitory factor possesses physiological functions (Hatano & Oosawa, 1966). However, it must be noted here that many aspects of the $Ca^{2+}$ control of *Physarum* cytoplasmic streaming, especially the role of phosphorylation and dephosphorylation of myosin and the function of actinotanglin, still remain unsolved.

# Evolution of the actin–myosin system

In the preceding sections, various biochemical and physiological studies on the actin–myosin regulatory protein system were discussed. In this final section, this system is discussed and summarized from an evolutionary point of view.

## Evolution of myosin

As has been described, myosin has been isolated not only from every eukaryotic cell that has been studied but also from a prokaryotic cell, *E. coli*. Unusual types of myosin have also been found. A typical example is *Acanthamoeba* myosin IA and IB, single-headed myosins. However, it is still uncertain how these myosins function in *Acanthamoeba* cells.

Though skeletal muscle is the most developed organ for rapid movement, skeletal muscle myosin and nonmuscle myosin have common characteristics, and both myosins have two heads with a long tail. On the other hand, the stability of thick filaments is clearly different in skeletal muscle and nonmuscle cells. In the case of nonmuscle cells, thick filament formation is regulated by phosphorylaton and dephosphorylation of myosin. This may be of significance in nonmuscle cells where the concentration of myosin is low.

## Evolution of actin

Actin can be considered a far more conservative molecule than myosin. This is possibly because of its physiological functions. The actin monomer, despite its relatively low molecular weight, must maintain such properties as self-polymerization, binding to myosin and providing binding sites for nucleotide and $Ca^{2+}$. The conservative nature of actin can easily be inferred from the fact that no significant difference is present among the characteristics of actins isolated from a wide variety of organisms ranging from amoebae or *Physarum* to skeletal muscle. Nevertheless, a variety of actin-binding proteins are present, possibly because each of these cells possesses various yet specific functions.

## $Ca^{2+}$ control system of cell motility

The $Ca^{2+}$ control system directly involved in the function of the actomyosin-ATPase is highly variable. With regard to the regulation of cell and muscle movements by $Ca^{2+}$, it is assumed that the regulatory system based on phosphorylation and dephosphorylation of myosin and the regulatory system linked to either myosin or actin filaments coexisted in cells at a considerably earlier stage of evolution. Both systems currently play important roles in evolutionarily lower organisms, with either one of the two systems having become the dominant system in higher organisms.

As has been described, the F-actin-activated ATPase activity of *Physarum* myosin increases when the heavy chains of myosin are phosphorylated, but the activity so increased is inhibited by the myosin-linked $Ca^{2+}$-dependent inhibitory factor (Ogihara *et al.*, 1983). In amoebae of *Acanthamoeba* and *Dictyostelium*, phosphorylation of myosin heavy chains is inhibitory to the ATPase activity. In *Acanthamoeba*, unlike the case of *Physarum*, the ATPase activity is regulated by the direct binding of $Ca^{2+}$ to myosin molecules (Collins & Korn, 1980). This type of regulation system might have evolved further into the regulatory system seen in mollusca. The $Ca^{2+}$-control system seen in *Physarum* and mollusca in which actomyosin ATPase is controlled by $Ca^{2+}$ binding to myosin may have evolved into the tropomyosin–troponin (TM–TN) system by changing the binding target from the thick filaments to the thin filaments. This system plays only a minor role in nonmuscle cells because the amount of TM–TN present in these cells is too small to regulate entire motile systems. Evolution in this direction has proceeded toward the current mode of regulation, in which the $Ca^{2+}$–TM–TN system plays a major role in the regulation of contraction in skeletal and cardiac muscles. Troponin might have specifically evolved from a calmodulin ancestor protein in striated muscle. The myosin phosphorylation system, on the other hand, must have evolved into motility control systems, such as those seen in smooth muscle and in platelets, by changing the phosphorylation target from the heavy chains to the light chains of myosin.

# References

Adelstein, R. S. & Eisenberg, E. (1980). Regulation and kinetics of the actin-myosin-ATP interaction. *Ann. Rev. Biochem.*, **49**, 921–56.

Allen, R. D. (1961). A new theory of amoeboid movement and protoplasmic streaming. *Exp. Cell Res.*, **8**, 13–31.

Allen, R. D. (1972). Biophysical aspects of pseudopodium formation and retraction. In *The Biology of Amoeba*, ed. K. W. Jeon, pp. 201–47. New York: Academic Press.

Allen, R. D. (1981). Motility. *J. Cell Biol.*, **91**, 148s–55s.

Allen, R. D. & Allen, N. S. (1978). Cytoplasmic streaming in amoeboid movement. *Ann. Rev. Biophys. Bioengin*, **7**, 469–95.

Chen, J. C. W. & Kamiya, N. (1975). Localization of myosin in the internodal cell of *Nitella* as suggested by differential treatment with N-ethylmaleimide. *Cell Struct. Funct.*, **1**, 1–9.

Clarke, M. & Spudich, M. A. (1977). Nonmuscle contractile proteins: the role of actin and myosin in cell motility and shape determination. *Ann. Rev. Biochem.*, **46**, 797–822.

Collins, J. H. & Korn, E. D. (1980). Actin activation of $Ca^{2+}$-sensitive $Mg^{2+}$-ATPase activity of *Acanthamoeba* myosin II is enhanced by dephosphorylation of its heavy chains. *J. Biol. Chem.*, **255**, 8011–14.

Collins, J. H. & Korn, E. D. (1981). Purification and characterization of actin-activatable, $Ca^{2+}$-sensitive myosin II from *Acanthamoeba*. *J. Biol. Chem.*, **256**, 2585–95.

Condeelis, J. S., Vahey, M., Carboni, J. M., Demey, J. & Ogihara, S. (1984). Properties of the 120 000 and 95 000-dalton actin-binding proteins from *Dictyostelium discoideum* and their possible functions in assembling the cytoplasmic matrix. *J. Cell Biol.*, **99**, 1195–265.

Craig, S. W. & Pollard, T. D. (1982). Actin-binding proteins. *Trends Biochem. Sci.*, **7**, 88–92.

Fujiwara, K. & Pollard, T. D. (1976). Fluorescent antibody localization of myosin in the cytoplasm, cleavage furrow, and mitotic spindle of human cells. *J. Cell Biol.*, **71**, 848–75.

Geiger, B., Tokuyasu, K. T., Dutton, A. H. & Singer, S. J. (1980). Vinculin, an intracellular protein localized at specialized sites where microfilaments terminate at cell membrane. *Proc. Natl. Acad. Sci (USA)*, **77**, 4127–31.

Goldman, R. D., Milsted, A., Schloss, J. A., Starger, J. & Yerna, M.-J. (1979). Cytoplasmic fibers in mammalian cells: cytoskeletal and contractile elements. *Ann. Rev. Physiol.*, **41**, 703–22.

Hasegawa, T., Takahashi, S., Hayashi, H. & Hatano, S. (1980). Fragmin: a calcium ion sensitive regulatory factor on the formation of actin filaments. *Biochemistry*, **19**, 2677–83.

Hatano, S. (1970). Specific effect of $Ca^{2+}$ on movement of plasmodial fragment obtained by caffeine treatment. *Exp. Cell Res.*, **61**, 199–203.

Hatano, S. & Oosawa, F. (1966). Isolation and characterization of plasmodium actin. *Biochim. Biophys. Acta*, **127**, 488–98.

Hatano, S. & Tazawa, M. (1968). Isolation, purification and characterization of myosin B from myxomycete plasmodium. *Biochim. Biophys. Acta*, **154**, 507–19.

Hellewell, S. & Taylor, L. D. (1979). The contractile basis of amoeboid movement. VI. The isolation-contraction-coupling hypothesis. *J. Cell Biol.*, **83**, 633–48.

Hitchcock, S. E. (1977). Regulation of motility in nonmuscle cells. *J. Cell Biol.*, **74**, 1–15.

Ishikawa, H., Bischoff, R. & Holtzer, H. (1969). Formation of arrowhead complexes with heavy meromyosin in a variety of cell types. *J. Cell Biol.*, **43**, 312–28.

Kamitsubo, E. (1972). A 'window technique' for detailed observation of characean cytoplasmic streaming. *Exp. Cell Res.*, **74**, 613–16.

Kamiya, N. (1959). Protoplasmic streaming. *Protoplasmatologia*, **8**(3a), 1–199.

Kamiya, N. (1973). Contractile characteristics of the myxomycete plasmodium. In *Proceedings of the Fourth International Biophysics Congress, Moscow*, ed. Academo of Science, USSR, pp. 447–494. Pushchino: International Union for Pure and Applied Biophysics.

Kamiya, N. (1981). Physical and chemical basis of cytoplasmic streaming. *Ann. Rev. Plant Physiol.*, **32**, 205–36.

Kamiya, N. & Kuroda, K. (1956). Velocity distribution of the protoplasmic streaming in *Nitella* cells. *Botanical Magazine (Tokyo)*, **69**, 544–54.

Kamiya, N. & Kuroda, K. (1958). Studies on the velocity distribution of the protoplasmic streaming in the myxomycete plasmodium. *Protoplasma*, **49**, 1–4.

Kato, T. & Tonomura, Y. (1977). Identification of myosin in *Nitella flexilis*. *J. Biochem.*, **82**, 777–82.

Kendrick-Jones, J. & Scholey, J. M. (1981). Myosin-linked regulatory systems. *J. Musc. Res. Cell Motil.*, **2**, 347–72.

Kersey, Y. M., Hepler, P. K., Palevitz, B. A. & Wessells, N. K. (1976). Polarity of actin filaments in characean algea. *Proc. Natl. Acad. Sci (USA)*, **73**, 165–7.

Kohama, K. (1981). Ca-dependent inhibitory factor for the myosin-actin-ATP interaction of *Physarum polycephalum*. *J. Biochem.*, **90**, 1829–32.

Komnick, H., Stockem, W. & Wohlfarth-Bottermann, K. E. (1973). Cell motility: mechanisms in protoplasmic streaming and amoeboid movement. *Int. Rev. Cytol.*, **34**, 169–249.

Korn, E. D. (1978). Biochemistry of actomyosin-dependent cell motility. *Proc. Natl. Acad. Sci. (USA)*, **75**, 588–99.

Korn, E. D. (1982). Actin polymerization and its regulation by proteins from nonmuscle cells. *Physiol. Rev.*, **62**, 672–737.

Lazarides, E. (1975). Immunofluorescence studes on the structure of actin filaments in tissue culture cells. *J. Histochem. Cytochem.*, **23**, 507–28.

Maruta, H., Baltes, W., Dieter, P., Marme, D. & Gerisch, G. (1983). Myosin heavy chain kinase inactivated by $Ca^{2+}$ calmodulin from aggregating cells of *Dictyostelium discoideum*. *EMBO J.*, **2**, 535–42.

Maruta, H. & Korn, E. D. (1977). *Acanthamoeba* cofactor protein is a heavy chain kinase required for actin activation of the $Mg^{2+}$-ATPase activity of Acanthamoeba myosin I. *J. Biol. Chem.*, **252**, 8329–32.

Mooseker, M. S. (1983). Actin-binding proteins of the brush border. *Cell.*, **35**, 11–13.

Mooseker, M. S. & Tilney, L. G. (1975). The organization of an actin filament-membrane complex: filaments polarity and membrane attachment in the microvilli of intestinal epithelial cells. *J. Cell Biol.*, **67**, 725–43.

Nagai, R. & Hayama, T. (1979). Ultrastructure of the endoplasmic factor responsible for the cytoplasmic streaming in *Chara* internodal cells. *J. Cell Sci.*, **36**, 121–36.

Nagai, R. & Hayma, T. (1979). Ultrastructural aspect of cytoplasmic streaming in Characean cells. In *Cell Motility: Molecules and Organization*, ed. S. Hatano, H. Ishikawa & H. Sato, pp. 321–37. Tokyo: University of Tokyo Press.

Nagai, R. & Rebhun, L. I. (1966). Cytoplasmic microfilaments in streaming *Notella* cells. *J. Ultrastruct. Res.*, **14**, 571–89.

Nagai, R., Yoshimoto, Y. & Kamiya, N. (1978). Cyclic production of tension force in the plasmodial strand of *Physarum polycephalum* and its relation to microfilament morphology. *J. Cell Sci.*, **33**, 205–25.

Nakamura, K. & Watanabe, S. (1978). Myosin-like protein and actin-like protein from *Escherichia coli K12 C600*. *J. Biochem.*, **83**, 1459–70.

Ogihara, S. (1982). Calcium and ATP regulation of the oscillatory torsional movement in a Triton model of *Physarum* plasmodial strands. *Exp. Cell Res.*, **138**, 377–84.

Ogihara, S. & Condeelis, J. S. (1983). Electron microscopic localization of myosin and 120-K actin-binding protein in the cortical actin matrix using IgG-gold. *J. Cell Biol.*, **97**, 270a.

Ogihara, S., Ikebe, M., Takahashi, K. & Tonomura, Y. (1983). Requirement of phosphorylation of *Physarum* myosin heavy chain for thick filament formation, actin activation of $Mg^{2+}$-ATPase activity, and $Ca^{2+}$-inhibitory superprecipitation. *J. Biochem.*, **93**, 205–23.

Ogihara, S. & Kuroda, K. (1979). Identification of a birefringent structure which appears and disappears in accordance with the shuttle streaming in *Physarum* plasmodia. *Protoplasma*, **100**, 167–77.

Ogihara, S. & Tonomura, Y. (1982). A novel 36,000-dalton actin-binding protein purified from microfilaments in *Physarum* plasmodia which aggregates actin filaments and blocks actin–myosin interaction. *J. Cell Biol.*, **93**, 604–14.

Palevitz, B. A. & Hepler, P. K. (1975). Identification of actin *in situ* at the ectoplasm–endoplasm interface of *Nitella*: Microfilament chloroplast association. *J. Cell Biol.*, **65**, 29–38.

Perry, S., Margareth, A. & Adelstein, R. (1977). *Contractile System in Nonmuscle Tissues*. Amsterdam: Elsevier.

Pollard, T. D. (1976). Cytoskeletal function of cytoplasmic contractile proteins. *J. Supramol. Struct.*, **5**, 317–34.

Pollard, T. D. & Korn, E. D. (1973). *Acanthamoeba* myosin. I. Isolation from *Acanthamoeba castellanii* of an enzyme similar to muscle myosin. *J. Biol. Chem.*, **248**, 4682–90.

Pollard, T. D. & Weihing, R. R. (1974). Actin and myosin in cell movement. *Crit. Rev. Biochem.*, **2**, 1–65.

Sanger, J. W. (1975). Changing patterns of actin localization during cell division. *Proc. Natl. Acad. Sci. (USA)*, **72**, 1913–16.

Schloss, J. A. & Goldman, R. D. (1980). Microfilaments and tropomyosin of cultured mammalian cells: isolation and characterization. *J. Cell Biol.*, **87**, 633–42.

Sheterline, P. (1983). *Mechanisms of Cell Motility: Molecular Aspects of Contractility*. New York: Academic Press.

Takahashi, K., Ogihara, S., Ikebe, M. & Tonomura, Y. (1983). Morphological aspects of thiophosphorylated and dephosphorylated myosin molecules from the plasmodium of *Physarum polycephalum*. *J. Biochem.*, **93**, 1175–83.

Taylor, D. L. & Condeelis, J. S. (1979). Cytoplasmic structure and contractility in amoeboid cells. *Int. Rev. Cytol.*, **56**, 57–144.

Tilney, L. G. (1975). Actin filaments in the acrosomal reaction of *Limulus* sperm: motility generated by alterations in the packing of the filaments. *J. Cell Biol.*, **64**, 289–310.

Tilney, L. G., DeRosier, D. J. & Mulroy, M. J. (1980). The organization of actin filaments in the stereocilia of cochlea hair cells. *J. Cell Biol.*, **86**, 244–59.

Tilney, L. G., Hatano, S., Ishikawa, H. & Mooseker, M. (1973). The polymerization of actin: its role in the generation of the acrosomal process of certain echinoderm sperm. *J. Cell Biol.*, **59**, 109–26.

Weeds, A. (1982). Actin-binding proteins – regulators of cell architecture and motility. *Nature*, **296**, 811–16.

Williamson, R. E. (1975). Cytoplasmic streaming in *Chara*: a cell model activated by ATP and inhibited by cytochalasin B. *J. Cell Sci.*, **17**, 655–68.

Wohlfarth-Bottermann, K. E. (1962). Weitreichende, fibrillare Protoplasma differenzierungen und ihre Bedeutung fur die Protoplasma stromung II: Lichtmikroscopische Darstellung. *Protoplasma*, **54**, 514–39.

Yoshimoto, Y., Matsumura, F. & Kamiya, N. (1981). Simultaneous oscillations in $Ca^{2+}$ efflux and tension generation in permealized plasmodial strand of Physarum. *Cell Motil.*, **1**, 433–43.

# 5    Dynein ATPase

Like the myosin–actin system, the dynein–tubulin system can also transduce the energy of ATP hydrolysis into mechanical work. Dynein and tubulin have functions similar to myosin and actin, respectively. Dynein is a huge molecule with an ATPase activity. It exists in eukaryotic cilia or flagella and functions in their bending movement. Tubulin molecules are normally polymerized and assembled into microtubules, which have outer diameters of 27–32 nm and lengths ranging from several micrometers to several millimeters. Microtubules are found in a wide variety of eukaryotes ranging from yeast and protozoa to higher plants and animals. Microtubules are believed to participate not only in the ciliary and flagellar movements of eukaryotes but also in many other motile functions, such as elongation of spindles during mitosis, movement of intracellular particles in the cytoplasm, movement of nerve substances and vesicles in nerve axons and secretion of hormones or other substances from cells. Among these systems, dynein has been isolated only from eukaryotic cilia and flagella. Microtubules also participate in the maintenance of cell shape. They form a network that covers the entire cytoplasm and serve as part of the cytoskelton (Chapter 4).

In this chapter, the molecular aspects of ciliary and flagellar movement by the dynein–tubulin system will be discussed in comparison with muscle contraction. For general characteristics of the dynein–tubulin system, several reviews have been published (Sleigh, 1974; Brokaw & Gibbons, 1975; Blum & Hines, 1979; Warner & Mitchell, 1980; Gibbons, 1981; Haimo & Rosenbaum, 1981).

# Cilia and flagella

## *Ultrastructure of cilia and flagella*

Cilia and flagella are organelles protruding from the cell surface of eukaryotes. They produce a flow of fluid and thus cause movement of the cell body as well as the movement of various substances along the cell surface. Prokaryotic flagella, in which the major structural protein is flagellin, move by a unique mechanism that is different from that of eukaryotic flagella, and ATP is not directly involved in their movement (see Berg, Manson & Conley, 1982).

The ultrastructures of cilia and flagella vary among organisms. Cilia and flagella having a 9 + 2 structure called an axoneme are used almost exclusively in the study of the mechanism of ciliary and flagellar movement (Gibbons, 1981). A simplified diagram of the transverse section of the 9 + 2 axoneme is shown in Fig. 5.1. In the center of the ciliary (or flagellar) axoneme, a pair of singlet microtubules (central pair tubules) are surrounded by nine pairs of doublet microtubules (outer doublets). One of the pair of microtubules which constitute an outer doublet forms a complete circle (A-tubule), while the other forms an incomplete C-shaped circle (B-tubule). These microtubules are hollow tubules with an outer diameter of 18–30 nm and are composed of tubulin molecules that are about $50\,000 \times 2$ in molecular weight. There are 13 tubulin molecules per cross section of central pair tubules and each A-tubule, while there are 11 for each B-tubule (Tilney *et al.*, 1973).

The central sheath projections protrude from the central pair tubules in a nearly circular form. They are arranged with a repeat of 14–16 nm along the longitudinal axis of the axoneme. Two arms extend from each A-tubule toward the B-tubule of the adjacent doublet. In electron micrographs of the transverse section of the axoneme, the inner arm is gently curved, while the outer arm has a hook-shaped structure. In electron micrographs of negatively stained outer doublets, the arms are arranged with a repeat of 23–24 nm along the A-tubule and are tilted toward the base of the axoneme (Sale & Satir, 1977). When cilia and flagella are viewed from base to tip, the arms always extend clockwise from the A-tubule. The outer doublets are numbered according to their position relative to the central pair tubules, as shown in Fig. 5.1. The effective stroke of cilia and flagella occurs in the direction defined by the plane perpendicular to the plane that contains the central pair tubules. Other substructures of the axoneme are the nexin link, an elastic bridge which joins the A-tubules of adjacent doublets, and the radial spoke, which extends from the A-tubule toward the central sheath projections. The radial spokes appear in sets of three in longitudinal sections. The three spokes in a given set are spaced unevenly but with regularity: the distance between the two spokes proximal to the tip of the axoneme is 21–24 nm and that between the two spokes proximal to the base of axoneme is 28–32 nm (Warner, 1976). The tip of the radial spoke is enlarged and forms a structure called the spoke head.

Fig. 5.1 The transverse section of the axoneme (9 + 2 structure) of eukaryotic cilia and flagella. The direction of view is from flagellar (ciliary) base to top. Numbers are the doublet numbers determined from their positions relative to the central pair tubules. The effective strokes in the ciliary beat cycle is toward doublets 5 and 6 (arrow).

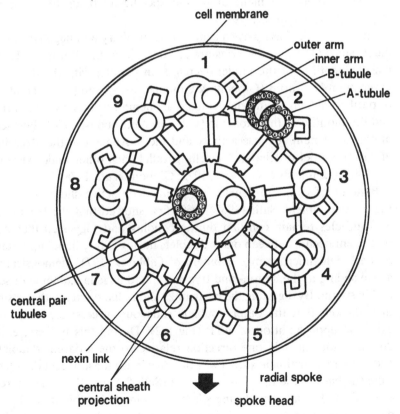

## Establishment of the sliding theory in ciliary and flagellar movement

Following the discovery by Engelhardt (1946) that the flagella of sperm have an ATPase activity and the finding by Hoffmann-Berling (1955) that the bending movement of the glycerine model of flagella is caused by ATP, it has been widely accepted that ciliary and flagellar movement requires ATP as an energy source.

Gibbons (1965, 1966) purified an ATPase from the cilia of *Tetrahymena* by treating it with EDTA at low ionic strength and named it dynein. Gibbons (1963) also observed, by electron microscopy, that the axoneme loses its arms upon extraction of dynein and that these lost arms can be reconstructed on the A-tubules by the addition of the purified dynein molecules in the presence of $Mg^{2+}$. Localization of dynein in the arms has also been confirmed by an observation that an antibody prepared against a tryptic fragment of dynein specifically binds to the arms (Ogawa, Mohri & Mohri, 1977).

The dynein ATPase plays a central role in ciliary and flagellar movement, since (1) when the arms are removed from the Triton model of flagella, motility of the model decreases in parallel with the decrease in the number of arms (Gibbons & Gibbons, 1973), and (2) the addition of purified dynein to an arm-depleted axoneme restores both the arms and the motility of the axoneme (Gibbons & Gibbons, 1976). Furthermore, the dynein ATPase activity and the movement in the glycerine or Triton model of cilia and flagella parallel each other under various conditions (Gibbons & Gibbons, 1972; Gibbons *et al.*, 1978).

Bending of cilia and flagella is explained by active sliding between the outer doublets. Satir (1965, 1968) first showed that the lengths of microtubules remain constant during bending and suggested that displacement occurs between outer doublets during ciliary bending. Later, by dark-field microscopy, Summers & Gibbons (1971) demonstrated active sliding in the Triton- and trypsin-treated flagella axoneme of sea urchin sperm. By digesting the nexin links and radial spokes with trypsin, they showed that the outer doublets slide out successively from the axoneme upon addition of ATP (Fig. 5.2). During this process, each doublet slides in only one direction relative to the adjacent doublet, and the arms function in such a way that they push the adjacent doublets from the base to the tip of the axoneme (Sale & Satir, 1977). The force exerted by dynein arms during sliding is estimated to be about 1 pN/dynein arm (Kamimura & Takahashi, 1981), which agrees well with

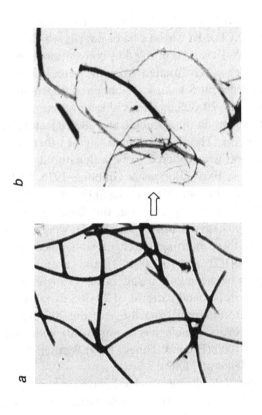

Fig. 5.2 Negatively stained trypsin-treated axonemes of *Tetrahymena* cilia before (*a*) and after (*b*) addition of ATP. Outer doublets are extruded from the axoneme by the addition of ATP (Takahashi & Tonomura, 1978).

the force of 2–5 pN estimated for the myosin crossbridges of muscle. This sliding movement between outer doublets constitutes the basic mechanism of the bending movement of cilia and flagella.

## Dynein ATPase

### *Structure of dynein ATPase*

Dynein can be extracted with a low concentration Tris-EDTA solution or 0.5–0.6 M NaCl or KCl from cilia or flagella which have been pretreated with Triton X-100 and is purified by conventional procedures (Bell *et al.*, 1982). Even when isolated from the same material, two types of dynein with different S values are obtained depending on the extraction conditions. The 20–30S dynein is able to rebind to the outer doublet, to restore the arms on the A-tubule and also to restore bending movement of the axoneme. The 10–15S dynein, on the other hand, does not restore the arms and is considered to be a denatured form of the 20–30S dynein (Gibbons, 1965; Gibbons & Gibbons, 1976, 1979; Blum & Hines, 1979; Gibbons & Fronk, 1979). Using SDS-gel electrophoresis, it has been shown that the 21S dynein from the flagella of sea urchin sperm is composed of two heavy chains with molecular weights of 330 000 (A$\alpha$) and 320 000 (A$\beta$); three intermediate chains with molecular weights of 122 000 (ICI), 90 000 (IC2) and 76 000 (IC3); and several light chains with molecular weights between 24 000 and 14 000 (Tang *et al.*, 1982). The SDS-gel electrophoretic band patterns of 20–30S dyneins prepared from other materials are different from those of the 20S dynein from sea urchin. However, every dynein has major bands with molecular weights in excess of 300 000 (Blum & Hines, 1979; Warner & Mitchell, 1980; Gibbons, 1981; Bell *et al.*, 1982).

The molecular weight of the dynein arm has been estimated by electron microscopy to be 1 300 000–1 820 000, which is higher than that of the major polypeptides of dynein (310 000–560 000) (Warner, Mitchell & Perkins, 1977; Gibbons & Fronk, 1979) and is nearly the same as that of 21S dynein (1 250 000). It has been reported that the 21S dynein can be dissociated into three kinds of subunit complexes: A$\alpha$, A$\beta$ + IC1 and IC2 + IC3. The 21S particle can be reconstructed by mixing isolated A$\alpha$ chain with the remaining chains (A$\beta$ + IC1 and IC2 + IC3). The reconstructed 21S particle contains A$\alpha$, A$\beta$ and IC1–3 in about the same proportions as found in the original dynein (Bell &

Gibbons, 1982; Tang *et al.*, 1982). In view of their molecular weights, it is almost certain that one 21S particle contains one mole each of A$\alpha$ and A$\beta$, but the amount of intermediate and light chains in the 21S particle remains obscure.

Although the reconstructed 21S particle is not able to restore motility to the arm-depleted axoneme, it is capable of inhibiting the restoration of movement of the axoneme by the original 21S dynein. Both A$\alpha$ and A$\beta$ possess ATPase activity with different characteristic properties and are thus expected to participate in different aspects of the function of the arm. In skeletal muscle, the two myosin heads that have ATPase activity are different from each other in both mechanism of ATPase reaction and physiological functions, as has been described in Chapter 2. In connection with this, further study on the differences between A$\alpha$ and A$\beta$ should be promising.

The function of A$\alpha$ and A$\beta$ has been studied by limited proteolytic digestion of dynein. The treatment of dynein with trypsin produces a subfragment (fragment A) that retains ATPase activity. Fragment A of dynein from sea urchin sperm flagella is composed of two polypeptides with molecular weights of 190 000 and 135 000. Fragment A has an ATPase activity but lacks the ability to bind to the outer doublet (Ogawa, 1973; Ogawa & Mohri, 1975). The 21S dynein loses its ability to bind to the A-tubule by more mild tryptic digestion. Upon this treatment, only A$\alpha$ is digested, and the molecular weight of A$\alpha$ decreases from 330 000 to 275 000. Thus, it is suggested that the binding site for the A-tubule resides on A$\alpha$. On the other hand, the arm's ATP-sensitive binding site for the B-tubule (discussed later in this chapter) has been speculated to reside on A$\beta$. This speculation is supported by the findings that the restoration of motility to the arm-depleted axoneme by 21S dynein is inhibited not only by the reconstructed 21S particle but by the A$\beta$/IC complex as well and that this inhibition is released by addition of ATP (Bell & Gibbons, 1982).

The inner and outer arms that form a pair on the A-tubule are distinctly different in their structure (Fig. 5.1). In addition, there have been many reports suggesting that these two arms are also different in their constituent proteins, including dynein (Kincaid, Gibbons & Gibbons, 1973; Ogawa, Negishi & Obika, 1982). On the other hand, it has been reported that the dynein isolated from the outer arm can bind to the binding site on the inner arm (Gibbons & Fronk, 1979). It has also been shown in the Triton model of flagella that, when the outer arms are selectively removed, only the frequency of movement decreases and the wave form remains constant (Gibbons & Gibbons, 1973). Moreover, it is shown that the inner arm alone can cause the sliding movement between doublets (Hata *et al.*, 1980). Thus, the functions of the two arms cannot be qualitatively distinguished.

## Basic properties of the dynein ATPase reaction

The dynein ATPase reaction has many properties in common with those of movement in the glycerine and Triton models of cilia and flagella (Blum & Hines, 1979). The dynein ATPase requires a divalent cation for its activity. Among the divalent cations, $Mg^{2+}$ is the most effective, followed by $Ca^{2+}$ and $Mn^{2+}$ in this order.

The dynein ATPase exhibits far stricter substrate specificity than the myosin ATPase: aside from ATP, only two compounds, 2'-deoxy-ATP and 3'-deoxy-ATP, are presently known to serve both as the substrate for the dynein ATPase reaction and as the inducer of movement in the glycerine and Triton models of flagella (Takahashi & Tonomura, 1979). These substrates are believed to function in the form of a $Mg^{2+}$ complex. Unhydrolyzable ATP analogs, AMPPNP and AMPPCP, inhibit both the ATPase activity and movement in the Triton model of flagella (Penningroth & Witman, 1978). Erythro-9-3-2(hydroxynonyl) adenine selectively inhibits the dynein ATPase activity (Penningroth *et al.*, 1982). In analogy with other ATPases, the dynein ATPase is noncompetitively inhibited by a several micromolar concentration of vanadate (Gibbons *et al.*, 1978).

The $K_m$ value for ATP in the dynein ATPase reaction varies among reports, ranging from 1 to 100 μM (Blum & Hines, 1979; Bell *et al.*, 1982). In the ATPase of the trypsin-treated axonemes of *Tetrahymena* cilia, the plot of $v^{-1}$ against $ATP^{-1}$ yields two straight lines and thus gives two $K_m$ values of about 1 μM and 10 μM (Takahashi & Tonomura, 1978). On the other hand, the movement in the Triton models of cilia and flagella occurs in the presence of micromolar concentrations of ATP, and the analysis of the relationship between the beat frequency and ATP concentration reveals apparent $K_m$ values of about 0.2 mM (Gibbons & Gibbons, 1972). However, the ATP-concentration dependence of motility and the ATPase activity cannot be compared directly, since the effects of various factors, e.g., viscosity of surrounding fluid, must be considered when analyzing motility.

The modification of SH groups of the 20–30S dynein results both in a transient increase of ATPase activity and in the loss of ability to form the arms. For these changes to occur, 30–40 mol of PCMB or NEM must bind per $10^5$ g of dynein molecules. However, no specific SH group that corresponds to either $SH_1$ or $SH_2$ of myosin has so far been identified in dynein (Shimizu & Kimura, 1977; Blum & Hunes, 1979; Gibbons & Fronk, 1979; Gibbons & Gibbons, 1979).

It has been reported that calmodulin can bind in a $Ca^{2+}$-dependent manner to the 14S and 30S dyneins extracted from *Tetrahymena* cilia and cause sixfold and twofold increases in the ATPase activity, respectively (Blum *et al.*, 1980). The physiological meaning of this phenomenon remains to be solved.

## Tubulin and microtubules

### Structure of tubulin

Tubulin, the major constituent protein of microtubules, is present in a wide variety of eukaryotes (Mohri, 1976; Luduena, 1979). From its assumed role in the sliding theory, tubulin is functionally analogous to actin in muscle. Tubulin generally exists as a dimer with a molecular weight of approximately 100 000 and is dissociated into monomers of 50 000 each in the presence of guanidine-HCl or SDS. There are two types of tubulin monomers with very similar molecular weights; the larger monomer is called $\alpha$ and the smaller monomer is called $\beta$. Since purified tubulin always contains $\alpha$ and $\beta$ in equal amounts and the majority of the dimers obtained by chemical crosslinking consists of both $\alpha$ and $\beta$ forms, the tubulin dimer is believed to be an $\alpha\beta$ heterodimer.

The primary structure has been determined for $\alpha$- and $\beta$-tubulins from chicken, porcine and rat brains (Krauhs *et al.*, 1981; Lemischka *et al.*, 1981; Ponstingl *et al.*, 1981; Valenzuela *et al.*, 1981). The chemical structures of $\alpha$- and $\beta$-tubulins are quite similar to each other, and homology is evident in 40–45% of the entire amino acid sequence. However, there is no homology between the primary structures of tubulin and actin.

The cytoplasmic microtubules, the central pair tubules, B-tubules and A-tubules of the axoneme are all different from each other in sensitivity to colchicine, cold treatment and $Ca^{2+}$ treatment. These four microtubules are different from each other in their amino acid compositions of both $\alpha$- and $\beta$-tubulins. Microheterogeneity in the tubulin of individual organisms has also been shown (Luduena, 1979).

## *Polymerization of tubulin*

Although microtubules are hardly assembled at all from purified tubulin alone *in vitro*, the addition of protein factors, either microtubule-associated proteins (MAPs) or factor $\tau$ (tau) markedly enhances the polymerization of the tubulin into microtubules (Kirshner, 1978; Sakai, 1979; Scheele & Borisy, 1979). The factor $\tau$ contains several polypeptides ranging in molecular weight from 58 000 to 65 000, while the MAPs are composed of $MAP_1$ and $MAP_2$ with molecular weights of about 350 000 and 270 000–290 000, respectively (Murphy, 1982). Both MAPs and factor $\tau$ are believed to be required for the formation of a nucleus that is indispensable to the formation and stabilization of microtubules. Thus, singlet microtubules can be assembled easily *in vitro*, but assembly of the doublet microtubules seen in cilia and flagella has not yet succeeded *in vitro*.

During polymerization of actin, 1 mol of ATP binds to each mole of G-actin and is hydrolyzed to free Pi and bound ADP, which is unexchangeable with surrounding ADP. In the case of tubulin, each dimer has 2 guanosine triphosphate (GTP)-binding sites. One is called the E-site, since its bound GTP is readily exchangeable with external GTP, and the other is called the N-site, since the GTP bound to it is nonexchangeable. During polymerization of tubulin, the GTP at the N-site remains intact, whereas the GTP at the E-site is split to GDP and becomes nonexchangeable with the external GTP. Polymerization of tubulin requires GTP on the E-site, and GDP cannot induce the polymerization (Jacobs, 1979).

## *Structure of microtubules*

Microtubules have a hollow, tubular structure with an outer diameter of 27–32 nm (Fig. 5.3; Roberts & Hyams, 1979; Sakai, Mohri & Borisy, 1982; Wilson, 1982). Since the bonds between the longitudinally adjacent tubulin molecules in the microtubule are stronger than those between cross-sectionally adjacent molecules, microtubules are thought to be made up of 13 protofilaments that parallel the long axis. The distance between tubulin monomers in a given protofilament is about 4 nm.

Microtubules have fixed polarity, as do actin filaments, and this structural orientation is important for the sliding movement between outer doublets. The polarity of outer doublet microtubules can be determined by examining the tilt angle of the bound dynein or the space between radial spokes (discussed earlier in this chapter; see Haimo, Telzer & Rosenbaum, 1979). When tubulin molecules are added to various types of microtubules in the presence of dimethyl sulfoxide at high ionic strength, a specific C-shaped hook that is caused by the excessive polymerization of tubulin appears on the wall of microtubules. The polarity of microtubules is also determined from the direction of the curvature of this hook examined in this section by electron microscopy (Heidemann & Euteneuer, 1982).

Fig. 5.3 Schematic model of a singlet microtubule. *a*, Cross-section. *b*, Side view. The tubulin monomer with a diameter of 4 nm is represented as a sphere.

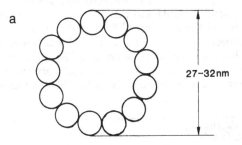

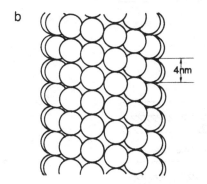

# Molecular mechanism of the sliding movement between outer doublets

## *Binding of dynein arms to microtubules and their dissociation by ATP*

The molecular mechanism of the sliding movement between outer doublets is considered to be analogous to that of muscle contraction (see Fig. 5.4). The sliding movement between outer doublets may include the following three basic processes: (1) Attachment of dynein arms to the adjacent B-tubules to form crossbridges. (2) Sliding movement induced by a change in the state of the crossbridges. (3) Detachment of the arms from the B-tubules, and thus a cycle is completed.

Earlier in this chapter, we described how the dynein arms are bound to specific sites on the A-tubule with a fixed periodicity (Fig. 5.1). However, using electron microscopy and traditional fixation of tissue, the crossbridge formation between adjacent B-tubules were not observed.

Fig. 5.4 The structural relationship between muscle fiber and the axoneme. The projections which forms crossbridges corresponds to the myosin head (HMM) and the dynein arm, respectively, in muscle fiber and the axoneme. The filament structure that supports the projection corresponds to the thick filament in muscle fiber and the A-tubule in the axoneme. The filament structure that moves in response to the crossbridge corresponds to the thin filament in muscle fiber and the B-tubule in the axoneme.

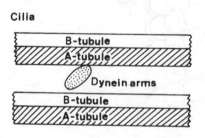

**Muscle**

Thick filament

Myosin head (HMM)

Thin filament

**Cilia**

B-tubule

A-tubule

Dynein arms

B-tubule

A-tubule

Later, Gibbons & Gibbons (1974) observed that, in the absence of ATP, the Triton model of flagella of sea urchin sperm can be set into a state called 'rigor wave', which is assumed to be homologous to the rigor state of muscle. They examined the axonemes in this state by electron microscopy using a different method of fixation and observed that the arms are bound to both the A- and B-tubules, thereby forming cross-bridges.

It has been shown that ATP induces the dissociation of dynein arms from B-tubules. Warner (1978) studied the effects of ATP on the distribution of crossbridges in the Triton model of cilia using electron microscopy. He showed that the number of crossbridges is reduced when the Triton model is reactivated with ATP. In addition, the extent of dissociation of crossbridges is largest between outer doublets 3 and 4 and between 7 and 8, where the largest displacement is thought to occur during axonemal bending. Dissociation of the crossbridges is almost negligible between outer doublets 8 and 9 and 1 and 2, and no displacement is thought to occur between 5 and 6.

The ATP-induced dissociation of the dynein arms from the B-tubules was more clearly demonstrated by electron microscopy using the negative-staining method (Takahashi & Tonomura, 1978). As shown in Fig. 5.5, when dynein is added to the purified outer doublets in the presence of $Mg^{2+}$, it binds to both the A- and B-tubules. When ATP is added, the arms that are bound to the B-tubules are selectively dissociated and rebind to the B-tubules after the hydrolysis of ATP. The $K_d$ value for ATP in the dissociation of arms was estimated to be $10^{-6}$ M by Takahashi & Tonomura (1978) and $1.3 \times 10^{-8}$ M by Mitchell & Warner (1980). It has also been reported that dynein binds to cytoplasmic singlet microtubules and forms arms in the absence of ATP and is dissociated from the microtubules by ATP (Haimo, Telzer & Rosenbaum, 1979).

All ATP analogs that can induce the sliding movement are capable of inducing the dissociation of dynein from the B-tubules (Takahashi & Tonomura, 1978). AMPPNP, a nonhydrolyzable ATP analog, also induces the dissociation of dynein arms. Vanadate ions, which inhibit the dynein ATPase, do not inhibit either dissociation or reassociation of the arms (Sale & Gibbons, 1979; Mitchell & Warner, 1980).

Takahashi & Tonomura (1978) showed with EM using negative staining that when dynein arms bind to the outer doublets in the absence of ATP, both arms have a tendency to point toward the base of the cilia (Figs 5.5 and 5.6). The angle is 48° for the dynein arms bound to A-tubules, while it is 135° for the dynein arms bound to B-tubules.

a

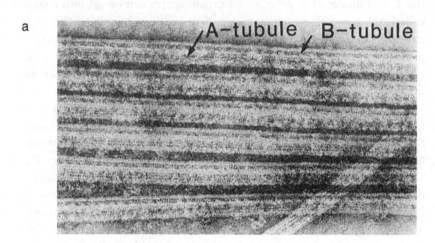

b

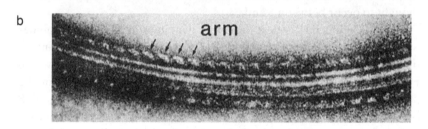

Fig. 5.5 ATP-induced dissociation of reconstituted dynein arms from the B-tubules of outer doublets isolated from *Tetrahymena* cilia. In this figure, all the B-tubules are the upper ones in each doublet. (*a*) Negatively contrasted EDTA-extracted outer doublets. All dynein arms are removed from the A- and B-tubules by the extraction. (*b*) A doublet mixed with purified 30S dynein of *Tetrahymena* cilia in the presence of $Mg^{2+}$. A row of arms appears along both A- and B-tubules, with a repeat of 23 nm each. The arms of both sides are tilted in the same direction. (*c*) The arm-decorated doublets after the addition of ATP. All arms along the B-tubule completely disappear. These arms reappeared after the hydrolysis of added ATP (*d*). (Reproduced from Takahashi & Tonomura, 1978).

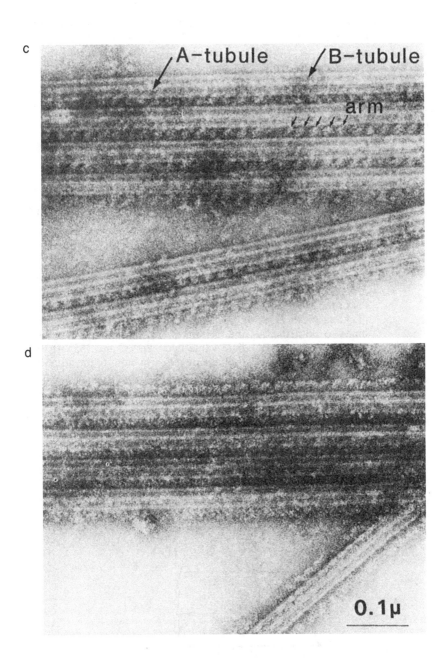

Therefore, when dynein arms are bound to both A- and B-tubules, after ATP hydrolysis the angle is assumed to become about 90°. Thus, ATP induces the change in the angle of the dynein arms through attachment and detachment of arms to the B-tubule. Goodenough & Heuser (1982) and Tsukita *et al.* (1983) also found using the quick-freeze, deep-etch technique that the arms have different configurations in the presence and absence of ATP and suggested that this change in configuration may induce the sliding movement between outer doublets.

## Mechanism of dynein ATPase reaction

To study how the ATPase reaction of dynein is coupled with the sliding between the outer doublets, it is necessary to understand the elementary steps of the dynein ATPase reaction.

It has been found that dynein ATPase has similar reaction interme-

Fig. 5.6 Schematic diagram of the variation of tilt angles of the dynein arm to the A-tubule. When the arm is bound to only the A-tubule (1) or the B-tubule (3), the tilt angle is 48° and 135°, respectively (Takahashi & Tonomura, 1978). Accordingly, when the arm binds to both the A- and B-tubules (2), the angle is assumed to take on an intermediate value between 48° and 135°. If this assumption is correct, when the dynein arm binds to the adjacent B-tubule, this tubule slides from axonemal base to tip. This direction agrees with the actual sliding direction shown by Sale & Satir (1977).

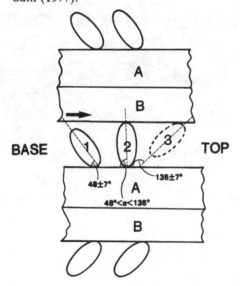

diates to those of myosin ATPase. Glycerine-treated sea urchin sperm catalyzes the $^{18}O$-exchange reaction between water and phosphate during an ATPase reaction, as observed using myosin ATPase (Barclay & Yount, 1972). This reaction occurs in myosin by the reversible step between E–ATP + $H_2O$ and E–P–ADP. Using dynein extracted from *Tetrahymena* cilia, Takahashi & Tonomura (1979) found that rapid liberation of Pi occurs at the initial phase of the ATPase reaction when the reaction is terminated by the addition of TCA (Fig. 5.7). Therefore, the following mechanism is given for the reaction of dynein ATPase.

$$E + ATP \rightleftharpoons E\text{-}ATP \rightleftharpoons E\text{-}P\text{-}ADP \dashrightarrow E + ADP + Pi$$

The existence of enzyme-bound ATP and ADP during the ATPase reaction was later shown by Terashita *et al.* (1983) using 21S dynein extracted from flagella of sea urchin sperm.

The sliding movement between doublets is coupled with ATP hydrolysis and dynein ATPase is thought to be activated by the sliding move-

Fig. 5.7  The initial burst of Pi liberation in the ATPase reaction of dynein extracted from *Tetrahymena* cilia with KCl. The extrapolated line of the plots does not pass through the origin, which suggests the rapid formation of a TCA-labile dynein phosphate complex as an intermediate in this ATPase reaction. Conditions: 0.1 mM ATP, 2.5 mM $MgSO_4$, 0.2 mM EDTA, 0.5 mM DTT, 30 mM Tris-HCl, pH 7.8, 0 °C. Open circle, 0.46 mg/ml dynein; filled circle, 0.49 mg/ml dynein.

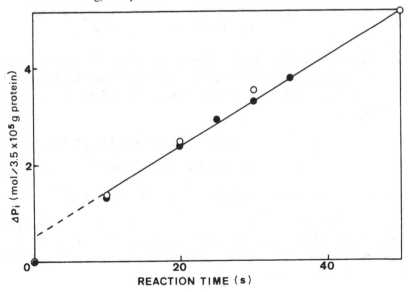

ment. In fact, it has been reported that the ATPase activity is reduced by 30–60% when movement in the Triton model of sea urchin sperm is repressed by removal of the sperm head of by increasing the viscosity of the surrounding fluid (Gibbons & Gibbons, 1972; Brokaw & Simonick, 1977).

There have been a number of reports showing that the addition of dynein-free doublets to isolated dynein results in a twofold to sixfold increase in the dynein ATPase activity (Gibbons & Fronk, 1979). This phenomenon is analogous to the activation of myosin ATPase by F-actin. Moreover, it has been reported that the ATPase function of the fragment A of dynein isolated from sea urchin sperm is strongly activated by the tubulin derived from the B-tubules but is far less effectively activated by the tubulin from the A-tubules (Ogawa, 1973).

Johnson & Porter (1982) measured the dissociation rate of dynein (E) from the singlet microtubules (MT) reconstituted from brain tubulin and showed that the dissociation of dynein arms occurs via two steps (MT–E + ATP → MT–E–ATP → MT + E–ATP). The rate constants for these steps were $2.4 \times 10^{-6}/\text{M/s}$ and about $100/\text{s}$, respectively.

At present very little is known about the mechanism of activation of dynein ATPase by microtubules. By combining the mechanism of dynein ATPase with that of the dissociation of dynein arms from microtubules the following mechanism is considered for the ATPase reaction of the dynein–tubulin system:

$$MT–E + ATP \longleftrightarrow MT–E–ATP \qquad\qquad MT–E + ADP + Pi$$
$$MT + E–ATP \longleftrightarrow MT + E–P–ADP$$

However, by analogy to the actomyosin ATPase (Chapter 3), it is considered that MT–E–ATP is in rapid equilibrium with the dissociated dynein (MT + E–ATP) and that ATP is hydrolyzed by the following two routes:

$$MT–E + ATP \longleftrightarrow MT–E–ATP \longleftrightarrow MT–E–ADP–P \longleftrightarrow MT + ADP + Pi$$
$$MT + E–ATP \longleftrightarrow MT + E–ADP–P$$

## Molecular mechanism of the sliding movement

Several models have been proposed for the molecular mechanism of the sliding movement (Sale & Gibbons, 1979; Goodenough & Heuser, 1982; Satir, 1982). However, these models are built on extremely little experimental evidence.

As described in the previous section, dynein arms bind to A- and B-tubules at different angles (48° and 135°, respectively), and when a dynein arm binds to both A- and B-tubules after ATP hydrolysis, the angle of the arm is assumed to be intermediate (90°). Although the mechanism of the dynein ATPase reaction and the relationship between the ATPase reaction and the binding of dynein arms with B-tubules have not yet been established, dynein arms are considered to have at least two states: (1) dynein arms dissociated from the B-tubule with angles of 48° and (2) dynein arms bound to both A- and B-tubules at an angle of 90° in the absence of ATP. Therefore, if we assume that dynein arms bind to B-tubules at an angle of 48° during ATP hydrolysis, the molecular mechanism of the sliding movement of dynein arms can be shown as in Fig. 5.8 (Takahashi & Tonomura, 1978). The dynein arms that form E–ATP or E–P–ADP and dissociate from B-tubules, rebind to the adjacent B-tubules at an angle of 48°. When MT–E–ATP or MT–E–P–ADP decomposes into MT–E and ADP + Pi, the angle of the attached arm changes from 48° to 90°, and the B-tubule slides past the A-tubule. By ATP binding to dynein ATPase, the dynein arm detaches from the B-tubule and the angle returns to 48°, and a cycle is completed.

Fig. 5.8 A molecular model for the active sliding between doublet microtubules. A, A-tubule; B, B-tubule. ATP-binding site shown on
the arm. Filled circle, ATP-sensitive binding site of the arm for
the adjacent B-tubule. In this model, the dynein arm is assumed
to show the two states with tilt angles of 48° and 90°, respectively.
For further explanation, see the text.

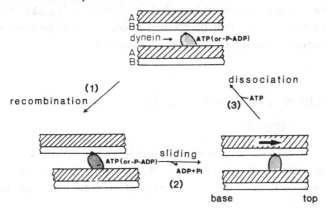

## Mechanism of the sliding–bending conversion

Conversion of the sliding movement to the bending movement is believed to require, at least (1) that the axoneme contains other structural components in addition to the outer doublets and the arms, and (2) that the on–off controls of the interdoublet sliding movement at individual sites along the axoneme are systematically integrated with time courses specific for each site.

It has been reported that the sliding movement does occur but the bending movement is not induced in axonemes in which the radial spokes, nexins or central pair microtubules have been removed by trypsin treatment (Summers & Gibbons, 1971) or mutation (Witman, Plummer & Sander, 1978). These observations suggest that structural components such as radial spokes, nexins and central pair microtubules play an important role in the conversion of sliding movement to bending movement.

Sugino and Naitoh (1982) have succeeded in drawing wave patterns identical to those of living cilia and flagella by computer simulation using a program in which the on–off controls of sliding are assumed to occur in an integrated manner along the length of the axoneme.

Furthermore, Shingyoji, Murakami & Takahashi (1977) showed that a local bending response occurred when the Triton model of sea urchin sperm flagella was locally reactivated by ionophoretic application of ATP. However, the mechanism that integrates these on–off controls of sliding that occur at various sites along the axoneme still remains obscure.

As described in Chapter 3, muscle contraction is regulated by $Ca^{2+}$. In ciliary and flagellar movement the orientation of effective strokes, the direction of wave propagation and the wave pattern are affected by micromolar concentrations of $Ca^{2+}$ (Naitoh & Kaneko, 1972; Blum & Hines, 1979). Further research is needed to elucidate the molecular mechanisms involved in this type of regulation.

## Dynein ATPase in other motile systems

### Mitosis

During eukaryotic cell division, chromosomes are separated and distributed to two daughter cells by a structure called the spindle. The spindle is made up of a bundle of microtubules, and its detailed structural and biochemical characteristics have been reviewed (Sakai, 1978; McIntosh, 1979; Petzelt, 1979; Inoue, 1981).

Involvement of a dynein-like ATPase in mitosis is strongly suggested by the findings that an antibody prepared against the fragment A of sea urchin sperm dynein binds to the spindle and also inhibits spindle movement (Sakai, 1978). The spindle movement is induced by $Mg^{2+}$ and ATP and is inhibited by inhibitors of dynein ATPase such as vanadate, erythro-9-[3-(2-hydroxynonyl)] adenine and sulfhydryl reagents (Sakai, 1978; Cande, 1982). Furthermore, the antibody prepared against myosin from sea urchin egg inhibits only cytokinesis and does not affect nuclear division (Mabuchi & Okuno, 1977). However, Cande (1982) has recently shown that movement of the spindle in early anaphase (anaphase A, when chromosomes are pulled toward the poles) does not require exogeneous ATP and is not affected by inhibitors of dynein ATPase. Only the movement in late anaphase (anaphase B, when the distance between the two poles increases and the spindles elongate) is caused by dynein ATPase. Therefore, it is still uncertain what causes chromosome movement during anaphase A.

### Movement of intracellular particles

As has been described earlier in this chapter, microtubules are believed to be involved in the movement of intracellular particles such as secretory vesicles, pigment granules and mitochondria (Hyams & Stebbings, 1979; Schwartz, 1979). Electron microscopy has revealed that microtubules are present along the paths of particle movement and that these particles are found linked to microtubules by crossbridges. In addition, the movement of granules is inhibited by the depolymerization of microtubules with colchicine. Involvement of a dynein-like ATPase in these movements is suggested by the result that the movement of pigment granules in pigment cells is blocked by vanadate or erythro-9-[3-(2-hydroxynonyl)] adenine (Beckerle & Porter, 1982; Stearns & Ochs, 1982).

## References

Barclay, R. & Yount, R. G. (1972). Evidence for myosin-like intermediates in the hydrolysis of adenosine triphosphate by sperm-tail flagella. *J. Biol. Chem.*, **247**, 4098–100.

*Dynein ATPase*

Beckerle, M. C. & Porter, K. (1982). Inhibitors of dynein activity block intracellular transport in erythrophores. *Nature*, **295**, 701–3.

Bell, C. W., Fraser, C. L., Sale, W. S., Tang, W. Y. & Gibbons, I. R. (1982). Preparation and purification of dynein. In *Methods in Enzymology*, vol. 85, ed. D. W. Grederiksen & L. W. Cunningham, pp. 450–74. New York: Academic Press.

Bell, C. W. & Gibbons, I. R. (1982). Structure of the dynein-1 outer arm in sea urchin sperm flagella. II. Analysis by proteolytic cleavage. *J. Biol. Chem.*, **257**, 516–22.

Berg, H. C., Manson, M. D. & Conley, M. P. (1982). Dynamics and energetics of flagellar rotation in bacteria. In *Prokaryotic and Eukaryotic Flagella, Symposia of the Society for Experimental Biology*, vol. 35, ed. W. B. Amos & J. G. Duckett, pp. 1–31. Cambridge University Press.

Blum, J. J., Hayes, A., Jamieson, G. A., Jr. & Varaman, T. C. (1980). Calmodulin confers calcium sensitivity on ciliary dynein ATPase. *J. Cell Biol.*, **87**, 386–97.

Blum, J. J. & Hines, M. (1979). Biophysics of flagellar motility. *Q. Rev. Biophys.*, **12**, 103–80.

Brokaw, C. J. & Gibbons, I. R. (1975). Mechanisms of movement in flagella and cilia. In *Swimming and Flying in Nature*, ed. T. Y. Wu, C. J. Brokaw & C. Brennen. pp. 89–126. New York & London: Plenum.

Brokaw, C. J. & Simonick, T. F. (1977). Mechanochemical coupling in flagella. V. Effects of viscosity on movement and ATP-dephosphorylation of Triton-demembranated sea-urchin spermatozoa. *J. Cell Sci.*, **23**, 227–41.

Cande, W. Z. (1982). Nucleotide requirements for anaphase chromosome movements in permeabilized mitotic cells: anaphase B but not anaphase A requires ATP. *Cell*, **28**, 15–22.

Engelhardt, V. A. (1946). Adenosinetriphosphatase properties of myosin. *Adv. Enzmol.*, **6**, 147–91.

Gibbons, B. H. & Gibbons, I. R. (1972). Flagellar movement and adenosine triphosphatase activity in sea urchin sperm extracted with Triton X-100. *J. Cell Biol.*, **54**, 75–97.

Gibbons, B. H. & Gibbons, I. R. (1973). The effect of partial extraction of dynein arms on the movement of reactivated sea urchin sperm. *J. Cell Biol.*, **13**, 337–57.

Gibbons, B. H. & Gibbons, I. R. (1974). Properties of flagellar 'rigor waves' formed by abrupt removal of adenosine triphosphate from actively swimming sea urchin sperm partially extracted with KCl. *J. Cell Biol.*, **63**, 970–85.

Gibbons, B. H. & Gibbons, I. R. (1976). Functional recombination of dynein 1 with demembranated sea urchin sperm partially extracted with KCl. *Biochem. Biophys. Res. Commun.*, **73**, 1–6.

Gibbons, B. H. & Gibbons, I. R. (1979). Relationship between the latent adenosine triphosphatase state of dynein 1 and its ability to recombine functionally with KCl-extracted sea urchin sperm flagella. *J. Biol. Chem.*, **254**, 197–201.

Gibbons, I. R. (1963). Studies on the protein components of cilia from *Tetrahymena pyriformis*. *Proc. Natl. Acad. Sci. (USA)*, **50**, 1002–10.

Gibbons, I. R. (1965). Chemical dissection of cilia. *Arch. Biol. (Liege)*, **76**, 317–52.

Gibbons, I. R. (1966). Studies on the adenosine triphosphatase activity of 14S and 30S dynein from cilia of *Tetrahymena*. *J. Biol. Chem.*, **241**, 5590–6.

Gibbons, I. R. (1981). Cilia and flagella of eukaryotes. *J. Cell Biol.*, **91**, 107s–24s.

Gibbons, I. R., Cosson, M. P., Evans, J. A., Gibbons, B. H., Houck, B., Martinson, K. H., Sale, W. S. & Tang, W. Y. (1978). Potent inhibition of dynein adenosine triphosphatase and of the motility of cilia and sperm flagella by vanadate. *Proc. Natl. Acad. Sci. (USA)*, **75**, 2220–4.

Gibbons, I. R. & Fronk, E. (1979). A latent adenosine triphosphatase form of dynein 1 from sea urchin sperm flagella. *J. Biol. Chem.*, **254**, 187–96.

Goodenough, V. W. & Heuser, J. E. (1982). Substructure of the outer dynein arm. *J. Cell Biol.*, **95**, 798–815.

Haimo, L. T. & Rosenbaum, J. L. (1981). Cilia, flagella, and microtubules. *J. Cell Biol.*, **91**, 125s–30s.

Haimo, L. T., Telzer, B. R. & Rosenbaum, J. L. (1979). Dynein binds to and crossbridges cytoplasmic microtubules. *Proc. Natl. Acad. Sci. (USA)*, **76**, 5759—63.

Hata, T., Yano, M., Mohri, H. & Miki-Noumura, T. (1980). ATP-driven tubule extrusion from axonemes without outer dynein arms of sea-urchin sperm flagella. *J. Cell Sci.*, **41**, 331–40.

Heidemann, S. R. & Euteneuer, U. I. (1982). Microtubule polarity determination based on conditions for tubulin assembly *in vitro*. *Meth. Cell Biol.*, **24**, 207–16.

Hoffmann-Berling, H. (1955). Geisselmodelle und adenosin-triphosphat (ATP). *Biochem. Biophys. Acta*, **16**, 146–54.

Hyams, J. S. & Stebbings, H. (1979). Microtubule associated cytoplasmic transport. In *Microtubules*, ed. K. Roberts & J. S. Hyams, pp. 487–530. New York: Academic Press.

Inoue, S. (1981). Cell division and the mitotic spindle. *J. Cell Biol.*, **91**, 131s–47s.

Jacobs, M. (1979). Tubulin and nucleotides. In *Microtubules*, ed. K. Roberts & J. S. Hyams, pp. 255–77. New York: Academic Press.

Johnson, K. A. & Porter, M. E. (1982). Transient state kinetic analysis of the dynein ATPase. *Cell Motil., Suppl.*, **1**, 101–6.

Kamimura, S. & Takahashi, K. (1981). Direct measurement of the force of microtubule sliding in flagella. *Nature*, **293**, 566–8.

Kincaid, H. L., Jr., Gibbons, B. H. & Gibbons, I. R. (1973). The salt-extractable fraction of dynein from sea urchin sperm flagella: an analysis by gel electrophoresis and by adenosine triphosphatase activity. *J. Supramol. Struct. Cell. Biochem.*, **1**, 461–70.

Kirschner, M. W. (1978). Microtubule assembly and nucleation. *Int. Rev. Cytol.*, **54**, 1–71.

Krauhs, E., Little, M., Kempf, T., Hofer-Warbinek, R., Ade, W. & Ponstingl, H. (1981). Complete amono acid sequence of β-tubulin from porcine brain. *Proc. Natl. Acad. Sci. (USA)*, **78**, 4156–60.

Lemischka, I., Farmer, S., Racaniello, V. R. & Sharp, P. A. (1981). Nucleotide sequence and evolution of a mammalian α-tubulin messenger RNA. *J. Mol. Biol.*, **151**, 101–20.

Luduena, R. F. (1979). Biochemistry of tubulin. In *Microtubules*, ed. K. Roberts & J. S. Hyams, pp. 65–116. New York: Academic Press.

Mabuchi, I. & Okuno, M. (1977). The effect of myosin antibody on the division of starfish blastomeres. *J. Cell Biol.*, **74**, 251–63.

McIntosh, J. R. (1979). Cell division. In *Microtubules*, ed. K. Roberts & J. S. Hyams, pp. 428–41. New York: Academic Press.

Mitchell, D. R. & Warner, F. D. (1980). Interactions of dynein arms with B subfibers of *Tetrahymena* cilia: quantitation of the effects of magnesium and adenosine triphosphate. *J. Cell Biol.*, **87**, 84–97.

Mohri, H. (1976). The function of tubulin in motile system. *Biochim. Biophys. Acta*, **456**, 85–127.

Murphy, D. B. (1982). Assembly-disassembly purification and characterization of microtubule protein without glycerol. *Meth. Cell Biol.*, **23**, 31–49.

Naitoh, Y. & Kaneko, H. (1972). Reactivated Triton-extracted models of *Paramecium*: modification of ciliary movement by calcium ions. *Science*, **176**, 523–4.

Ogawa, K. (1973). Studies on flagellar ATPase from sea urchin spermatozoa. II. Effect of trypsin digestion on the enzyme. *Biochim. Biophys. Acta*, **293**, 514–25.

Ogawa, K. & Mohri, H. (1975). Preparation of antiserum against a tryptic fragment (fragment A) of dynein and an immunological approach to the subunit composition of dynein. *J. Biol. Chem.*, **250**, 6476–83.

Ogawa, K., Mohri, T. & Mohri, H. (1977). Identification of dynein as the outer arms of sea urchin sperm axonemes. *Proc. Natl. Acad. Sci. (USA)*, **74**, 5006–10.

Ogawa, K., Negishi, S. & Obika, M. (1982). Immunological dissimilarity in protein component (dynein 1) between outer and inner arms within sea urchins sperm axonemes. *J. Cell Biol.*, **92**, 706–13.

Penningroth, S. M. & Witman, G. B. (1978). Effects of adenyl imidodiphosphate, a nonhydrolyzable adenosine triphosphate analog, on reactivated and rigor wave sea urchin sperm. *J. Cell Biol.*, **79**, 827–32.

Penningroth, S. M., Cheung, A., Bouchard, P., Gagnon, C. & Bardin, W. (1982). Dynein ATPase is inhibited selectively *in vitro* by erythro-9-3-2-(hydroxynonyl) adenine. *Biochem. Biophys. Res. Commun.*, **104**, 234–40.

Petzelt, C. (1979). Biochemistry of the mitotic spindle. *Int. Rev. Cytol.*, **60**, 53–92.

Ponstingl, H., Krauhs, E., Litlle, M. & Kemph, T. (1981). Complete amino acid sequence of α-tubulin from porcine brain. *Proc. Natl. Acad. Sci. (USA)*, **78**, 2757–61.

Roberts, K. & Hyams, J. S. (1979). *Microtubules*. New York: Academic Press.

Sakai, H. (1978). The isolated mitotic apparatus and chromosome motion. *Int. Rev. Cytol.*, **55**, 23–48.

Sakai, H., Mohri, H. & Borisy, G. G. (1982). *Biological Functions of Microtubules and Related Structures*. New York: Academic Press.

Sale, W. S. & Gibbons, I. R. (1979). Study of the mechanism of vanadate inhibition of the dynein cross-bridge cycle in sea urchin sprem flagella. *J. Cell Biol.*, **82**, 291–8.

Sale, W. S. & Satir, P. (1977). Direction of active sliding of microtubules in *Tetrahymena* cilia. *Proc. Natl. Acad. Sci. (USA)*, **74**, 2045–9.

Satir, P. (1965). Studies on cilia. II. Examination of the distal region of the ciliary shaft and the role of the filaments in motility. *J. Cell Biol.*, **26**, 805–34.

Satir, P. (1968). Studies on Cilia. III. Further studies on the cilium tip and a 'sliding filament' model of ciliary motility. *J. Cell Biol.*, **39**, 77–94.

Satir, P. (1982). Mechanisms and controls of microtubule sliding in cilia. In *Prokaryotic and Eukaryotic Flagella, Symposium of the Society for Experimental Biology*, vol. 35, ed. W. B. Amos & J. G. Duckett, pp. 179–201. Cambridge University Press.

Scheele, R. B. & Borisy, G. G. (1979). *In vitro* assembly of microtubules. In *Microtubules*, ed. K. Roberts & J. S. Hyams, pp. 175–254. New York: Academic Press.

Schwartz, J. H. (1979). Axonal transport: components, mechanisms, and specificity. *Ann. Rev. Neuroscience*, **2**, 467–504.

Shimizu, T. & Kimura, I. (1977). Effects of adenosine triphosphate on *N*-ethylmaleimide-induced modification of 30S dynein from *Tetrahymena* cilia. *J. Biochem.*, **82**, 165–73.

Shingyoji, C., Murakami, A. & Takahashi, K. (1977). Local reactivation of Triton-extracted flagella by iontophoretic application of ATP. *Nature*, **265**, 269–70.

Sleigh, M. A. (1974). *Cilia and Flagella*. New York: Academic Press.

Stearns, M. E. & Ochs, R. L. (1982). A functional *in vitro* model for studies of intracellular motility in digitonin-permeabilized erythrophores. *J. Cell Biol.*, **94**, 727–39.

Sugino. K. & Naitoh, Y. (1982). Simulated cross-bridge patterns corresponding to ciliary beating in *Paramecium*. *Nature*, **295**, 609–11.

Summers, K. E. & Gibbons, I. R. (1971). Adenosine triphosphate-induced sliding of tubules in trypsin-treated flagella of sea-urchin sperm. *Proc. Natl. Acad. Sci. (USA)*, **68**, 3092–6.

Takahashi, M. & Tonomura, Y. (1978). Binding of 30S dynein with the B-tubule of the outer doublet of axoneme from *Tetrahymena pyriformis* and adenosine triphosphate-induced dissociation of the complex. *J. Biochem.*, **84**, 1339–56.

Takahashi, M. & Tonomura, Y. (1979). Kinetic properties of dynein ATPase from *Tetrahymena pyriformis*. The initial phosphate burst of dynein

ATPase and its interaction with ATP analogs. *J. Biochem.*, **86**, 413–23.

Tang, W. Y., Bell, C. W., Sale, W. S. & Gibbons, I. R. (1982). Structure of the dynein-1 outer arm in sea urchin sperm flagella. I. Analysis by separation of subunits. *J. Biol. Chem.*, **257**, 508–15.

Terashita, S., Kato, T., Sato, H. & Tonomura, Y. (1983). Reaction mechanism of 21S dynein ATPase from sea urchin sperm. II. Formation of reaction intermediates. *J. Biochem.*, **93**, 1575–81.

Tilney, L. G., Bryan, J., Bush, D. J., Fujiwara, K., Mooseker, M. S., Murphy, D. B. & Snyder, D. H. (1973). Microtubules: evidence for 13 protofilaments. *J. Cell Biol.*, **59**, 267–75.

Tsukita, S., Tsukita, S., Usukura, J. & Ishikawa, H. (1983). ATP-dependent structural changes of the outer dynein arm in *Tetrahymena* cilia: a freeze-etch replica study. *J. Cell Biol.*, **96**, 1480–5.

Valenzuela, P., Quiroga, M., Zaldivar, J., Rutter, W. J., Kirschner, M. W. & Cleceland, D. W. (1981). Nucleotide and corresponding amino acid sequences encoded by α- and β-tubulin mRNAs. *Nature*, **289**, 650–5.

Warner, F. D. (1976). Cross-bridge mechanism in ciliary motility: the sliding-bending conversion. In *Cell Motility*, ed. R. Goldman, T. Pollard & J. L. Rosenbaum, pp. 891–914. Cold Spring Harbor, N.Y.: Cold Spring Harbor Laboratory.

Warner, F. D. (1978). Cation-induced attachment of ciliary dynein cross bridge. *J. Cell Biol.*, **77**, R19–R26.

Warner, F. D. & Mitchell, D. R. (1980). Dynein: the mechanochemical coupling adenoine triphosphatase of microtubule-based sliding filament mechanisms. *Int. Rev. Cytol.*, **66**, 1–43.

Warner, F. D., Mitchell, D. R. & Perkins, C. R. (1977). Structural conformation of the ciliary ATPase dynein. *J. Mol. Biol.*, **114**, 367–84.

Wilson, L. (1982). The cytoskeleton. Part A. Cytoskeletal proteins, isolation and characterization. *Methods in Cell Biology*, vol. 24. New York: Academic Press.

Witman, G. B., Plummer, J. & Sander, G. (1978). *Chlamydomonas* flagellar mutants lacking radial spokes and central tubules. Structure, composition, and function of specific axonemal components. *J. Cell Biol.*, **76**, 729–47.

# 6  $F_1$-ATPase

In chloroplasts of green plants and plasma membranes of photosynthetic bacteria, the energy of light is captured by pigments such as chlorophyll and released during the passage of electrons through an electron transfer chain, which results in the synthesis of ATP (photophosphorylation). In eukaryotic mitochondria and bacterial plasma membranes, on the other hand, the energy required for this function is furnished in the form of free energy liberated by the oxidation of sugars and fatty acids (oxidative phosphorylation). The mechanism by which electron transfer and ATP synthesis are coupled has been studied for many years by numerous investigators (Boyer *et al.*, 1977; Hinkle & McCarty, 1978).

Two major contributions have emerged from research on the mechanism of energy transduction in photophosphorylation and oxidative phosphorylation. The first was the discovery of 'coupling factors'. Vigorous sonication of mitochondria in the presence of $Mg^{2+}$ yields submitochondrial particles (SMP) in which the original mitochondrial membranes are turned inside out (Figs 6.1 and 6.2). SMP are capable of synthesizing ATP in an electron transfer-dependent manner. This ATP-synthesizing activity is absent in SMP prepared in the presence of EDTA. However, the addition of the supernatant fraction obtained in the preparation of the SMP restored the ATP-synthesizing activity. Racker and co-workers isolated from this supernatant fraction a protein capable of ATP hydrolysis and named it coupling factor 1 or $F_1$ (Penefsky *et al.*, 1960). Following this discovery, extensive research has been carried out on the structure and function of $F_1$ and on its binding factor, $F_0$, which is embedded in the membrane. These efforts led to the finding that $F_1$ and $F_0$ constitute an enzyme complex that functions as a $H^+$-translocating ATPase or, in reverse, as an ATP synthetase (for reviews, see Senior, 1973; Nelson, 1976; Racker, 1976; Criddle, Johnston & Stack, 1979;

Kagawa *et al.*, 1979; Penefsky, 1979; Futai & Kanazawa, 1980; Shavit, 1980; Fillingame, 1981).

The other major contribution was the formulation of the chemiosmotic hypothesis for the coupling mechanism of ATP synthesis. Mitchell (1961) proposed that the high-energy intermediate in ATP synthesis is the electrochemical potential of $H^+$ formed across the membrane. He suggested that the electron-transfer system embedded in the membrane uses the energy stored in NADH during substrate oxidation to establish a $H^+$ gradient across the membrane. The $F_1$–$F_0$ complex present in the same membrane translocates $H^+$ down this $H^+$ gradient and uses the energy in ATP synthesis (Fig. 6.2). This hypothesis has led to many important experiments and its validity is now widely accepted (for review, see Boyer *et al.*, 1977; McCarty, 1978*b*).

Thus, the framework of energy transduction in ATP synthesis has been established. The problem of utmost importance at present is to understand the molecular mechanism that couples ATP synthesis in the $F_1$–$F_0$ complex to $H^+$ transport. This chapter discusses this problem, placing emphasis on the structure and function of the $F_1$–$F_0$ ATPase.

Fig. 6.1 Electron micrograph of mitochondria isolated from parietal cells of mole gastric gland. (Photograph by courtesy of T. Kanaseki).

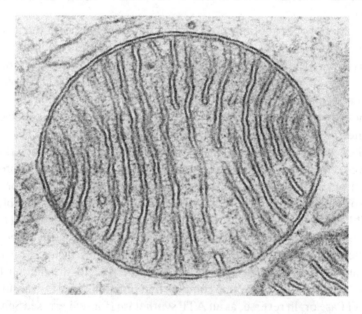

0.5 um

Fig. 6.2 Schematic depiction of mitochondria and the chemiosmotic hypothesis.

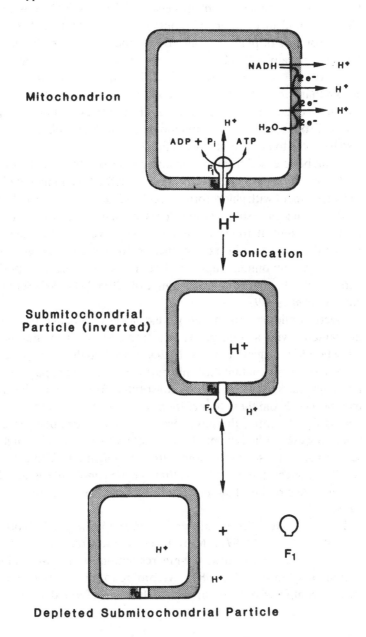

## Structure of the $F_1$–$F_0$ complex

### *Size and shape of $F_1$*

$F_1$-ATPase ($F_1$) is present in the mitochondrial inner membranes of all animal and plant cells. ATPases homologous to $F_1$ can also be found in other membranes: $CF_1$ in the thylakoid membranes of chloroplasts of plant cells and $BF_1$ in the plasma membranes of bacteria, specifically called $EF_1$ in *E. coli* and $TF_1$ in thermophilic bacterium PS3 (Avron, 1963; Kagawa *et al.*, 1979). In addition, it was recently shown that chromaffin granules present in chromaffin cells of the adrenal medulla contain $H^+$-translocating ATPase (Njus, Knoh & Zallakian, 1981). From structural and functional similarities, this $H^+$ ATPase is also believed to be homologous to $F_1$.

$F_1$ can be released from the membrane in a soluble form by sonication in the presence of high concentrations of EDTA (Pullman *et al.*, 1960) or by treatment with chloroform (Beechey *et al.*, 1975). Since the soluble $F_1$ dissociates into subunits and loses its activity at low temperature, it must be kept at room temperature. However, $F_1$ is protected from this cold inactivation in the presence of 50% glycerol or when precipitated with ammonium sulphate. The molecular weight of purified $F_1$ varies somewhat but is within the range of 320 000–400 000 (Senior, 1973; Kagawa *et al.*, 1979).

Electron micrographs of mitochondria show beadlike particles 9 nm in diameter with a stalk protruding from the inner membrane of the mitochondria. Kagawa & Racker (1966*b*) showed that $F_1$-depleted membranes no longer contain the 9-nm particles and that the particles appear again on the surface of SMP reconstituted from $F_1$ and the depleted membranes. From these observations they concluded that this 9-nm particle is $F_1$. Recently, the physicochemical properties of $F_1$ in aqueous solution were studied in detail using methods such as small-angle X-ray scattering, light scattering and ultracentrifugation (Table 6.1). The results of such studies indicate that the $CF_1$ molecule is an ellipsoid with a Stokes radius of about 6 nm (Paradies, Zimmermann & Schmidt, 1978).

$F_1$ and $TF_1$ were first crystallized in 1977 (Spitsberg & Haworth, 1977; Wakabayashi *et al.*, 1977). More recently, the structure of the rat liver $F_1$ crystal was analyzed at 0.9-nm resolution by an X-ray diffraction technique (Amzel *et al.*, 1982). According to this analysis, the $F_1$ molecule is composed of two equivalent halves, each formed by three regions of approximately equal size.

Table 6.1 *Physicochemical properties of $CF_1$ in aqueous solution*
*(from Paradies et al., 1978)*

| | |
|---|---|
| Molecular weight | |
| from small-angle X-ray scattering | $320\,000 \pm 5000$ |
| from light scattering | $330\,000 \pm 5000$ |
| from sedimentation equilibrium | $325\,000 \pm 1000$ |
| Sedimentation coefficient $s_{20,w}^0$ | $13.2 \pm 0.5$ |
| Diffusion coefficient $D_{20,w}^0 \times 10^{-7}\,cm^2/s$ | $3.75 \pm 0.05$ |
| Stokes radius (m) | 5.82 |
| Gyration radius (nm) | 4.25 |
| Axial ratio | 1.5–2.0 |
| Volume ($\times 10^2\,nm^3$) | 7.01 |

## Subunit structure of $F_1$

$F_1$ is composed of five subunits: in order of decreasing molecular weights, two major subunits, $\alpha$ and $\beta$, and three minor subunits $\gamma$, $\delta$ and $\varepsilon$ (Table 6.2). However, a four-subunit $EF_1$ that lacks a subunit but functions normally in ATP hydrolysis was isolated from a strain of *E. coli* (Futai, Sternweis & Heppel, 1974). Furthermore, $F_1$ from mitochondria contains a sixth subunit, a peptide with a molecular weight of 10 000–12 000, that inhibits $F_1$-ATPase activity (Pedersen, Schwerzmann & Cintron, 1981). Usually, however, this inhibitory peptide dissociates from the $F_1$ during purification procedures.

The molecular weight of $F_1$ is between 320 000 and 400 000 whereas the molecular weights for both the $\alpha$ and $\beta$ subunits, which are believed to comprise at least 80% of the $F_1$ molecule, are between 50 000 and 60 000. Thus, more than one of each of the $\alpha$ and $\beta$ subunits must be present in each $F_1$ molecule. Accordingly, the subunit stoichiometry of $F_1$ has been studied by various experimental procedures, including titration of SH groups present in the $\alpha$ and $\beta$ subunits, measurement of binding between $F_1$ and aurovertin, an inhibitor that binds to the $\beta$ subunit in the ratio of 1:1, chemical crosslinking and analysis of incorporation of [14]C-labelled amino acids into $F_1$ (Kagawa *et al.*, 1979). At present a stoichiometry of $\alpha_3\beta_3\gamma$ is estimated for $EF_1$ (Foster & Fillingame, 1982) and $TF_1$ (Kagawa *et al.*, 1979). For $F_1$ and $CF_1$, however, two subunit compositions have been reported, $\alpha_3\beta_3\gamma$ (Esch & Allison, 1979; Merchant *et al.*, 1983) and $\alpha_2\beta_2\gamma_{1-2}$ (Baird & Hammes, 1976; Nelson, 1976; Verschoor, Van der Sluis & Slater, 1977). The exact

Table 6.2 *F₁ subunit structures*

| Designation | Source | m.w. of subunits (in thousands) | | | | | Deduced subunit stoichiometry |
|---|---|---|---|---|---|---|---|
| | | $\alpha$ | $\beta$ | $\gamma$ | $\delta$ | $\varepsilon$ | |
| $F_1{}^a$ | mitochondria | 54 | 50 | 33 | 17.5 | 7.5 | $\alpha_2\beta_2\gamma_2\delta\varepsilon$ or $\alpha_3\beta_3\gamma\delta\varepsilon$ |
| $CF_1{}^b$ | chloroplast | 59 | 56 | 37 | 17.5 | 13 | $\alpha_2\beta_2\gamma\delta\varepsilon_2$ or $\alpha_3\beta_3\gamma\varepsilon$ |
| $EF_1{}^c$ | E. coli | 55 | 50 | 32 | 19 | 14 | $\alpha_3\beta_3\gamma\delta\varepsilon$ |
| $TF_1{}^d$ | thermophilic bacterium PS3 | 54 | 51 | 30 | 21 | 16 | $\alpha_3\beta_3\gamma\delta\varepsilon$ |
| $(F_1)^e$ | chromaffin granule | 51 | 50 | 28 | — | — | — |

[a] Knowles & Penefsky, 1972a,b.
[b] Nelson, 1976.
[c] Foster & Fillingame, 1982.
[d] Yoshida et al., 1979.
[e] Apps & Glover, 1978.

stoichiometry is difficult to determine for the $\delta$ and $\varepsilon$ subunits because these subunits readily dissociate from $F_1$ during purification procedures.

The arrangement of subunits in the $F_1$ molecule has been studied by the reconstitution method, which will be described later, along with the use of chemical crosslinking reagents, but no definite conclusion has yet been reached. Two types of arrangements have so far been proposed: one is that the $\alpha$ and $\beta$ subunits are arranged alternately and thus, the $F_1$ molecule has an axial symmetry (Todd & Douglas, 1981), and the second is that two to three molecules of each of the $\alpha$ and $\beta$ subunits are clustered and thus, the $F_1$ molecule is not axially symmetrical (Baird & Hammes, 1976, 1977; Enns & Criddle, 1977). However, when the relationship of the $\alpha$ and $\beta$ subunits to the rest of the subunits is taken into consideration, even the former proposition does not affirm that the $F_1$ molecule has perfect axial symmetry.

For the present, reconstitution of $F_1$ from independently isolated subunits is possible only in $TF_1$ and $EF_1$. Five subunits, $\alpha$ to $\varepsilon$, have been obtained by treatment of $TF_1$ with guanidine-HCl, followed by ion-exchange chromatography in the presence of urea (Yoshida *et al.*, 1977*b*). $EF_1$ can be dissociated into subunits at low temperature and high ionic strength. The $\alpha$, $\beta$ and $\gamma$ subunits can be isolated from these dissociated subunits by hydrophobic chromatography. The $\delta$ and $\varepsilon$ subunits can be isolated by gel filtration of the supernatant obtained by the treatment of $EF_1$ with 50% pyridine (Dunn & Futai, 1980).

None of the isolated subunits of $EF_1$ or $TF_1$ or $TF_1$ alone has ATPase activity. However, the ATPase activity can be reconstituted from $\alpha$, $\beta$ and $\gamma$ subunits (Futai, 1977). When the $\delta$ and $\varepsilon$ subunits are present, the reconstituted complex can bind to membranes containing $F_0$ and form an $F_1$–$F_0$ complex capable of energy conversion (Fig. 6.3; Yoshida *et al.*, 1977*a*; Dunn & Futai, 1980).

The isolated $\alpha$ subunit contains a high-affinity or tight nucleotide-binding site ($K_d = 0.1$–$20\,\mu$M), and it has been reported that the binding of ATP to this site causes a drastic change in the conformation of the $\alpha$ subunit (Dunn, 1980). The isolated $\beta$ subunit, on the other hand, contains a low-affinity or loose nucleotide-binding site (Ohta *et al.*, 1980). The function of these nucleotide-binding sites will be discussed in detail in the next section. No nucleotide-binding site has yet been found in the isolated $\gamma$ subunit. It has been suggested from reconstruction experiments using $TF_1$ that the $\gamma$ subunit functions in the $F_1$–$F_0$ complex as a gate for the $H^+$ channel (Yoshida *et al.*, 1977*a*).

Because of its slender shape in the isolated form (axial ratio of 4–5) and its high content of α-helix (Schmidt & Paradies, 1977a), the δ subunit is thought to correspond to the stalk portion of the $F_1$–$F_0$ (TF$_1$–F$_0$ or EF$_1$–F$_0$) complex observed in electron microscopy (Fig. 6.4). The δ subunit is also assumed to bind to the headpiece of $F_1$ through the amino terminus of the α subunit (Dunn, Heppel & Fullmer, 1980). Mitochondrial $F_1$–$F_0$ complex contains another component, oligomycin sensitivity-conferring protein (OSCP), which is reported to be located in the stalk portion (MacLennan & Asai, 1968; MacLennan & Tzagoloff, 1968). The ε subunits of EF$_1$ and CF$_1$ are inhibitory to F$_1$-ATPase activity (Sternweis & Smith, 1980), while the ε subunits of TF$_1$ and mitochondrial F$_1$ are not. In addition, as has been described, an inhibitory peptide that is different from the ε subunit has already been isolated from mitochondria. Thus, whether the ε subunits isolated from different sources are homologous to each other is quite questionable. The δ and ε subunits of EF$_1$ may be structurally homologous to the OSCP and the sixth, inhibitory subunit of mitochondrial F$_1$, respectively, as suggested by amino acid sequence analysis (Walker, Runswick & Saraste, 1982).

Fig. 6.3 Dissociation and reconstitution of $F_1$. (Schematic depiction originally drawn by Kanazawa & Futai, 1981).

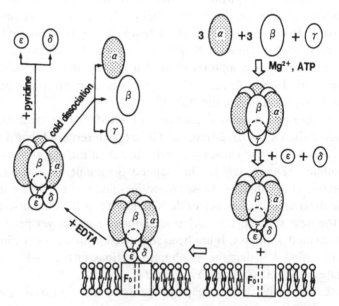

Fig. 6.4 Electron micrograph of the $F_1$–$F_0$ complex isolated from rat liver mitochondria. The tripartite structure of the $F_1$–$F_0$ complex (headpiece–stalk–basepiece) can be clearly observed. (From Soper, Decker & Pedersen, 1979.)

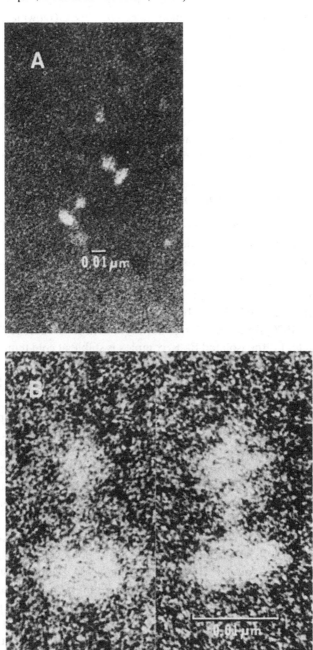

*F₁-ATPase*

Recently, a gene cluster coding for the $F_1$–$F_0$ complex, the *unc* operon (Kanazawa *et al.* argue that the term *pap* operon is better suited for this operon), has been identified in *E. coli* DNA (Downie, Gibson & Cox, 1979; Kanazawa *et al.*, 1980). From the analysis of the primary structure of this gene cluster, the amino acid sequence has been deduced for each subunit of $EF_1$ (Gay & Walker, 1981*a*,*b*; Kanazawa *et al.*, 1981*a*,*b*, 1982; Mabuchi *et al.*, 1981). Mutant strains deficient in $EF_1$ function have been isolated in *E. coli* (Kanazawa, Saito & Futai, 1978; Kanazawa *et al.*, 1980). The analysis of the primary structure of the *unc* genes from these mutant strains should continue greatly to further understanding the relationship between structure and function of $F_1$.

## Size and shape of the $F_1$–$F_0$ complex

The $F_1$–$F_0$ complex consisting of the water-soluble $F_1$-ATPase and the membrane-embedded, hydrophobic protein complex, $F_0$, has been solubilized from the host membranes (e.g., the mitochondrial inner membrane) by treatment with detergent (e.g., octylglucoside, cholic acid or deoxycholic acid) and purified (Kagawa & Racker, 1966*a*; Tzagoloff, Byingtron & MacLennan, 1968; Sone *et al.*, 1975; Serrano, Kanner & Racker, 1976; Foster & Fillingame, 1979; Pick & Racker, 1979). Unlike the singlet $F_1$, the isolated $F_1$–$F_0$ complex is stable at cold temperatures.

The molecular weight of *E. coli* $EF_1$–$F_0$ complex has been estimated to be 448 000–460 000 by small-angle X-ray scattering and laser light scattering (Paradies *et al.*, 1981). Electron micrographs of the mitochondrial $F_1$–$F_0$ complex are shown in Fig. 6.4. The $F_1$–$F_0$ complex shows a structure in which the headpiece and basepiece are connected by a stalk. The diameter of the headpiece is about 9 nm, which corresponds well to that of the knob observed in electron microscopy of SMP, i.e., $F_1$. The basepiece, which is believed to derive from $F_0$, has a thickness of about 6 nm, enough to penetrate membrane. The isolated $TF_0$ and $EF_0$ contain three hydrophobic subunits, a, b and c, with molecular weights of 24 000, 19 000 and 8000, respectively (Negrin, Foster & Fillingame, 1980), while the mitochondrial $F_0$ contains more than three subunits. The complete amino acid sequences of all $F_0$ subunits have been determined from the analysis of the primary structure of the gene cluster, the *unc* operon, that codes for *E. coli* $EF_1$–$F_0$. In particular, the amino acid sequence of subunit c has been determined, not only for *E. coli*, but also for mitochondria, chloroplasts and the thermophilic bacterium PS3 (Sebald & Wachter, 1980). The amino acid sequences of subunit c are very similar among these sources. $F_0$ is believed to be made up of several copies of each of the three subunits, but their stoichiometry has not yet been determined. Subunit c is also called the DCCD (dicyclohexylcarbodimide)-binding protein. This is because DCCD binds to a specific aspartic acid or glutamic acid residue of subunit c and inhibits the ATPase activity of the $F_1$–$F_0$ complex and the $H^+$-transport activity of $F_0$. That $F_0$ is a $H^+$ channel was first proven by reconstitution experiments in thermophilic bacterium PS3 and *E. coli* (Yoshida *et al.*, 1977*a*; Negrin *et al.*, 1980). These experiments showed that liposomes embedded with $F_0$ transport $H^+$, depending on the diffusion potential of $K^+$ that is generated when valinomycin is added. This $H^+$ transport is inhibited by low concentrations of DCCD. Furthermore, $TF_1$ binds to the $F_0$ liposomes and inhibits $H^+$ transport. The reconstituted $TF_1$–$F_0$ liposomes are capable of catalyzing both the $[^{32}P]Pi$–ATP exchange reaction and ATP synthesis driven by an acid–base transition (Sone *et al.*, 1977). These experimental results suggest that the $F_1$–$F_0$ complex synthesizes ATP in a manner directly coupled to $H^+$ transport.

## Reaction mechanism of $F_1$-ATPase

One approach to understanding the molecular mechanism of ATP synthesis by the $F_1$–$F_0$ complex is to elucidate the mechanism of its reverse reaction, the $F_1$–ATPase reaction. However, many aspects of the reaction mechanism of $F_1$–$F_0$ ATPase still remain unsolved, despite many investigations. This is because the $F_1$-ATPase, like myosin ATPase (Chapter 2), contains a reaction pathway for the uncoupled ATPase, which has a very high turnover rate, in addition to the pathway for the reverse reaction of ATP synthesis (Leimgruber & Senior, 1976*a,b*; Kumar, Kalra & Brodie, 1979; Matsuoka, Watanabe & Tonomura, 1981). Only recently have the elementary steps of the ATPase reaction begun to be analyzed by new techniques in terms of their functions as the reverse reactions of ATP synthesis (Grubmeyer & Penefsky, 1981*a,b*; Matsuoka *et al.*, 1981).

The first half of this section discusses the general characteristics of $F_1$-ATPase in a steady state. The discussion is extended to the effects of ligand binding and chemical modifications involved in the ATPase activity and then to the structural properties of the active sites of the $F_1$ molecule. The second half of this section deals with the kinetic properties of the elementary steps and reaction intermediates of $F_1$-ATPase.

## Catalytic sites and regulatory sites of F₁-ATPase

F₁-ATPase shows a hydrolytic activity of 70–120 μmol Pi/mg/ min with 5 mM Mg²⁺-ATPase as the substrate at 30 °C and an optimum pH of 8.0 (Pennin, Godinot & Gautheron, 1979). The ATPase activity of CF₁ is latent and therefore requires activation by heat treatment or by limited proteolysis with trypsin.

The ATPase activity of the isolated F₁–F₀ complex is only a fraction of the F₁-ATPase activity. Furthermore, the F₁–F₀ complex catalyzes a [³²P]Pi–ATP exchange reaction which is thought to be a partial reaction of ATP synthesis (Serrano *et al.*, 1976), whereas F₁ alone is not capable of this reaction.

ATP is thought to be not only the substrate for F₁-ATPase but also a regulatory factor for the ATPase. Ebel & Lardy (1975*a*) and Pedersen (1976) examined the relationship between the initial reaction velocity and the ATP concentration during ATP hydrolysis by F₁ and found that a bend is present in the Lineweaver–Burk plot or Eadie–Hofstee plot of the data (Fig. 6.5). The maximum velocity ($V_{max}$) and Michaelis

Fig. 6.5 Steady-state ATPase activity of F₁. Kinetic data are presented in an Eadie–Hofstee plot. Two pairs of $K_m$ (0.06 and 0.98 mM) and $V_{max}$ (11 and 24 μmol/mg/min) were obtained (Pedersen, 1976).

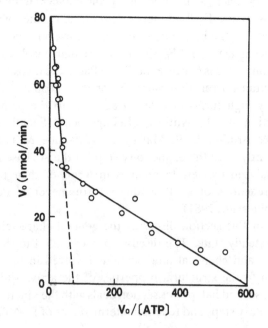

constant ($K_m$) are far higher when ATP is present at high concentrations than when it is present at low concentrations. Moreover, this bend is specific to ATP and is absent when ITP or GTP are used as substrates. These results were interpreted as an indication that the catalytic site and regulatory site of F₁-ATPase exist as nucleotide-binding sites that are different from each other. The catalytic site is of relatively high affinity ($K_d$ of 20–70 μM or lower) and provides the site for the hydrolysis of nucleotides such as ATP, GTP and ITP, whereas the regulatory site is of relatively low affinity ($K_d$ of 0.2–1 mM) with high specificity for ATP. The binding of the nucleotide to the regulatory site stimulates the hydrolysis of a nucleotide at the catalytic site. The nonhydrolizable ATP analog AMPPNP competitively inhibits the hydrolysis of high concentrations of ATP but noncompetitively inhibits the hydrolysis of GTP and ITP. These phenomena can be explained by assuming that AMPPNP binds only to the regulatory site in competition with ATP, thereby inducing the competitive inhibition. On the other hand, when GTP or ITP are used as substrates, neither can bind to the regulatory site and, therefore, hydrolysis is noncompetitively inhibited by the AMPPNP that binds to the regulatory site (Schuster, Ebel & Lardy, 1975).

## *Binding of nucleotides and Pi to F₁*

*Tightly bound nucleotides (very slowly exchangeable sites)*     One mole F₁ contains 3 mol of 'tightly bound nucleotides' with apparent dissociation constants being estimated at about $10^{-10}$ M (Slater *et al.*, 1979). These consist of ATP and ADP at various molar ratios. Neither of these nucleotides dissociates from F₁ during purification procedures. It is not known which subunit of F₁ contains these tightly bound nucleotides.

The tightly bound nucleotides can be removed from F₁ by treatment with 50% glycerol, by mild treatment with trypsin or by gel filtration in the presence of EDTA at high ionic strength. F₁ molecules from which the tightly bound nucleotides have been removed retain nearly normal ATPase activity and the ability to rebind to the membrane but lack the ability to synthesize ATP (Garrett & Penefsky, 1975; Leimgruber & Senior, 1976*a,b*). These observations indicate that the tightly bound nucleotides do not play any role in the function of the uncoupled F₁-ATPase, at least under normal conditions. However, it is highly possible that these nucleotides are involved in the ATP synthesis reaction.

*F₁-ATPase*

*Exchangeable binding sites for nucleotides*   In addition to tightly bound nucleotides, 2–3 mol of adenine nucleotides can bind reversibly/mol of $F_1$ (Harris, 1978). These 'exchangeable' sites are of at least two types, one with high affinity ($K_d$ about $10^{-7}$ M) and the other with low affinity ($K_d$ about $10^{-5}$ M) for nucleotides. Furthermore, the nucleotides that are bound to these sites dissociate at two different rates, a relatively fast rate ($t_{\frac{1}{2}} \ll 1$ min) and a slow rate ($t_{\frac{1}{2}}$ about 1 min). Up to 3 mol of the nonhydrolyzable ATP analog AMPPNP can bind to the exchange sites/mol of $F_1$, but binding of the first mole almost completely abolishes $F_1$-ATPase activity (Cross & Nalin, 1982). Thus it is assumed that all or part of these exchangeable sites are catalytic sites of the $F_1$-ATPase. If it is assumed that one exchange site is present per pair of $\alpha$ and $\beta$ subunits, then the fact that the properties of these sites are not identical indicates the presence of cooperativity between $\alpha$–$\beta$ pairs.

*Pi-binding site*   One mole of $F_1$ contains 1 mol of high-affinity Pi-binding sites (Penefsky, 1977; Kasahara & Penefsky, 1978). The binding of Pi to $F_1$ requires divalent cations and is inhibited by ATP, ADP or AMPPNP. Analyses using photoaffinity analogs of Pi have suggested that the high-affinity binding site for Pi is located on the $\beta$ subunit of $F_1$. More than 1 mol of low-affinity Pi-binding sites are also present in addition to the high-affinity sites.

## Modifiers of F₁-ATPase and chemical modification

*Inhibitors of F₁-ATPase*   The effects of many specific inhibitors have been studied to gain insights into the structure of $F_1$ and its ATPase reaction mechanism.

An inhibitory peptide that can be regarded as the sixth subunit is present in mitochondrial $F_1$ (Pullman & Monroy, 1963; Pedersen *et al.*, 1981). This inhibitory peptide dissociates from $F_1$ upon conformational changes of $F_1$ induced by membrane energization (discussed later in this chapter) and is now believed to inhibit not only the ATPase activity but ATP synthesis as well.

Aurovertin, an antibiotic isolated from the mold *Calcarisporium arbuscula*, binds to $F_1$ at a ratio of 3 mol/mol of $F_1$ and inhibits both $F_1$-ATPase activity and ATP synthesis (Ebel & Lardy, 1975b; Issartel *et al.*, 1983). The binding site for aurovertin is located on the $\beta$ subunit of $F_1$ and is different from the binding sites for nucleotides (Verschoor,

154

Van der Sluis & Slater, 1977). When bound to $F_1$, aurovertin exhibits strong fluorescence, with its intensity being strongly affected by nucleotides and by $Mg^{2+}$. Thus it provides a useful probe for detecting conformational changes of $F_1$ (to be discussed later in this chapter).

A peptide antibiotic, efrapeptin (A23187), binds to $F_1$ with high affinity at a ratio of $1 mol/mol$ of $F_1$ and inhibits both ATPase activity and ATP synthesis (Cross & Kohlbrenner, 1978).

*Chemical modification of F₁*  Chemical structures related to the functions of particular subunits of $F_1$ have been studied using chemical modifiers specific to various amino acid residues and using photoaffinity nucleotide analogs. Many chemical modifiers bind covalently to the $\beta$ subunit and inactivate the ATPase activity (Table 6.3). The simplest explanation for this is that the catalytic sites for ATP hydrolysis reside on the $\beta$ subunit. However, since modification of the regulatory sites can also cause the inhibition of ATPase activity, an accurate interpretation must await detailed elucidation of the reaction mechanism of $F_1$-ATPase.

7-Chloro-4-nitrobenzo-2-oxa-1,3-diazide binds to $F_1$ at about $1 mol/mol$ and causes complete inactivation of the ATPase. This, together with the inhibitory effects of AMPPNP and efrapeptin described above, is considered to be evidence of the presence of cooperativity among the $\alpha-\beta$ subunit pairs, two to three copies of which are present in each $F_1$.

The effects photoaffinity nucleotide analogs have on $F_1$ are summarized in Table 6.3. The following facts are unique: in the absence of $Mg^{2+}$, i.e., under conditions where no ATPase activity occurs, these analogs selectively bind to the $\beta$ subunit, whereas in the presence of $Mg^{2+}$, i.e., under the conditions where ATPase activity occurs, they bind equally to both the $\alpha$ and $\beta$ subunits or selectively to the $\alpha$ subunit. From these facts Williams & Coleman (1982) speculated that one catalytic site is formed in the interfacial cleft of an $\alpha-\beta$ subunit pair and that, in the absence of $Mg^{2+}$, ATP binds only to the $\beta$ subunit side, whereas in the presence of $Mg^{2+}$, ATP is hydrolyzed to ADP and then transferred to the $\alpha$ subunit side.

## Phosphorylated intermediates of F₁-ATPase reaction

As has been described in Chapter 2, if the hydrolysis of ATP occurs in a single step, the stereo structure of the Pi liberated in the

Table 6.3 *Chemical modifcation of $F_1$*

| Compound | Type of $F_1$ | Sites/$F_1$ | Subunit specificity | Inhibition of ATPase | Reference |
|---|---|---|---|---|---|
| DCCD[a] | $F_1$ | 2–3 | $\beta$ | yes | Pougeois, Satre & Vignais (1979); Esch *et al.* (1981) |
| DCCD | EF$_1$ | 1 | $\beta$ | yes | Satre *et al.* (1979) |
| DCCD | CF$_1$ | 2 | $\beta$ | yes | Shoshan & Selman (1980) |
| NBD-Cl[b] | F1 | | $1\beta$ | yes | Ferguson *et al.* (1975) |
| Dial-ATP[c] | MF1[d] | 1 | $\alpha$ | yes | Kumar, Kalra & Brodie (1979) |
| Dial-ADP | MF1[d] | 3 | $\alpha > \beta$ (2:1) | no[e] | Kumar, Kalra & Brodie (1979) |
| 8-Azido-ATP | $F_1$ | 2 | $\alpha < \beta$ ($-Mg^{2+}$)<br>$\alpha > \beta$ ($+Mg^{2+}$) | yes<br>yes | Wagenvoord, Kemp & Slater (1980) |
| 8-Azido-ADP | $F_1$ | 2 | $\alpha = \beta$ ($-Mg^{2+}$)<br>$\alpha$ ($+Mg^{2+}$) | yes | Wagenvoord, van der Kraan & Kemp (1979) |
| FSBA[f] | $F_1$ | 4 | $\alpha < \beta$ | yes | Esch & Allison (1978) |
| BzATP[g] | $F_1$ | | $\beta$ ($-Mg^{2+}$)<br>$\alpha < \beta$ (1:2, $+Mg^{2+}$) | yes | Williams & Coleman (1982) |

[a] DCCD, dicyclohexylcarbodiimide.
[b] NBD-Cl, 7-chloro-4-nitrobenzo-2-oxa-1,3-diazole.
[c] Dial-ATP(ADP), 2',3'-dialdehyde derivatives of ATP(ADP).
[d] MF$_1$, F$_1$ from *Mycobacterium phlei*.
[e] However, the ATP synthesis on the membrane is inhibited.
[f] FSBA, *p*-fluorosulfonylbenzoyl-5'-adenosine.
[g] BzATP, 3'-*O*-(4-benzoyl)benzoyl ATP.

ATPase reaction should be an inversion type, relative to the stereo structure of the $\gamma$-phosphoryl group of the substrate ATP. On the other hand, if the hydrolysis occurs in a two-step reaction involving phosphate transfer between ATP and enzyme, as seen in the phosphorylated intermediates, the stereo structure of the Pi should be a retention type. Webb *et al.* (1980) analyzed stereochemically the $F_1$-drive hydrolysis of ATP$\gamma$S that has its $\gamma$-phosphoryl group labeled with $^{18}O$, in $[^{17}O]$water (they used the same method previously for the study of myosin ATPase), and showed that the stereo structure of the thiophosphate formed by this reaction was an inversion type. This finding strongly argues that the $F_1$-ATPase reaction occurs without the formation of a covalent bond-containing intermediate such as a phosphorylated enzyme.

## Presteady-state analysis of $F_1$-ATPase

*Analysis using ATP analogs* Recently, Matsuoka *et al.* (1981) and Grubmeyer & Penefsky (1981*a,b*) obtained interesting results from the analysis of the $F_1$-ATPase reaction using the fluorescent ATP analogs 2'-(5-dimethyl-amino-naphthalene-1-sulfonyl)amino-2'-deoxy ATP (DNS-ATP) and 2',3'-$O$-(2,4,6-trinitrophenyl)-ATP (TNP-ATP), respectively. These analogs both have much slower rates of hydrolysis by $F_1$ than ATP has. This makes it easy to analyze the elementary steps in their hydrolysis and, since changes in fluorescence intensity or in absorption occur when these analogs bind to $F_1$, quantitative as well as time-course analysis of their binding to $F_1$ can be measured.

A reaction mechanism (Fig. 6.6) has been proposed from the analysis of the hydrolysis of DNS-ATP by $F_1$. $F_1$ contains two equivalent, yet independent, high-affinity binding sites for DNS-ATP. DNS-ATP shows an enhancement of fluorescence intensity when it binds to these sites. The initial rate of fluorescence increase is dependent on the concentration of DNS-ATP, and this dependency follows the Michaelis–Menten equation. From these observations it is proposed that the complex ($E^{*DNS\text{-}ATP}$), with an enhanced fluorescence intensity, is formed through the $F_1$–substrate complex ($E^{DNS\text{-}ATP}$), which is in rapid equilibrium with the DNS-ATP in the medium and has a low fluorescence intensity. The appearance of $E^{*DNS\text{-}ATP}$ is followed by the formation of the intermediate ($E_p^{*DNS\text{-}ADP}$ in which DNS-ATP has been cleaved. The release of free Pi from $F_1$ then ensues. Since the intensity of fluorescence remains enhanced throughout this process, it is assumed that the $F_1$–DNS-ADP complex ($E^{*DNS\text{-}ADP}$) is formed and that the release of DNS-ADP from

$F_1$-ATPase

this intermediate serves as the rate-limiting step for the overall reaction. Thus the reaction mechanism of $F_1$–DNS-ATPase closely resembles that of myosin ATPase described in Chapter 2.

Another important finding obtained with DNS-ATP is that the three elementary steps in the $F_1$–DNS-ATPase reaction, i.e., cleavage of the γ-phosphate bond of DNS-ATP (Fig. 6.7a) and release of DNS-ADP (Fig 6.7b), are markedly accelerated by high concentrations of ATP or other nucleotides. This can be explained by assuming that a nucleotide-binding site (regulatory site), in addition to the catalytic site to which DNS-ATP is bound, is present on $F_1$ and that the binding of ATP or other nucleotides to the former site causes a conformational change in $F_1$, thus resulting in the acceleration of the above three steps. The relationship of the catalytic and regulatory sites, proposed from the kinetic analysis of the reaction mechanism, to the nucleotide-binding sites located on the α and β subunits of $F_1$, provides an interesting research subject. DCCD, which binds to the β subunit of $EF_1$, does not affect the binding of DNS-ATP to $EF_1$ and the liberation of TCA-labile Pi during a single turnover, whereas it almost completely inhibits the $EF_1$–DNS-ATPase activity in steady state and the acceleration of the release of DNS-ADP by a high concentration of ATP. Moreover, DNS-ATP binds with high affinity only to the isolated α subunit and not to the isolated β subunit of $EF_1$ (Matsuoka *et al.*, 1982). These results seem to suggest that a high-affinity catalytic site is present on the α subunit and a low-affinity regulatory site on the β subunit of $EF_1$. However, since the isolated α subunit is not capable of hydrolyzing DNS-

Fig. 6.6 Reaction mechanism of $F_1$-DNS-ATPase. Although two identical catalytic sites are postulated, only one of them is indicated here (Matsuoka *et al.*, 1981).

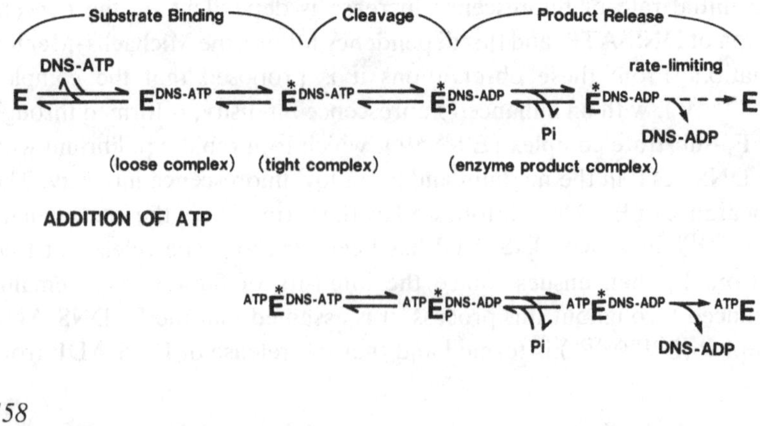

ATP alone, the possibility must be considered that between the α and β subunits a catalytic site is formed that is capable of complete function through interactions between these subunits, as described above.

The reaction mechanism shown in Fig. 6.8 has been proposed from the analysis of the $F_1$-driven hydrolysis of TNP-ATP (Grubmeyer & Penefsky, 1981a,b). This mechanism is clearly different from the mechanism described above for DNS-ATP hydrolysis. TNP-ATP is rapidly

Fig. 6.7 Acceleration of the $F_1$-DNS-ATPase reaction by high concentrations of ATP. (A) ATP acceleration of TCA-labile Pi liberation (open circles) and Pi release from $F_1$ (open squares). Dashed line indicates the amount of DNS-nucleotide bound to $F_1$. 0.2 mM ATP was added at the arrow. (B) ATP acceleration of DNS-ADP release from $F_1$-DNS-ATPase reaction was started, the fluorescence intensity decreased rapidly (solid line) to the level of free DNS-ADP (0), and a corresponding release of DNS-ADP (open circles) was observed (Matsuoka *et al.*, 1981). ΔFL, decrease in fluorescence intensity.

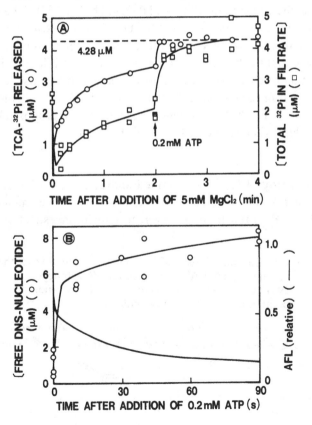

hydrolyzed only when there is cooperativity between the two catalytic sites, whereas DNS-ATP is independently hydrolyzed at each of the two catalytic sites on $F_1$. When TNP-ATP is bound to these sites the complex undergoes absorption changes and a fluorescence increase. In Fig. 6.8 the high-affinity binding of TNP-ATP is characterized by $<(K_d \ll 10\,\text{nM})$ and the low-affinity binding by $\cdot (K_d = 20\text{–}80\,\text{nM})$. An important feature here is that these sites are not independent of each other and thus are interchangeable.

As shown in Fig. 6.9, the cleavage of TNP-ATP is very slow when it occupies only one of the two sites on $F_1$ ($\text{E} < \text{TNP-ATP} \xrightarrow{\text{slow}} \text{E} {\leq}_{\text{Pi}}^{\text{TNP-ADP}}$). On the other hand, when both of the two sites are occupied by TNP-ATP to form TNP-ATP $\cdot$ E $<$ TNP-ATP, only one TNP-ATP is rapidly cleaved (TNP-ATP $\cdot$ E $<$ TNP-ATP $\rightarrow$ TNP-ATP $>$ E $:_{\text{Pi}}^{\text{TNP-ADP}}$) and the next cleavage of TNP-ATP at the remaining site is very slow. However, no difference is observed between the hydrolysis rate of TNP-$[\gamma\text{-}^{32}\text{P}]$ATP bound at the first site (TNP-AT$^{32}$P $\cdot$ E $<$ TNP-ATP) and that bound at the second site (TNP-ATP $\cdot$ E $<$ TNP-AT$^{32}$P). These results can be explained by assuming that the two sites are intrinsically equivalent catalytic sites and become different only after the binding of substrate, and that the binding of substrate to the second (loose) site causes acceleration of cleavage of the substrate bound to the first (tight) site, resulting in transformation of the loose site to the tight site. This assumption agrees well with the binding-change mechanism proposed by Boyer (1979), which will be discussed in the next section.

*Analysis using ATP*   As described above, the use of ATP analogs has for the first time opened the way to analysis of the elementary steps of the $F_1$-ATPase reaction. However, this line of investigation requires

Fig. 6.8 Reaction mechanism of $F_1$-ATPase deduced from the analysis of the $F_1$-TNP-ATPase reaction (Grubmeyer & Penefsky, 1981*b*). Arrowhead indicates high-affinity binding and a raised point indicates low-affinity binding.

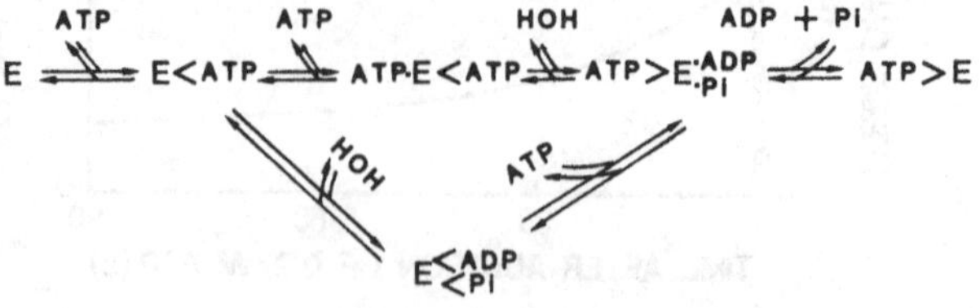

further, more-direct analysis using ATP itself. Grubmeyer, Cross & Penefsky (1982) analyzed the elementary steps of ATP hydrolysis occurring at a single catalytic site on $F_1$, using the rapid-mixing method and the simplified binding assay technique developed by Penefsky.

When $[\gamma\text{-}^{32}P]$ATP is added to an excess of $F_1$, it can be assumed to bind only to a single catalytic site on $F_1$. As shown in Fig. 6.10 (acid quench), the turnover rate under such conditions is very slow $(10^{-4}/s)$. When the hydrolysis of $[\gamma\text{-}^{32}P]$ATP is measured after the addition of a high concentration of cold ATP to the reaction mixture (Fig. 6.10, cold chase), the rate is enhanced, as is the case using ATP analogs described previously. Since enzyme-bound, but not free, $[\gamma\text{-}^{32}P]$ATP is hydrolyzed rapidly upon adding cold ATP, the time course of the binding of $[\gamma\text{-}^{32}P]$ATP to the first site can be estimated from the time course of the cold ATP chase. Furthermore, it has been concluded that

Fig. 6.9 Promoted activity of $F_1$-TNP-ATPase. The rate and extent of TNP-ATP hydrolysis by $F_1$. When only site 1 was filled with TNP-$[\gamma\text{-}^{32}P]$ATP (1 mol/mol $F_1$ was added for curve 3), the initial rate of hydrolysis was very slow. When both site 1 and site 2 were filled with TNP-$[\gamma\text{-}^{32}P]$ATP (2 mol/mol $F_1$ and 4.9 mol/mol $F_1$ were added for curve 2 and curve 1, respectively), the initial rate of hydrolysis was very fast. However, 1 mol of TNP-$[\gamma\text{-}^{32}P]$ATP/mol $F_1$ remained unhydrolyzed even over the longer reaction time (Grubmeyer & Penefsky, 1981b). PCA, perchloric acid.

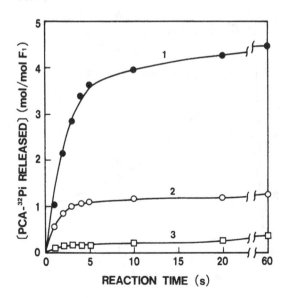

Pi is released before ADP from the $F_1^{ADP}$ complex and that the release of ADP serves as the rate-limiting step for ATP hydrolysis, since the amount of bound $^{32}P$ in $[\gamma\text{-}^{32}P]ATP$ (see Fig. 6.11, $E^{ATP} + E_{Pi}^{ADP}$) decreases faster ($2.7 \times 10^{-3}$/s) than the amount of bound $^3H$ in $[^3H]ATP$ (see Fig. 6.11, $E^{ATP} + E_{Pi}^{ADP} + E^{ADP}$) decreases ($3.6 \times 10^{-4}$/s).

As a result, the reaction mechanism shown in Fig. 6.11 has been proposed for the hydrolysis of ATP at a single catalytic site on $F_1$. This reaction mechanism agrees well with the mechanism described previously for the hydrolysis of DNS-ATP and, as expected, closely resembles the reaction mechanism of myosin ATPase (Chapter 2). Grubmeyer $et\ al.$ (1982) also analyzed in detail the binding of the substrate to and the release of the products from $F_1$ to estimate the forward and reverse rate constants as well as the equilibrium constants. The results revealed additional interesting features: the binding of ATP to the first site on $F_1$ occurs with a very low dissociation constant ($K_1 \geqslant 10^{12}\,\text{M}^{-1}$); and an equilibrium is established between ATP and ADP + Pi bound to this

Fig. 6.10 Elementary steps of the $F_1$-ATPase reaction. See text for details. The binding and hydrolysis of $[\gamma\text{-}^{32}P]ATP$ by a molar excess of $F_1$. $^{32}Pi$ (cold-ATP chase), $^{32}Pi$ acid quench and bound $^{32}P$ indicate amounts of $F_1$–ADP–Pi + $F_1$–ADP, $F_1$–ADP–Pi + $F_1$–ADP and $F_1$–ATP + $F_1$–ADP–Pi, respectively (cf. Fig. 6.11) (Grubmeyer $et\ al.$, 1982).

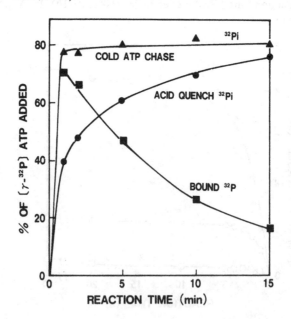

site (catalytic step) with an equilibrium constant $(K_2)$ of 0.5. These studies suggest that free energy input is required for the release of bound ATP and not for the formation of bound ATP from bound ADP and Pi in ATP synthesis catalyzed by the $F_1$–$F_0$ complex.

Figure 6.11 shows a scheme in which only a single catalytic site is involved in the reaction. When ATP binds to the second site $(K_d = 3 \times 10^{-5} M)$, the turnover rate of ATP hydrolysis increases to $300/s$ (Cross, Grubmeyer & Penefsky, 1982). This indicates that the values for $k_3$ and $k_4$ in Fig. 6.11 increase at least $10^5$- and $10^6$-fold, respectively, on binding of ATP to the second site. From analyses using TNP-ATP as described above, it has been concluded that the second ATP-binding site described here is a cooperative catalytic site that is intrinsically identical to the first site. The turnover rate of ATP hydrolysis increases an additional two-fold when the concentration of ATP is further increased. Although this increase is likely to be due to the binding of ATP to a third site, it is not known whether this third site is a cooperative catalytic site similar to the first and second sites or a low-affinity site having a regulatory function. In this connection, Gresser, Myers and Boyer (1982) have proposed a model that argues that the three cooperative catalytic sites on $F_1$ undergo interconversion in response to the binding changes. This model will be discussed in the next section.

## *Synthesis of F₁-bound ATP from ADP and Pi*

A large amount of free energy is required for ATP synthesis in an aqueous solution, so it has been assumed for many years that

Fig. 6.11 Rate constants of the elementary steps of ATP hydrolysis by $F_1$ (Grubmeyer *et al.*, 1982).

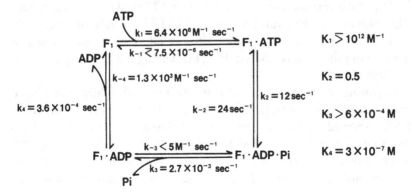

free energy input is also necessary for the formation of ATP from ADP and Pi at the catalytic site of $F_1$. However, it has already been confirmed in the reverse reaction of myosin ATPase that a free energy input from outside is not required for the dehydration condensation of ADP and Pi to synthesize ATP bound tightly to the enzyme. It has also been confirmed by analysis using the ATP analog DNS-ATP that two types of ATP are present during the $F_1$-ATPase reaction, one tightly bound and the other loosely bound to the enzyme. Thus, by analogy with myosin ATPase, the synthesis of tightly bound ATP can be expected to occur even in the absence of a $H^+$ gradient when an excess amount of ADP + Pi is added to $F_1$ or $CF_1$.

This possibility was first proven by Feldman & Sigman (1982), who showed that the addition of 20–100 mM [$^{32}$P]Pi to $CF_1$, which contains 1 mol of intrinsic, tightly bound ADP/mol $CF_1$, results in the reversible synthesis of [$\gamma$-$^{32}$P]ATP with a maximum ratio of 0.25 mol ATP/mol $CF_1$ and $t$ of 20 s. The ATP synthesized was tightly bound to $CF_1$ and could not be released from it by treatment of the enzyme with hexokinase in the presence of glucose. Furthermore, ADP in solution could not serve as the substrate for this ATP synthetic reaction. Accordingly, they speculated that the synthesis of ATP tightly bound to $CF_1$ occurs by the following mechanism:

$$\text{E} < \text{ADP} + \text{Pi} \underset{k_{-1}}{\overset{k_1}{\rightleftharpoons}} \text{E} < \text{ADP} \cdot \text{Pi} \underset{k_{-2}}{\overset{k_2}{\rightleftharpoons}} \text{E} < \text{ATP}$$

where the arrowheads indicate tight binding. The dissociation constant for Pi binding is 60–90 mM at pH 6, a figure considerably greater than the Michaelis constant for Pi in the steady-state synthesis of ATP, and the equilibrium constant for the covalent bond-formation step ($K_2 = k_{-2}/k_2$) is 2.5

They also made the interesting observation that synthesized bound [$\gamma$-$^{32}$P]ATP is very rapidly hydrolyzed when ATP is added to the medium. This implies that the hydrolysis of [$\gamma$-$^{32}$P]ATP is stimulated when ATP in the medium binds to a binding site (an alternating catalytic site or a regulatory site) that is different from the binding site for the synthesized [$\gamma$-$^{32}$P]ATP.

Following the success with $CF_1$, the synthesis of tightly bound ATP has recently been accomplished in beef heart $F_1$. Sakamoto & Tonomura (1983) showed that a maximum of 0.4–0.6 mol of tightly bound [$\gamma$-$^{32}$P]ATP/mol of $F_1$ is formed from ADP in the medium and [$^{32}$P]Pi when dimethylsulfoxide is present in the reaction mixture at concentrations higher than 30%. From kinetic measurements, the following reaction mechanism has been proposed:

$$E + ADP + Pi \underset{\substack{\text{rapid}\\ \text{equilibrium}}}{\rightleftharpoons} E \cdot {\overset{ADP}{Pi}} \rightleftharpoons E \cdot {\overset{<ADP}{Pi}} \underset{\substack{\text{rapid}\\ \text{equilibrium}}}{\rightleftharpoons} E$$

where the dissociation constants for substrate ADP and Pi have been estimated to be 3 μM and 0.5 mM, respectively. However, when dimethylsulfoxide is not present in the reaction mixture, the apparent dissociation constant for Pi increases to more than 400 mM. This implies that the role of dimethylsulfoxide is to cause conformational changes in $F_1$ and, as a result, to increase its affinity for Pi. These observations also suggest that in the $F_1$-$F_0$ complex on the membrane the free energy input from outside causes an increase in the affinity for Pi.

Thus, the fact that tightly bound ATP can be formed from ADP and Pi without outside free energy input supports the mechanism proposed by Boyer (1979). This mechanism assumes that the energy generated by electron transport is used by both $F_1$ and $CF_1$ mainly for changes in their affinities for substrates and products.

# Molecular mechanism of ATP synthesis by the $F_1$-$F_0$ complex

## *Primary acceptor in ATP synthesis*

The overall reaction of ATP synthesis is the formation of ATP from ADP and Pi. However, there had been difference of opinion concerning whether the primary acceptor of Pi is ADP or AMP. This discrepancy was caused by the finding that $\beta$-labeled ADP appears upon short incubation of thylakoid membranes with AMP and $[^{32}P]Pi$, implying that the AMP that is bound to $F_1$ is the primary acceptor of Pi (Tiefert & Moudrianakis, 1979). This finding has been interpreted as follows: the energy-dependent reaction between $[^{32}P]Pi$ and AMP that is bound to $CF_1$ gives rise to bound $[\beta\text{-}^{32}P]ADP$, which in turn undergoes an adenylate kinase-like reaction with ADP that is bound to the other site on $CF_1$ and forms bound AMP and free $[\gamma\text{-}^{32}P]ATP$, thus accomplishing a cycle of ATP synthesis.

On the other hand, the following experimental results suggest that ADP is the primary acceptor of Pi in ATP synthesis. (1) $[\beta\text{-}^{32}P]ADP$ does not appear when the above experiment is carried out in the presence of sufficient amounts of hexokinase and glucose (McCarty, 1978*a*). Since hexokinase utilizes only ATP for its substrate, this observation suggests that [γ-$^{32}$P]ATP is synthesized by photophosphorylation from ADP tightly bound to CF$_1$ and [$^{32}$P]Pi and is subsequently released from CF$_1$. [β-$^{32}$P]ADP is then synthesized from this free [γ-$^{32}$P]ATP by adenylate kinase which is independent of CF$_1$. (2) During the initial phase of ATP synthesis on thylakoid membranes driven by an acid–base transition or by light, the rate of appearance of [β-$^{32}$P]ADP is much slower than that of [γ-$^{32}$P]ATP (Yamamoto & Tonomura, 1975). (3) As has been described in the preceding section, the stereo structure of thiophosphate generated by the F$_1$-driven hydrolysis of ATPγS is an inversion type, indicating that phosphate transfer occurs only once. If AMP is the primary acceptor of Pi, phosphate transfer must occur twice during the course of ATP synthesis. Therefore, assuming that the hydrolysis of ATPγS by F$_1$ occurs by the same pathway as does the reverse reaction of ATP synthesis, this observation suggests that ADP and not AMP is the primary acceptor of Pi.

## Kinetic analysis of ATP synthesis in the steady state

The sequence of ADP and Pi binding during the formation of the Michaelis intermediate ($F_{1Pi}^{ADP}$) has been studied by examination of the dependence of the initial rate of ATP synthesis on the substrate concentration. As a result, several models have been proposed: A ternary complex composed of enzyme, ADP and Pi ($F_{1Pi}^{ADP}$) is formed in a random sequence; a ternary complex is formed in an ordered sequence, with ADP being the first to bind; or Pi binds first to form the complex. In addition, a broad range of Michaelis constants have been reported: 2–85 μM for ADP and 0.1–20 mM for Pi. As has been described in the preceding section, the presence of cooperativity between catalytic sites has been implied for the F$_1$-ATPase reaction. Thus, since a similar situation is expected for ATP synthesis driven by the F$_1$–F$_0$ complex, it is difficult to obtain valid insights into the reaction mechanism from only a kinetic analysis of the steady state.

The relationship between ATP synthesis and H$^+$ translocation has been studied by examining the dependence of their initial rates on pH, in both the acid and base phases when ATP synthesis is driven by an acid–base transition (Yamamoto & Tonomura, 1975; Dewey & Hammes, 1981). These studies have revealed that ionic groups with pK values of about 6 and 8 are involved in the acid phase and base phase, respectively. In addition, the order of dependence (Hill coefficient) on

$H^+$ concentration has been found to be 2–3. This result is consistent with the report that the molar ratio of $H^+$ that must be transported for ATP synthesis in the steady state, the $H^+/ATP$ ratio, is 3 (Alexandre, Reynafaje & Lehninger, 1978).

## Exchange reaction catalyzed by the $F_1$–$F_0$ complex

To elucidate the properties of the partial reactions required for ATP synthesis, Boyer and his co-workers have long been involved in the analysis of various exchange reactions that occur during the synthesis and hydrolysis of ATP (Rosing, Kayalar & Boyer, 1977; Kayalar, Rosing & Boyer, 1977; Choate, Hutton & Boyer, 1979). Figure 6.12 summarizes the exchange reactions they have studied, along with the

Fig. 6.12 (*a*) Schematic representation of the minimal steps of substrate binding and release and ATP formation and cleavage. (*b*) The required reaction steps for oxygen exchanges and $Pi \rightleftharpoons ATP$ exchange in ATP synthesis and hydrolysis (Rosing *et al.*, 1977).

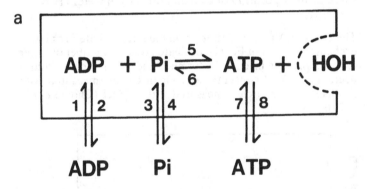

a

b

| Exchange reaction | Minimal required steps | | | | | | |
|---|---|---|---|---|---|---|---|
| | ADP bind-ing (Step 1) | ADP release (Step 2) | $P_i$ binding (Step 3) | $P_i$ release (Step 4) | Substrate intercon-version (Steps 5 & 6) | ATP release (Step 7) | ATP bind-ing (Step 8) |
| ADP $\rightleftharpoons$ ATP[a] | + | + | | | + | + | + |
| $P_i \rightleftharpoons$ ATP[b] | | | + | + | + | + | + |
| Medium $P_i \rightleftharpoons$ HOH | | | + | + | + | | |
| Intermediate $P_i \rightleftharpoons$ HOH | | | | + | + | | + |
| Medium ATP $\rightleftharpoons$ HOH | | | | | + | + | + |
| Intermediate[b] ATP $\rightleftharpoons$ HOH | | | + | | + | + | |

[a] Requires presence of bound $P_i$ derived either from medium $P_i$ or from ATP cleavage.
[b] Requires presence of bound ADP derived either from medium ADP or from ATP cleavage.

necessary partial reactions. Among the results obtained from the analyses of these exchange reactions, those of particular importance are described below.

(1) Incorporation of oxygen atoms from water ($H^{18}OH$) occurs at a ratio of more than 1 mol/mol of Pi liberated upon the hydrolysis of ATP by SMP. This so-called intermediate $Pi \rightleftharpoons HOH$ exchange does not disappear in the presence of uncouplers and occurs in solubilized $F_1$. These results suggest that the conversion of bound ADP and Pi to bound ATP (step 5 in Fig. 6.12) occurs without outside energy input during ATP hydrolysis. The intermediate $P_i \rightleftharpoons HOH$ exchange is markedly depressed when the concentration of ATP is increased during ATP hydrolysis by SMP or solubilized $F_1$ (Fig. 6.13). This result suggests that the dissociation of ADP and Pi formed at the catalytic site is stimulated by the binding of ATP to the other site, the regulatory site or the alternating site, as discussed earlier in this chapter.

(2) In SMP undergoing ATP synthesis, step 5 can also be measured as incorporation of oxygen atoms from water ($H^{18}OH$)

Fig. 6.13 The effect of ATP concentration on the intermediate $Pi \rightleftharpoons HOH$ exchange reaction of $F_1$. The ordinate gives the number of water oxygens present in each Pi formed, and this increases with a decrease in ATP concentration. Open circles, measured with $[^{18}O]ATP$; filled circles, measured with $H^{18}OH$ (Choate *et al.*, 1979).

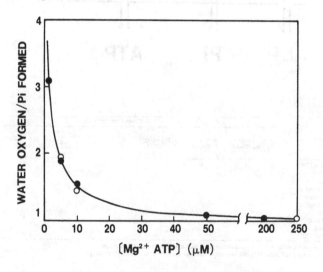

into ATP (the intermediate ATP $\rightleftharpoons$ HOH exchange). A pronounced enhancement of this exchange reaction occurs when the concentrations of substrate ADP and Pi are lowered, suggesting that the dissociation of ATP formed at the catalytic site is also promoted by the binding of ADP and Pi to the other site, the regulatory site or the alternating site.

(3) In SMP, ATP $\rightleftharpoons$ Pi and ATP $\rightleftharpoons$ ADP exchanges are strongly inhibited by uncouplers, in contrast to Pi $\rightleftharpoons$ HOH exchange. These exchange reactions do not occur in solubilized $F_1$. Furthermore, both reactions require the dissociation of ATP from $F_1$ (step 7 in Fig. 6.12). These findings suggest that free energy input is required for the dissociation of ATP from $F_1$.

Thus it has been suggested from the analyses of exchange reactions that the dissociation of ATP from $F_1$ is the reaction step that requires outside energy input. Boyer (1979) has proposed a binding-change mechanism which assumes the presence of cooperativity between binding sites. In this mechanism the binding of substrate or product to one catalytic site causes a change in the tightness of the binding of the substrate or product at the other site. This mechanism is described below.

## *Reaction scheme of ATP synthesis*

As described above, the partial reactions of ATP synthesis driven by the $F_1$–$F_0$ complex have been studied by analysis of the exchange reactions. These results agree well with results obtained from kinetic analysis of the elementary steps of the $F_1$-ATPase described in the previous section in the following two aspects. First, tightly bound ATP is spontaneously formed upon addition of ADP and Pi to $CF_1$ or $F_1$, and the step in which tightly bound ATP is released through a loosely bound state requires free energy input. The second aspect is that the kinetic properties of nucleotides and Pi bound to a catalytic site(s) drastically change upon binding of nucleotide and/or Pi to the other site. Accordingly, several types of reaction schemes can be considered for ATP synthesis driven by the $F_1$–$F_0$ complex, depending on how the cooperativity between catalytic sites is specified and the degree to which the role of the regulatory site is considered important.

The binding-change mechanism proposed by Boyer (1979) contains two features in addition to the two features described above. The first

is that the free energy input is used not only for promotion of the dissociation of ATP tightly bound to $F_1$, but also for promotion of the binding
of ADP and Pi to $F_1$. The second feature is that the changes in affinity
of $F_1$ for substrates and products are due to the cooperative alterations
of the catalytic sites, which in turn are caused by the conformational
changes of $F_1$ coupled to the energy input to the $F_1$–$F_0$ complex on the
membrane.

Figure 6.14a is a schematic representation of the binding-change
mechanism. $F_1$ contains two equivalent and interconvertible catalytic
sites, site 1 and site 2. First, ADP and Pi bind loosely to site 2 on
molecule of $F_1$ containing an ATP bound tightly to site 1 (ATP > E),
thus giving rise to $\text{ATP} > \text{E}:^{\text{ADP}}_{\text{Pi}}$) (step 1). A binding change then occurs
in response to the free energy input, and the ADP and Pi that have
been loosely bound to site 2 now become tightly bound. At the same
time, the ATP tightly bound to site 1 undergoes a binding change, form-
ing a loose complex, $\text{ATP} \cdot \text{E*}^{<\text{ADP}}_{\leq\text{Pi}}$ (step 2), and subsequently dissociates
from $F_1$, giving rise to $\text{E*}^{<\text{ADP}}_{\leq\text{Pi}}$ (step 3). Although not directly mentioned
in this scheme, the free energy is thought to be supplied to $F_1$ through
$F_0$ by the $H^+$ gradient. The conversion of $\text{E*}^{<\text{ADP}}_{<\text{Pi}}$ to E < ATP (step

Fig. 6.14 The binding-change mechanism of ATP synthesis and hydrolysis.
Scheme A describes an earlier version of the model that assumed
catalytic cooperativity between two alternating sites. For details
see text. Scheme B is a current, expanded version that assumes
there are three alternating sites on each of the three catalytic
subunits of $F_1$ (Kayalar et al., 1977; Gresser et al., 1982).

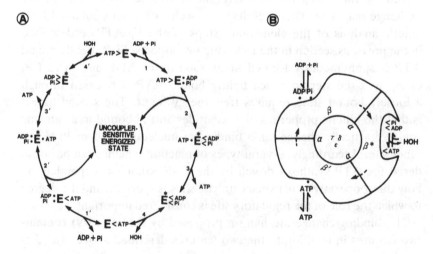

4) occurs without outside free energy input. Finally, the above reactions proceed by replacing site 1 and site 2 with each other (steps 1'–4'), and $F_1$ returns to the original state, ATP > E.

Boyer and his colleagues and other investigators have obtained experimental results that are consistent with or supportive of the binding-change mechanism. Kayalar *et al.* (1976) have shown that the Michaelis constants for ADP and Pi increase markedly when the $H^+$ gradient is cancelled by uncouplers during SMP-driven oxidative phosphorylation. This finding is consistent with the binding-change mechanism, which argues that the $H^+$ gradient causes conformational changes of $F_1$ and, as a result, promotes the binding of ADP and Pi.

Thylakoid membranes with ongoing ATP synthesis contain ATP that cannot be removed by hexokinase at a ratio of 1 mol/mol of $CF_1$. This amount of bound ATP remains constant even under conditions where the concentration of Pi is well below the Michaelis constant. Moreover, as shown in Fig. 6.15, a large amount of Pi is bound to $CF_1$ during the initial phase of ATP synthesis driven by the thylakoid membranes.

Fig. 6.15 Occurrence of transitorily tightly bound ATP and Pi in thylakoid membranes during ATP synthesis. After a steady-state rate of ATP synthesis from ADP and [$^{32}$P]Pi by preilluminated thylakoid was obtained, [$^{32}$P]Pi was diluted with unlabeled Pi. After the [$^{32}$P]Pi dilution, about 0.3 mol of [$^{32}$P]ATP/mol $CF_1$ was synthesized. This amount of [$^{32}$P]ATP indicates the presence of transitorily tightly bound [$^{32}$P]Pi (Rosen *et al.*, 1979).

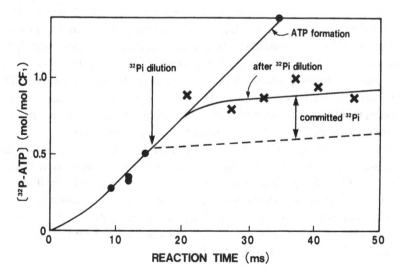

This Pi cannot readily be replaced by a chase reaction. These findings are consistent with the binding-change mechanism, which argues that 2 mol or more of catalytic sites are present/mol of $CF_1$ and that the ATP that is tightly bound to one site becomes loosely bound when ADP and Pi bind tightly to the other site.

The fact that the nonhydrolyzable ATP analog AMPPNP inhibits ATP hydrolysis but not ATP synthesis has been regarded as evidence for the argument that the reaction pathways for ATP synthesis and ATP hydrolysis are different from each other (Penefsky, 1974). However, this effect of AMPPNP can also be explained nicely in the above two models, which argue that both the synthesis and hydrolysis of ATP occur by the same pathway: AMPPNP binds to $F_1$ and forms a tight complex. Under conditions where no energy input from outside is available, it remains bound to $F_1$, consequently blocking ATP hydrolysis. Under conditions where energy input is available from a source such as oxidative phosphorylation, on the other hand, bound AMPPNP, like bound ATP, dissociates from $F_1$ upon its affinity change caused by free energy input from outside and thus fails to inhibit ATP synthesis. However, as has been described in the preceding section, a considerable portion of the ATP hydrolysis catalyzed by isolated $F_1$ does not occur through the reverse reaction of ATP synthesis driven by the $F_1$–$F_0$ complex. We believe that this fact makes it difficult to infer the reaction mechanism of ATP synthesis from the previous studies on the reaction mechanism of $F_1$-ATPase.

As shown in Fig. 6.14a, two alternating catalytic sites are depicted for each $F_1$ in an early version of the scheme proposed by Boyer (1979). However, this scheme is not limited to two catalytic sites and can easily be expanded to accommodate the presence of three alternating catalytic sites (Cross, 1981). Gresser, Myers & Boyer (1982) recently argued that the alternating three-site model shown in Fig. 6.14b better explains ATP hydrolysis driven by beef heart $F_1$. They argued this on the basis of results that they had obtained in binding measurements: ATP at concentrations of about 1 μM or less saturates the first catalytic site on $F_1$. About 10–20 μM ATP then saturates the second catalytic site, promotes the release of the product from the first site and causes a reduction in the Pi ⇌ $H_2O$ exchange. The concentration of ATP required for half maximal velocity of ATP hydrolysis is higher than these concentrations. These results indicate that $F_1$-driven ATP hydrolysis is controlled by ATP at three different concentrations and that at least three ATP-binding sites are, therefore, involved in ATP hydrolysis.

If it is assumed that the subunit stoichiometry of $F_1$ is $\alpha_3\beta_3\gamma\delta\varepsilon$ and that the $F_1$ ternary structure is axially symmetrical, the alternating three-site model proposed by Gresser *et al.* (1982) can link the structure and the catalytic mechanism of $F_1$: if the behavior of the three catalytic subunit pairs, each composed of $\alpha$ and $\beta$ subunits, is controlled by a control subunit core composed of $\gamma$, $\delta$ and $\varepsilon$ subunits, only one catalytic subunit pair can interact with the control subunit core at any stage of the catalytic cycle, though the three pairs should be equally capable of such interaction during the steady-state reaction. It is presumed that this interaction is driven by the energy input to the control subunit core through $F_0$, and that the catalytic subunit pairs rotate around the central subunit core.

## Conformational changes of $F_1$ during ATP hydrolysis and ATP synthesis

As has been described earlier in this chapter, the activity of $F_1$-ATPase is stimulated by ATP. Thus $F_1$ can be expected to undergo conformational changes upon its binding with ligands, regardless of whether the presence of a regulatory site is assumed or the presence of cooperative catalytic sites is assumed. In addition, the binding-change mechanism postulates that energization of the membrane causes conformational changes of $F_1$ and thus, changes in the binding of ligands. Boyer (1979) has suggested that the conformational change of $F_1$ serves as a key interaction in the $F_1$-driven energy transduction at the molecular level. A large number of studies have been carried out on the conformational change of $F_1$ and its subunits induced by ligand binding, and the conformational change of $F_1$ in the $F_1$–$F_0$ complex induced by membrane energization (Penefsky, 1979).

### Conformational changes of $F_1$ induced by ligand binding

As described earlier in this chapter, the fluorescence intensity of aurovertin increases when it binds to the $\beta$ subunit of $F_1$. This fluorescence intensity is enhanced by ADP and diminished by ATP. When the $\beta$ subunit of $F_1$ is blocked by DCCD, the binding of aurovertin occurs normally, but the fluorescence changes induced by ADP and ATP become very small (Pougeois, Satre & Vignais, 1979). This finding suggests that the inhibition of ATPase activity by DCCD is closely related to the inhibition of the conformational change of $\beta$ subunit induced by nucleotide binding.

Evidence exists for the involvement of an $\alpha$ subunit in the conformational change of a $\beta$ subunit induced by nucleotide binding. In EF₁ the fluorescence intensity of aurovertin bound to the $\beta$ subunit is enhanced by the high-affinity binding of ADP ($K_d = 4 \, \mu M$). In the mutant EF₁ that has normal $\beta$ subunits and altered $\alpha$ subunits, however, ADP cannot enhance the fluorescence intensity of bound aurovertin, although it is still capable of binding to the mutant EF₁ with high affinity (Wise, Latchney & Senior, 1981).

The conformational change of F₁ is induced not only by the binding of nucleotide, but by the binding of Pi as well. Matsuoka *et al* (1981) observed that the fluorescence intensity of DNS-ADP bound to the catalytic site of F₁ is further enhanced by the addition of Pi and concluded that the environment of the catalytic site of F₁ becomes more hydrophobic upon binding of Pi.

## *Conformational changes of F₁ induced by membrane energization*

Ryrie & Jagendorf (1971, 1972) showed that hydrogen atoms of the CF₁ molecule undergo an exchange with tritium in the medium upon illumination of the thylakoid membranes. This result constitutes the first example of the occurrence of an illumination-dependent conformational change in CF₁. McCarty & Fagan (1973) showed that illumination of thylakoids in the presence of *N*-ethylmaleimide leads to chemical modification of the $\gamma$ subunit of CF₁ and inhibition of ATP synthesis. This result suggests that the SH groups of the $\gamma$ subunit involved in the catalytic activity are exposed upon illumination.

It was shown in experiments using fluorescent probes that the conformational changes of F₁ induced by membrane energization occur rapidly enough to be comparable to the onset of ATP synthesis. When respiration is initiated by the addition of oxygen to SMP, the fluorescence intensity of aurovertin bound to the $\beta$ subunit of F₁ increases with a half time of 36 ms (Chang & Penefsky, 1974).

As has been pointed out repeatedly, one of the basic features of the theory proposed by Boyer (1979) is the fact that the conformational changes of $F_1$ induced by membrane energization lead to changes in the properties of the binding of nucleotide and Pi to $F_1$. Harris & Slater (1975) showed that when thylakoid membranes are exposed to light, the total amount of nucleotide bound to $CF_1$ remains constant but the rate of nucleotide exchange with the medium increases. It also was reported that ADP tightly bound to thylakoids is released within 20 ms when an acid–base transition is used to energize thylakoids (Rosing *et al.*, 1976). In this case, the release of ADP occurs faster than the synthesis of $[\gamma^{32}P]$ATP from exogenously added ADP and $[^{32}P]$Pi.

Thus, conformational changes of $F_1$ have been shown to occur upon membrane energization. However, many difficult problems remain to be solved for the elucidation of the relationship between these conformational changes and ATP synthesis. For example, it is not clear at present whether the observed conformational changes are indispensable to any particular elementary steps of energy transduction driven by the $F_1$–$F_0$ complex or are mere secondary phenomena. If the former is the case, then additional problems arise: how is the transport of $H^+$, mediated by $F_0$ in the presence of an electrochemical potential gradient of $H^+$, converted to the conformational changes of $F_1$ and what role does a local $H^+$ gradient, generated in the vicinity of the $F_1$–$F_0$ complex (Williams, 1983), play during this process? Extensive studies are needed for the future resolution of these problems and for an understanding of the true molecular mechanism of energy transduction in oxidative phosphorylation and photophosphorylation.

# References

Alexandre, A., Reynafaje, B. & Lehninger, A. L. (1978). Stoichiometry of vectorial $H^+$ movements coupled to electron transport and to ATP synthesis in mitochondria. *Proc. Natl. Acad. Sci. (USA)*, **75**, 5296–300.

Amzel, L. M., Mickinney, M., Narayanan, P. & Pedersen, P. I. (1982). Structure of the mitochondrial $F_1$-ATPase at 9–1 resolution. *Proc. Natl. Acad. Sci. (USA)*, **79**, 5852–6.

Apps, D. K. & Glover, L. A. (1978). Isolation and characterization of magnesium adenosinetriphosphatase from the chromaffin granule membrane. *FEBS Lett.*, **85**, 254–8.

Avron, M. (1963). A coupling factor in photophosphorylation. *Biochim. Biophys. Acta*, **77**, 699–702.

Baird, B. A. & Hammes, G. G. (1976). Chemical cross-linking studies of chloroplast coupling factor 1. *J. Biol. Chem.*, **251**, 6953–62..

Baird, B. A. & Hammes, G. G. (1977). Chemical cross-linking studies of beef heart mitochondrial coupling factor 1. *J. Biol. Chem.*, **252**, 4743–8.

Beechey, R. B., Hubbard, S. A., Linnett, P. E., Mitchell, A. D. & Munn, E. A. (1975). A simple and rapid method for the preparation of adenosine triphosphatase from submitochondrial particles. *Biochem. J.*, **148**, 533–7.

Boyer, P. D. (1979). Binding-change mechanism of ATP synthesis. In *Membrane Bioenergetics*, ed. C. P. Lee, G. Schatz & L. Ernster, pp. 461–79. Reading, MA: Addison-Wesley Publishing Co.

Boyer, P. D., Chance, B., Ernster, L., Mitchell, P., Racker, E. & Slater, E. C. (1977). Oxidative phosphorylation and photophosphorylation. *Ann. Rev. Biochem.*, **46**, 955–1026.

Chang, T. & Penefsky, H. S. (1974). Energy-dependent enhancement of aurovertin fluorescence. *J. Biol. Chem.*, **249**, 1090–8.

Choate, G. L., Hutton, R. L. & Boyer, P. D. (1979). Occurrence and significance of oxygen exchange reactions catalyzed by mitochondrial adenosine triphosphatase preparations. *J. Biol. Chem.*, **254**, 286–90.

Criddle, R. S., Johnston, R. F. & Stack, R. J. (1979). Mitochondrial ATPases. *Curr. Top. Bioenerg.*, **89–145.**

Cross, R. L. (1981). The mechanism and regulation of ATP synthesis by F$_1$-ATPases. *Ann. Rev. Biochem.*, **50**, 681–714.

Cross, R. L., Grubmeyer, C. & Penefsky, H. S. (1982). Mechanism of ATP hydrolysis by beef heart mitochondrial ATPase. *J. Biol. Chem.*, **257**, 12101–5.

Cross, R. L. & Kohlbrenner, W. E. (1978). The mode of inhibition of oxidative phosphorylation by efrapeptin (A23871). *J. Biol. Chem.*, **253**, 4865–73.

Cross, R. L. & Nalin, C. M. (1982). Adenine nucleotide binding sites on beef heart F$_1$-ATPase. *J. Biol. Chem.*, **257**, 2874–81.

Dewey, T. G. & Hammes, G. G. (1981). Steady-state kinetics of ATP synthesis and hydrolysis catalyzed by reconstituted chloroplast coupling factor. *J. Biol. Chem.*, **256**, 8941–6.

Downie, J. A., Gibson, F. & Cox, G. B. (1979). Membrane adenosine triphosphatases of prokaryotic cells. *Ann. Rev. Biochem.*, **48**, 103–31.

Dunn, S. D. (1980). ATP causes a large change in the conformation of the isolated alpha subunit of *Escherichia coli* F$_1$-ATPase. *J. Biol. Chem.*, **255**, 11857–60.

Dunn, S. D. & Futai, M. (1980). Reconstitution of a functional coupling factor from the isolated subunits of *Escherichia coli* F$_1$-ATPase. *J. Biol. Chem.*, **255**, 113–18.

Dunn, S. D., Heppel, L. A. & Fullmer, C. S. (1980). The NH$_2$-terminal portion of the alpha subunit of *Escherichia coli* F$_1$-ATPase is required for binding the delta subunit. *J. Biol. Chem.*, **255**, 6891–6.

Ebel, R. E. & Lardy, H. A. (1975a). Stimulation of rat liver mitochondrial adenosine triphosphatase by anions. *J. Biol. Chem.*, **250**, 191–6.

Ebel, R. E. & Lardy, H. A. (1975b). Influence of aurovertin on mitochondrial ATPase activity. *J. Biol. Chem.*, **250**, 4992–5.

Enns, R. & Criddle, R. S. (1977). Investigation of the structural arrangement of the protein subunits of mitochondrial ATPase. *Arch. Biochem. Biophys.*, **183**, 742–52.

Esch, F. S. & Allison, W. S. (1978). Identification of a tyrosine residue at a nucleotide binding site in the beta subunit of the mitochondrial ATPase with p-fluorosulfonyl[$^{14}$C]-benzoyl-5'-adenosine. *J. Biol. Chem.*, **253**, 6100–6.

Esch, F. S. & Allison, W. S. (1979). On the subunit stoichiometry of the $F_1$-ATPase and the sites in it that react specifically with p-fluorosulfonylbenzoyl-5'-adenosine. *J. Biol. Chem.*, **254**, 10740–6.

Esch, F. S., Bohlen, P., Ohtsuka, A. S., Yoshida, M. & Allison, W. S. (1981). Inactivation of the bovine mitochondrial $F_1$-ATPase with dicyclohexyl[$^{14}$C]carbodiimide leads to the modification of a specific glutamic acid residue in the beta subunit. *J. Biol. Chem.*, **256**, 9084–9.

Feldman, R. I. & Sigman, D. S. (1982). The synthesis of enzyme-bound ATP by soluble chloroplast coupling factor 1. *J. Biol. Chem.*, **257**, 1676–83.

Feldman, R. I. & Sigman, D. S. (1982). The synthesis of enzyme-bound ATP by soluble chloroplast coupling factor 1. *J. Biol. Chem.*, **257**, 1676–83.

Ferguson, S. J., Lloyd, W. J., Lyons, M. H. & Radda, G. K. (1975). The mitochondrial ATPase: evidence for a single essential tyrosine residue. *Eur. J. Biochem.*, **54**, 117–26.

Fillingame, R. H. (1981). Biochemistry and genetics of bacterial H$^+$-translocating ATPases. *Curr. Top. Bioenerg.*, **11**, 35–106.

Foster, D. I. & Fillingame, R. H. (1979). Energy-transducing H$^+$-ATPase of *Escherichia coli*. *J. Biol. Chem.*, **254**, 8230–6.

Foster, D. L. & Fillingame, R. H. (1982). Stoichiometry of subunits in the H$^+$-ATPase complex of *Escherichia coli*. *J. Biol. Chem.*, **257**, 2009–15.

Futai, M. (1977). Reconstitution of ATPase activity from the isolated $\alpha$, $\beta$, $\gamma$ and $\delta$ subunits of the coupling factor, $F_1$, of *Escherichia coli*. *Biochem. Biophys. Res. Commun.*, **79**, 1231–7.

Futai, M. & Kanazawa, H. (1980). Role of subunits in proton-translocating ATPase ($F_0$–$F_1$). *Curr. Top. Bioenerg.*, **10**, 181–215.

Futai, M., Sternweis, P. C. & Heppel, L. A. (1974). Purification and properties of reconstitutively active and inactive adenosinetriphosphatase from *Escherichia coli*. *Proc. Natl. Acad. Sci. (USA)*, **71**, 2725–9.

Garrett, N. E. & Penefsky, H. S. (1975). Interaction of adenine nucleotides with multiple binding sites on beef heart mitochondrial adenosine triphosphatase. *J. Biol. Chem.*, **250**, 6640–7.

Gay, N. J. & Walker, J. E. (1981a). The atp operon: nucleotide sequence of the region encoding the $\alpha$-subunit of *Escherichia coli* ATP-synthase. *Nucleic Acids Res.*, **9**, 2187–94.

Gay, N. J. & Walker, J. E. (1981b). The atp operon: nucleotide sequence of the promotor and the genes for the membrane proteins, and the δ subunit of *Escherichia coli* ATP-synthase. *Nucleic Acids Res.*, **9**, 3919–26.

Gresser, M. J., Myers, J. A. & Boyer, P. D. (1982). Catalytic site cooperativity of beef heart mitochondrial $F_1$ adenosine triphosphatase. *J. Biol. Chem.*, **257**, 12030–8.

Grubmeyer, C., Cross, R. L. & Penefsky, H. S. (1982). Mechanism of ATP hydrolysis by beef heart mitochondrial ATPase. *J. Biol. Chem.*, **257**, 12092–100.

Grubmeyer, C. & Penefsky, H. S. (1981*a*). The presence of two hydrolytic sites on beef heart mitochondrial adenosine triphosphatase. *J. Biol. Chem.*, **256**, 3718–27.

Grubmeyer, C. & Penefsky, H. S. (1981*b*). Cooperativity between catalytic sites in the mechanism of action of beef heart mitochondrial adenosine triphosphatase. *J. Biol. Chem.*, **256**, 3728–34.

Harris, D. A. (1978). The interactions of coupling ATPase with nucleotides. *Biochim. Biophys. Acta*, **463**, 245–73.

Harris, D. A. & Slater, E. C. (1975). Tightly bound nucleotides of the energy-transducing ATPase of chloroplasts and their role in photophosphorylation. *Biochim. Biophys. Acta*, **387**, 335–48.

Hinkle, P. C. & McCarty, R. E. (1978). How cells make ATP. *Sci. Am.*, **238**, 104–23.

Issartel, J. P., Klein, G., Satre, M. & Vignais, P. V. (1983). Aurovertin binding sites on beef heart mitochondrial $F_1$-ATPase. Study with $^{14}$C aurovertin D of the binding stoichiometry and of the interaction between aurovertin and the natural ATPase inhibitor for binding to $F_1$. *J. Biol. Chem.*, **22**, 3492–7.

Kagawa, Y. & Racker, E. (1966*a*). Partial resolution of the enzymes catalyzing oxidative phosphorylation. IX. Reconstitution of oligomycin-sensitive adenosine triphosphatase. *J. Biol. Chem.*, **241**, 2467–74.

Kagawa, Y. & Racker, E. (1966*b*). Partial resolution of the enzymes catalyzing oxidative phosphorylation. X. Correlation of morphology and function in submitochondrial particles. *J. Biol. Chem.*, **241**, 2475–82.

Kagawa, Y., Sone, N., Hirata, H. & Yoshida, M. (1979). Structure and function of $H^+$-ATPase. *J. Bioenerg. Biomem.*, **11**, 39–78.

Kanazawa, H. & Futai, M. (1981). Proton-translocating ATPase of *Escherichia coli*. *Protein, Nucleic Acid and Enzyme*, **26**, 1999–2013.

Kanazawa, H., Kayano, T., Kiyasu, T. & Futai, M. (1982). Nucleotide sequence of the genes coding for $\beta$ and $\varepsilon$ subunits of proton-translocating ATPase from *E. coli*. *Biochem. Biophys. Res. Commun.*, **105**, 1257–64.

Kanazawa, H., Kayano, T., Mabuchi, K. & Futai, M. (1981*a*). Nucleotide sequence of the genes coding for $\alpha$, $\beta$ and $\gamma$ subunits of the proton-translocating ATPase of *E. coli*. *Biochem. Biophys. Res. Commun.*, **103**, 604–12.

Kanazawa, H., Mabuchi, K., Kayano, T., Noumi, T., Sekiya, T. & Futai, M. (1981*b*). Nucleotide sequence of the genes for $F_0$ components of the proton-translocating ATPase of *E. coli. Biochem. Biophys. Res. Commun.*, **103**, 613–20.

Kanazawa, H., Mabuchi, K., Kayano, T., Tamura, F. & Futai, M. (1981). Nucleotide sequence of genes coding for dicyclohexylcarbodiimide-binding protein and the α subunit of proton-translocating ATPase of *Escherichia coli. Biochem. Biophys. Res. Commun.*, **100**, 219–25.

Kanazawa, H. Saito, S. & Futai, M. (1978). Coupling factor ATPase from *Escherichia coli*: an UncA mutant (uncA401) with defective subunit. *J. Biochem.*, **84**, 1513–17.

Kanazawa, H., Tamura, F., Mabuchi, K., Miki, T. & Futai, M. (1980). Organization of *unc* gene cluster of *Escherichia coli* coding for proton-translocating ATPase of oxidative phosphorylation. *Proc. Natl. Acad. Sci. (USA)*, **77**, 7005–0.

Kasahara, M. & Penefsky, H. S. (1978). High affinity binding of monovalent $P_i$ by beef heart mitochondrial adenosine triphosphatase. *J. Biol. Chem.*, **253**, 4180–7.

Kayalar, C., Rosing, J. & Boyer, P. D. (1976). 2,4-Dinitro phenol causes a marked increase in the apparent Km of $P_i$ and of ADP for oxidative phosphorylation. *Biochem. Biophys. Res. Commun.*, **72**, 1153–9.

Kayalar, C., Rosing, J. & Boyer, P. D. (1977). An alternating site sequence for oxidative phosphorylation suggested by measurement of substrate binding patterns and exchange reaction inhibitions. *J. Biol. Chem.*, **252**, 2486–91.

Knowles, A. F. & Penefsky, H. S. (1972a). The subunit structure of beef heart mitochondrial adenosine triphosphatase: isolation procedures. *J. Biol. Chem.*, **247**, 6617–23.

Knowles, A. F. & Penefsky, H. S. (1972b). The subunit structure of beef heart mitochondrial adenosine triphosphatase. *J. Biol. Chem.*, **247**, 6624–30.

Kumar, G., Kalra, V. K. & Brodie, A. F. (1979). Affinity labeling of coupling factor-latent ATPase from *Mycobacterium phlei* with 2′,3′-dialdehyde derivatives of adenosine 5′-triphosphate and adenosine 5′diphosphate. *J. Biol. Chem.*, **254**, 1964–71.

Leimgruber, R. M. & Senior, A. E. (1976a). Removal of 'tightly bound' nucleotides from soluble mitochondrial adenosine triphosphatase ($F_1$). *J. Biol. Chem.*, **251**, 7103–9.

Leimgruber, R. M. & Senior, A. E. (1976b). Removal of 'tightly bound' nucleotides from phosphorylating submitochondrial particles. *J. Biol. Chem.*, **251**, 7110–3.

Mabuchi, K., Kanazawa, H., Kayano, T. & Futai, M. (1981). Nucleotide sequence of the gene coding for the δ subunit of proton-translocating ATPase of *Escherichia coli. Biochem. Biophys. Res. Commun.*, **102**, 172–9.

McCarty, R. E. (1978a). AMP is converted to ADP and ATP in the medium before it is bound to coupling factor 1 in illuminated spinach chloroplast thylakoids. *FEBS Lett.*, **95**, 299–302.

McCarty, R. E. (1978b). The ATPase complex of chloroplasts and chromatophores. Curr. Top. Bioenerg., 7, 245–77.

McCarty, R. E. & Fagan, J. (1975). Light-stimulated incorporation of N-ethylmaleimide into coupling factor 1 in spinach chloroplasts. Biochemistry, 12, 1503–7.

MacLennan, E. H. & Asai, J. (1968). Studies on the mitochondrial adenosine triphosphatase system. V. Localization of the oligomycin-sensitivity conferring protein. Biochem. Biophys. Res. Commun., 33, 441–7.

MacLennan, D. H. & Tzagoloff, A. (1968). Studies on the mitochondrial adenosine triphosphatase system. IV. Purification and characterization of the oligomycin sensitivity conferring protein. Biochemistry, 7, 1603–10.

Matsuoka, I., Takeda, K., Futai, M. & Tonomura, Y. (1982). Reactions of fluorescent ATP analog, 2'-(5-dimethyl-amino-naphthalene-1-sulfonyl)amino-2'-deoxyATP, with E. coli $F_1$-ATPase and its subunits. J. Biochem., 92, 1383–98.

Matsuoka, I., Watanabe, T. & Tonomura, Y. (1981). Reaction mechanism of the ATPase activity of mitochondrial $F_1$ studied by using a fluorescent ATP analog, 2'-(5-dimethyl-aminonaphthalene-1-sulfonyl) amino-2'-deoxyATP. J. Biochem., 90, 967–89.

Merchant, S., Shaner, S. L. & Selman, B. R. (1983). Molecular weight and subunit stoichiometry of the chloroplast coupling factor 1 from Chlamydomononas reinhardi. J. Biol. Chem., 258, 1026–31.

Mitchell, P. (1961). Coupling of phosphorylation to electron and hydrogen transfer by a chemi-osmotic type of mechanism. Nature, 191, 144–8.

Negrin, R. S., Foster, D. L. & Fillingame, R. H. (1980). Energy-transducing $H^+$-ATPase of Escherichia coli. J. biol. Chem., 255, 5643–8.

Nelson, N. (1976). Structure and function of chloroplast ATPase. Biochim. Biophys. Acta, 456, 314–38.

Njus, D., Knoh, J. & Zallakian, M. (1981). Proton-linked transport in chromaffin granules. Curr. Top. Bioenerg., 11, 107–47.

Ohta, S., Tsuboi, M., Ohshima, T., Yoshida, M. & Kagawa, Y. (1980). Nucleotide binding to isolated $\alpha$ and $\beta$ subunits of proton translocating adenosine triphosphatase studied with circular dichroism. J. Biochem., 87, 1609–17.

Paradies, H. H., Mertens, G., Schmid, R., Schneider, E. & Altendorf, K. (1981). Molecular properties of the ATP synthetase from Escherichia coli. Biochem. Biophys. Res. Commun., 98, 595–606.

Paradies, H. H., Zimmermann, J. & Schmidt, U. D. (1978). The conformation of chloroplast coupling factor 1 from spinach in solution. J. Biol. Chem., 253, 8972–9.

Pedersen, P. L. (1976). ATP-dependent reactions catalyzed by inner membrane vesicles of rat liver mitochondria. J. Biol. Chem., 251, 934–40.

Pedersen, P. L., Schwerzmann, K. & Cintron, N. (1981). Regulation of the synthesis and hydrolysis of ATP in biological systems. Role of peptide inhibitors of $H^+$-ATPases. Curr. Top. Bioenerg., 11, 149–99.

Penefsky, H. S. (1974). Differential effects of adenyl imidodiphosphate on adenosine triphosphate synthesis and the partial reactions of oxidative phosphorylation. *J. Biol. Chem.*, **249**, 3579–85.

Penefsky, H. S. (1977). Reversible binding of $P_i$ by beef heart mitochondrial adenosine triphosphatase. *J. Biol. Chem.*, **252**, 2891–9.

Penefsky, H. S. (1979). Mitochondrial ATPase. *Adv. Enzymol.*, **49**, 223–80.

Penefsky, H. S., Pullman, M. E., Datta, A. & Racker, E. (1960). Partial resolution of the enzymes catalyzing oxidative phosphorylation. II. Participation of a soluble adenosine triphosphatase in oxidative phosphorylation. *J. Biol. Chem.*, **235**, 3330–6.

Pennin, F., Godinot, C. & Gautheron, D. C. (1979). Optimization of the purification of mitochondrial $F_1$-adenosine triphosphatase. *Biochim. Biophys. Acta*, **548**, 63–71.

Pick, U. & Racker, E. (1979). Purification and reconstitution of N,N'-dicyclohexylcarbodimide sensitive ATPase complex from spinach chloroplasts. *J. Biol. Chem.*, **254**, 2793–9.

Pougeois, R., Satre, M. & Vignais, P. V. (1979). Reactivity of mitochondrial $F_1$-ATPase to dicyclohexylcarbodiimide. *Biochemistry*, **18**, 1408–13.

Pullman, M. E. & Monroy, G. C. (1963). A naturally occurring inhibitor of mitochondrial adenosine triphosphatase. *J. Biol. Chem.*, **238**, 3762–9.

Pullman, M. E., Penefsky, H. S., Datta, A. & Racker, E. (1960). Partial resolution of the enzymes catalyzing oxidative phosphorylation. I. Purification and properties of soluble dinitrophenol-stimulated adenosine triphosphatase. *J. Biol. Chem.*, **235**, 3322–9.

Racker, E. (1976). *A New Look at Mechanisms in Bioenergetics*. New York: Academic Press.

Rosen, G., Gresser, M., Vinkler, C. & Boyer, P. D. (1979). Assessment of total catalytic sites and the nature of bound nucleotide participation in photophosphorylation. *J. Biol. Chem.*, **254**, 10654–61.

Rosing, J., Kayalar, C. & Boyer, P. D. (1977). Evidence for energy-dependent change in phosphate binding for mitochondrial oxidative phosphorylation based on measurements of medium and intermediate phosphate–water exchanges. *J. Biol. Chem.*, **252**, 2478–85.

Rosing, J., Smith, D. J., Kayalar, C. & Boyer, P. D. (1976). Medium ADP and not ADP already tightly bound to thylakoid membranes forms the initial ATP in chloroplast phosphorylation. *Biochem. Biophys. Res. Commun.*, **72**, 1–8.

Ryrie, I. J. & Jagendorf, A. T. (1971). An energy-linked conformational change in the coupling factor protein in chloroplasts. *J. Biol. Chem.*, **246**, 3771–4.

Ryrie, I. J. & Jagendorf, A. T. (1972). Correlation between a conformational change in the coupling factor protein and the high energy state in chloroplasts. *J. Biol. Chem.*, **247**, 4453–9.

Sakamoto, J. & Tonomura, Y. (1983). Synthesis of enzyme bound ATP by mitochondrial soluble $F_1$-ATPase in the presence of dimethylsulfoxide. *J. Biochem.*, **93**, 1601–14.

Satre, M., Lundardi, J., Pougeois, R. & Vignais, V. (1979). Inactivation of *Escherichia coli* BF₁-ATPase by dicyclohexylcarbodiimide. Chemical modification of the β subunit. *Biochemistry*, **18**, 3134–40.

Schmidt, U. D. & Paradies, H. H. (1977a). The structure of the epsilon-subunit from the chloroplast coupling factor (CF₁) studied by means of small angle X-ray scattering and inelastic light scattering. *Biochem. Biophys. Res. Commun.*, **78**, 383–92.

Schuster, S. M., Ebel, R. E. & Lardy, H. A. (1975). Kinetic studies on rat liver and beef heart mitochondrial ATPase. *J. Biol. Chem.*, **250**, 7848–53.

Sebald, W. & Wachter, E. (1980). Amino acid sequence of the proteolipid subunit of the ATP synthase from spinach chloroplasts. *FEBS Lett.*, **122**, 307–11.

Senior, A. E. (1973). The structure of mitochondrial ATPase. *Biochim. Biophys. Acta*, **301**, 249–77.

Serrano, R., Kanner, B. I. & Racker, E. (1976). Purification and properties of the proton-translocating adenosine triphosphatase complex of bovine heart mitochondria. *J. Biol. Chem.*, **251**, 2453–61.

Shavit, N. (1980). Energy transduction in chloroplasts: structure and function of the ATPase complex. *Ann. Rev. Biochem.*, **49**, 111–38.

Shoshan, V. & Selman, B. R. (1980). The interaction of *N,N'*-dicyclohexylcarbodiimide with chloroplast coupling factor 1. *J. Biol. Chem.*, **255**, 384–9.

Slater, E. C., Kemp, A., Van der Kraan, I., Muller, J. L. M., Roveri, O. A., Vershoor, G. J., Wagenvoord, R. J. & Wielders, J. P. M. (1979). The ATP- and ADP-binding sites in mitochondrial coupling factor F₁ and their possible role in oxidative phosphorylation. *FEBS Lett.*, **103**, 7–11.

Sone, N., Yoshida, M., Hirata, H. & Kagawa, Y. (1975). Purification and properties of a dicyclohexylcarbodiimide-sensitive adenosine triphosphatase from a thermophilic bacterium. *J. Biol. Chem.*, **250**, 7917–23.

Sone, N., Yoshida, M., Hirata, H. & Kagawa, Y. (1977). Adenosine triphosphate synthesis by electrochemical proton gradient in vesicles reconstituted from purified adenosine triphosphatase and phospholipids of thermophilic bacterium. *J. Biol. Chem.*, **252**, 2956–60.

Soper, J. W., Decker, G. L. & Pedersen, P. L. (1979). Mitochondrial ATPase complex. *J. Biol. Chem.*, **254**, 11170–6.

Spitsberg, V. & Haworth, R. (1977). The crystalization of beef heart mitochondrial adenosine triphosphatase. *Biochim. Biophys. Acta.*, **492**, 237–40.

Sternweis, P. C. & Smith, J. B. (1980). Characterization of the inhibitory (ε) subunit of the proton-translocating adenosine triphosphatase from *Escherichia coli. Biochemistry*, **19**, 526–31.

Tiefert, M. A. & Moudrianakis, E. N. (1979). Role of AMP in photophosphorylation by spinach chloroplasts. *J. Biol. Chem.*, **254**, 9500–8.

Todd, R. D. & Douglas, M. G. (1981). A model for the structure of the yeast mitochondrial adenosine triphosphatase complex. *J. Biol. Chem.*, **256**, 6984–9.

Tzagoloff, A., Byington, K. H. & MacLennan, D. H. (1968). Studies on the mitochondrial adenosine triphosphatase system. II. The isolation and characterization of an oligomycin-sensitive adenosine triphosphatase from bovine heart mitochondria. *J. Biol. Chem.*, **243**, 2405–12.

Verschoor, G. J., Van der Sluis, P. R. & Slater, E. C. (1977). The binding of aurovertin to isolated $\beta$ subunit of $F_1$ (mitochondrial ATPase). *Biochim. Biophys. Acta*, **462**, 438–49.

Wagenvoord, R. J., Kemp, A. & Slater, E. C. (1980). The number and localization of adenine nucleotide-binding sites in beef-heart mitochondrial ATPase ($F_1$) determined by photolabelling with 8-azido-ATP and 8-azido-ADP. *Biochim. Biophys. Acta*, **593**, 204–11.

Wagenvoord, R. J., van der Kraan, I. & Kemp, A. (1979). Localization of adenine nucleotide-binding sites on beef heart mitochondrial ATPase by photolabelling with 8-azido-ADP and 8-azido-ATP. *Biochim. Biophys. Acta*, **548**, 85–95.

Wakabayashi, T., Kubota, M., Yoshida, M. & Kagawa, Y. (1977). Structure of ATPase (coupling factor $TF_1$) from a thermophilic bacterium. *J. Mol. Biol.*, **117**, 515–19.

Walker, J. E., Runswick, M. J. & Saraste, M. (1982). Subunit equivalence in *Escherichia coli* and bovine heart mitochondrial $F_1F_0$ ATPases. *FEBS Lett.*, **146**, 393–6.

Webb, M. R., Grubmeyer, C., Penefsky, H. S. & Trentham, D. R. (1980). The stereochemical course of phosphoric residue transfer catalyzed by beef heart mitochondrial ATPase. *J. Biol. Chem.*, **255**, 11637–9.

Williams, N. & Coleman, P. S. (1982). Exploring the adenine nucleotide binding sites on mitochondrial $F_1$-ATPase with a new photoaffinity probe, 3'-O-(4-benzoyl)benzoyl adenosine 5'triphosphate. *J. Biol. Chem.*, **257**, 2834–41.

Williams, R. J. P. (1983). Mitochondrial compartments and chemi-osmosis. *Trends Biochem. Sci.*, **8**, 48.

Wise, J. G., Latchney, L. R. & Senior, A. E. (1981). The defective proton-ATPase of *uncA* mutants of *Escherichia coli*. *J. Biol. Chem.*, **256**, **10383–9.**

Yamamoto, T. & Tonomura, Y. (1975). pH jump-induced phosphorylation of adenosine diphosphate in thylakoidal membrane. *J. Biochem.*, **77**, 137–46.

Yoshida, M., Okamoto, H., Sone, N., Hirata, H. & Kagawa, Y. (1977a). Reconstitution of thermostable ATPase capable of energy coupling from its purified subunits. *Proc. Natl. Acad. Sci. (USA)*, **74**, 936–40.

Yoshida, M., Sone, N., Hirata, H. & Kagawa, Y. (1977b). Reconstitution of adenosine triphosphatase of thermophilic bacterium from purified individual subunits. *J. Biol. Chem.*, **252**, 3480–5.

Yoshida, M., Sone, N., Hirata, H., Kagawa, Y. & Ui, N. (1979). Subunit structure of adenosine triphosphatase. *J. Biol. Chem.*, **254**, 9525–44.

# 7    The Sarcoplasmic
       Reticulum ATPase

## Sarcoplasmic reticulum and Ca$^{2+}$ transport

### Active transport of cation

Active transport across biomembranes, like muscle contraction, oxidative phosphorylation and photophosphorylation, is an
advanced area within the field of bioenergetics. Singer & Nicolson (1972)
proposed a model of the fundamental structure of a biomembrane, showing that protein molecules are embedded in a mosaic manner in a lipid
bilayer with high fluidity. This model has now been widely accepted.
Ion movement across a biomembrane, from one side with high electrochemical potential to the other side with low electrochemical potential,
is referred to as passive ion transport. Organisms also have the ability
to transport ions across a membrane against an electrochemical gradient
and this transport is accompanied by an increase in free energy. When
this free energy is supplied metabolically, this ion movement is called
active transport. In many cases the free energy for active transport is
supplied from ATP hydrolysis catalyzed by membrane-bound ATPases.
When ion transport against an electrochemical gradient is directly
coupled to ATP hydrolysis, it is called primary active transport. When
an electrochemical gradient of Na$^+$ or H$^+$ is established by primary active
transport and then the gradient drives the transport of another substance
against its electrochemical gradient, this transport is called secondary
or ion-coupled active transport.

Since this monograph is mainly concerned with energy-transducing
ATPases, we will focus upon primary active transport in this and the
following chapters. The Ca$^{2+}$,Mg$^{2+}$-ATPase of the sarcoplasmic reticulum (SR) is one of the simplest of the active transport ATPases and
will be discussed in this chapter. Chapter 8 deals with the Na$^+$,K$^+$-
ATPase, which has been as well studied as the Ca$^{2+}$,Mg$^{2+}$-ATPase of

the SR. We will also discuss the remarkable similarities between these transport ATPases in both structure and reaction mechanism.

## *The role of the sarcoplasmic reticulum in excitation–contraction coupling*

The intracellular Ca$^{2+}$ concentration is usually maintained at less than 1 µM; this is $10^3$–$10^4$ times lower than the extracellular Ca$^{2+}$ concentration. This allows intracellular Ca$^{2+}$ to play an essential role as a trigger for many important physiological processes and, therefore, its concentration must be very finely regulated. Regulation of the intracellular Ca$^{2+}$ concentration is carried out by Ca$^{2+}$ pumps present in the plasma membrane and in the intracellular membrane system.

As described in Chapter 2, the cycle of contraction and relaxation in skeletal muscle fiber is induced by the rapid increase and the subsequent decrease in the intracellular Ca$^{2+}$ concentration. A considerable amount of Ca$^{2+}$ is required for the induction of the maximum contraction of a muscle fiber because the concentration of troponin, a Ca$^{2+}$-receptor protein in the skeletal muscle fiber, is rather high. For this physiological requirement, the sarcoplasmic reticulum, an intracellular membrane system in the muscle fiber, functions as a large-capacity Ca$^{2+}$ storehouse, releasing Ca$^{2+}$ to induce contraction and taking it back up to bring about relaxation. As shown in Fig. 7.1, the membrane system that controls Ca$^{2+}$ concentration in muscle cells consists of the plasma membrane with its tubular infoldings (the T-system) running transversely to the fiber axis and a reticular structure (the SR) that forms a network surrounding the myofibrils (Porter & Palade, 1957; Peachey, 1965; Peachey & Schild, 1968; Franzini-Armstrong, Landmesser & Pilar, 1975). The reticular membranes of the SR anastomose at periodic intervals (at the Z-line in amphibian muscle or at the A–I junction in mammalian muscle) to form cisternae that are attached to the T-system but not continuous with it. The T system is open to the external milieu (Endo, 1964; Huxley, 1964; Franzini-Armstrong *et al.*, 1975), but the SR is a closed system in the cell. Electron microscopic observations suggest that a junction with a characteristic structure exists between the T-system and the cisternae of SR (Rayns, Devine & Sutherland, 1975).

Huxley & Taylor (1958) suggested that the T-system may mediate the transmission of electrical stimuli from the surface membrane to the interior of the muscle fiber. A rapid increase in intracellular Ca$^{2+}$ concentration, immediately after electrical stimulation of the muscle fiber, has been observed to precede muscle contraction (Jobsis & O'Conner, 1966; Ashley, Moisescu & Rose, 1974). In addition, Somlyo *et al.* (1981) showed with electron probe analysis that the $Ca^{2+}$ concentration decreased in the lumen of SR and increased conversely in the cytoplasm after a muscle fiber was tetanized. On the basis of these and many other observations, the physiological role of the SR in the contraction–relaxation cycle of muscle is explained as follows. During rest, muscle cell $Ca^{2+}$ accumulates in the SR and the concentration of cytoplasmic $Ca^{2+}$ is maintained at extremely low levels. When stimuli are transmitted from the nerve to the SR through the T-system, $Ca^{2+}$ is rapidly released from the SR into the cytoplasm and, as a result, induces muscle contraction. After excitation of the muscle fiber ceases, $Ca^{2+}$ is taken up by the SR and the concentration of cytoplasmic $Ca^{2+}$ is again reduced to a low level. Thus the muscle fiber returns to the initial resting state.

Fig. 7.1 The sarcoplasmic reticulum and the transverse tubules. The sleevelike, flattened membrane system surrounding each myofibril is called the sarcoplasmic reticulum. The terminal cisterna of the membrane system lies at the level of the Z-line of the myofibril in this diagram of a frog muscle. Pairs of parallel terminal cisternae run across the myofibrils in close contact with the transverse tubules. From Peachey (1965).

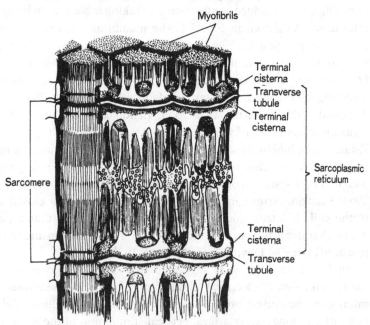

The mechanism by which Ca$^{2+}$ is released from the SR following muscle excitation is still unknown. Endo, Tanaka & Ogawa (1970) and Ford & Podolsky (1972) used skinned muscle fibers to show that Ca$^{2+}$ release from the SR can be induced by a small amount of Ca$^{2+}$ in the presence of adenine nucleotides. A similar Ca$^{2+}$-induced Ca$^{2+}$ release was found with a preparation of isolated SR (Ogawa & Ebashi, 1976; Miyamoto & Racker, 1982; Morii & Tonomura, 1983). It is expected that the study of Ca$^{2+}$ release will further develop in the near future using preparations of isolated SR. It has also been shown with isolated SR that Ca$^{2+}$ uptake by SR is an active process involving the Ca$^{2+}$,Mg$^{2+}$-ATPase of the SR membrane (Hasselbach & Makinose, 1961; Ebashi & Lipmann, 1962). The reaction mechanism of this Ca$^{2+}$,Mg$^{2+}$-ATPase has been clarified in detail using isolated SR. In the following sections, the mechanism of coupling between active Ca$^{2+}$ transport and the ATPase reaction in SR will be considered in detail, but other topics of the SR will be discussed only briefly. The reader is referred to the reviews by Tonomura (1972), Martonosi (1975), Tada, Yamamoto & Tonomura (1978), Hasselbach (1979) and Yamamoto, Takisawa & Tonomura (1979) for general studies on SR, and to those by MacLennan & Holland (1975), Inesi (1979) and Møller, Andersen & le Maire (1982) for the structure of the SR membrane. The reader is also referred to the reviews by Tada *et al.* (1978) for the regulation of Ca$^{2+}$ transport, by De Meis (1981) for the reverse reaction of Ca$^{2+}$,Mg$^{2+}$-ATPase and by Berman (1982) for thermodynamic studies. This monograph also does not deal with Ca$^{2+}$ release from the SR, and the excellent reviews by Endo (1977) and Fabiato & Fabiato (1978) are recommended.

## Chemical structure of the Ca$^{2+}$,Mg$^{2+}$-ATPase

### *Protein components of fragmented sarcoplasmic reticulum (FSR)*

The major protein components of FSR membrane consist of the Ca$^{2+}$,Mg$^{2+}$-ATPase with a molecular weight of about 100 000 (MacLennan, 1970; Meissner, Conner & Fleischer, 1973), calsequestrin (MacLennan & Wong, 1971) with a molecular weight of about 65 000 and a glycoprotein with a molecular weight of about 55 000 (Campbell & MacLennan, 1981). Calsequestrin is an extrinsic protein readily removed from the membrane by treatment with a low concentration of detergent. This protein may be involved in the storage of $Ca^{2+}$ inside the SR lumen. In addition to these proteins, a proteolipid with a molecular weight of about 6000 is tightly bound to the ATPase (MacLennan *et al.*, 1973), but its physiological function is still unknown.

FSR can be separated into light and heavy fractions by sucrose density gradient centrifugation. The heavy fraction is derived mainly from the cisternae of SR, and the content of the calsequestrin in this fraction is rather high. By contrast, the light fraction is derived from the longitudinal SR and contains scarcely any calsequestrin (Meissner, 1975).

## *Size and shape of the ATPase molecule*

An electron microscopic picture of negatively stained FSR is shown in Fig. 7.2. Projections with 4–5-nm-wide particles attached to 2-nm-long stalks are arranged regularly on the surface of the SR membrane. In many experiments in which trypsin digestion (Migala, Agostini & Hasselbach, 1973; Stewart & MacLennan, 1974; Scott & Shamoo, 1982), antibody-binding reactions (Stewart, MacLennan & Shamoo, 1976; Sumida & Sasaki, 1975) and chemical modification (Hasselbach & Elfvin, 1967) were performed, it was demonstrated that these projections represent part of the $Ca^{2+},Mg^{2+}$-ATPase.

The shape of the ATPase has been investigated after solubilizing the ATPase with detergent. Electron micrographs of the negatively stained, solubilized ATPase show the protein to be a cylindrical molecule 3.5 nm in diameter and 11 nm in length (Hardwicke & Green, 1974). Further

Fig. 7.2 Surface particles of the ATPase molecules visualized by negative staining of vesicles formed from the purified $Ca^{2+}$ ATPase and phospholipid. Courtesy of D. H. MacLennan.

50 nm

electron microscopic observations by freeze-fracture of the ATPase dispersed in the high concentration of a nonionic detergent such as octaethyleneglycolmonododecylether, $C_{12}E_8$, revealed that the ATPase molecule has a cylindrical shape 11 nm in length and 7.5 nm in diameter (le Maire, Møller & Gulik-Krzywicki, 1981). On the basis of low-angle X-ray scattering and sedimentation equilibrium of the ATPase solubilized with deoxycholate (DOC), le Maire *et al.* (1981) have proposed a double cylindrical model. In this model, the ATPase molecule consists of a head 7.5 nm in diameter and 3 nm in height and a tail 4 nm in diameter and 7 nm in height, and lipid or DOC are bound to the lateral side of the tail. Further investigations will be required to determine whether the shape of the detergent-solubilized ATPase monomer is identical to that of the ATPase in the native SR membrane.

## Primary structure of the ATPase protein

When SR was exposed briefly to trypsin, the ATPase protein with a molecular weight of about 100 000 was degraded into two fragments with molecular weights of about 55 000 and 45 000 (Migala *et al.*, 1973; Thorley-Lawson & Green, 1973; Inesi & Scales, 1974; Stewart & MacLennan, 1974). Further digestion of FSR with trypsin degraded the m.w. 55 000 subfragment into two fragments with molecular weights of about 30 000 and 20 000 but did not affect the m.w. 45 000 fragment (Thorley-Lawson & Green, 1975; Stewart *et al.*, 1976). After the second cleavage, the transport activity disappeared but the phosphorylation activity was retained (Scott & Shamoo, 1982).

Recently, the amino acid sequence of the $Ca^{2+}$, $Mg^{2+}$-ATPase protein was extensively investigated and about 70% of the whole sequence was determined (Allen, Trinnaman & Green, 1980; Klip, Reithmeier & MacLennan, 1980; Tong, 1977, 1980). From these studies, it was found that the tryptic fragment with a molecular weight of 20 000 contains the N-terminal of the ATPase polypeptide and the tryptic fragment with a molecular weight of 45 000 contains the C-terminal. Thus, the polypeptides with molecular weights of 20 000, 30 000 and 45 000 were successively linked from the N-terminal to the C-terminal in the amino acid sequence of the ATPase. It is known that the subfragment with a molecular weight of 30 000 contains an aspartyl residue in which the carboxyl group is phosphorylated by ATP. The amino acid sequence in the vicinity of the aspartyl residue has been also clarified. The lysine residue located at position 160 from the aspartyl residue toward the C-terminal can be specifically modified by fluorescein isothiocyanate (FITC), and the ATP-binding capacity disappears following this modification (Mitchinson *et al.*, 1982).

Shamoo *et al.* (1976) reported that the tryptic fragment with a molecular weight of 20000 displays a $Ca^{2+}$-specific ionophore activity. This is an interesting observation in relation to the result obtained by Allen *et al.* (1980) that indicated that $Ca^{2+}$ may be transported through a pore formed by the intramembranal portion of the ATPase peptide. However, further information on the $Ca^{2+}$-binding site and the $Ca^{2+}$-transport site have not been obtained. It has been suggested by Pick & Racker (1979) that the binding site of dicyclohexylcarbodiimide (DCCD), an inhibitor of $Ca^{2+}$ uptake, is located on the subfragment with a molecular weight of 20000. An antibody against the subfragment with a molecular weight of 30000 shows an intense immunoprecipitation with SR vesicles, but an antibody against the subfragment with a molecular weight of 20000 reacts with the vesicles very weakly. Stewart *et al.* (1976) showed that an antibody against the subfragment with a molecular weight of 45000 scarcely reacts with the vesicles and suggested that the subfragment with a molecular weight of 30000 is mostly exposed on the outer surface of the membrane, whereas subfragments with a molecular weight of 20000 and 45000 are partially exposed on the outer surface, as schematically illustrated in Fig. 7.3.

## Lipid component of FSR membrane

FSR membrane contains about 100 mol of lipid/mol of ATPase. Phospholipids make up approximately 80% of the total lipid and the remainder is neutral lipid, of which more than 50% is cholesterol. The phospholipids consist of phosphatidylcholine (80%), phosphatidylethanolamine (9%), phosphatidylserine (2%) and cardiolipin (0.2%) (Martonosi, Donley & Halpin, 1968; Meissner & Fleischer, 1971).

The SR lipid can be divided into two different groups: the bilayer lipid, which forms the basic framework of the membrane, and the boundary lipid, which is directly attached to the ATPase.

Warren *et al.* (1974) found that when SR membranes were solubilized and delipidated by increasing the concentration of cholate, full ATPase activity was maintained at a molar ratio of phospholipid to protein above about 30. Below this ratio the ATPase activity decreased and had a negligible level at about 15 lipid molecules/molecule of ATPase. These 30 mol of lipid/mol of ATPase protein, which have been called the 'annular lipid' (Warren *et al.*, 1975; Bennet, McGill & Warren, 1980), coat the enzyme protein as the first bilayer shell of lipid molecules. Experimental observations using spin-labelled lipids have demonstrated that the same number of lipid molecules are immobilized, relative to the bulk bilayer lipids, as are attributed to direct interaction with the ATPase protein (Nakamura & Ohnishi, 1975; Montecucco *et al.*, 1977). Hesketh *et al.* (1976) estimated that the exchange rate of the annular lipid into the bulk bilayer lipid was lower than $10^6$/s by ESR techniques. Chapman, Gomez-Fernandez & Goni (1979) suggested that a lower limit for the exchange rate is about $10^5$/s on the basis of $^2$H-NMR analysis. Therefore, the mobility of annular lipids appears to be restricted only by comparison with that of the fluid bulk lipid ($10^8$/s).

Recently, Møller *et al.* (1982) demonstrated that the mobility of the spin-labelled fatty acid was strongly restricted when it was covalently

Fig. 7.3 Diagrammatic representation of the folding of a single polypeptide of ATPase in the lipid bilayer of sarcoplasmic reticulum membrane. According to this representation, the first and the second tryptic cleavage sites and the phosphorylation site are exposed on the outer surface of the membrane. Tryptic subfragments: hatched area, m.w. 20000; open area, m.w. 30000; and stippled area, m.w. 45000. Dashed lines indicate peptides remaining insoluble after extensive proteolysis.

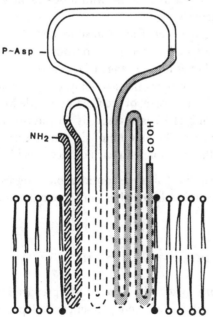

bound to the ATPase protein and that part of the strongly immobilized label was converted into a slightly restricted form when $C_{12}E_8$ was added to the membrane to disperse the aggregated ATPase molecules. They interpreted the strong immobilization of lipid to result from lipid–protein contact.

## Reaction mechanism of $Ca^{2+}$,$Mg^{2+}$-ATPase

Since the $Ca^{2+}$,$Mg^{2+}$-ATPase in the SR membrane serves as a carrier of $Ca^{2+}$ as well as an energy transducer in $Ca^{2+}$ transport, kinetic studies of the ATPase reaction are important for obtaining insight into the molecular basis of active $Ca^{2+}$ transport. FSR preparations have the following advantages over other transport membranes in the study of active transport of cations:

(1) The $Ca^{2+}$,$Mg^{2+}$-ATPase is a major component of skeletal muscle SR and accounts for up to 90% of the total protein (Meissner, 1975).

(2) FSR are tightly sealed vesicles oriented to pump $Ca^{2+}$ to their luminal spaces. The membrane orientation can be eliminated by solubilizing the membrane with a detergent without loss of ATPase activity.

(3) The coupling between $Ca^{2+}$ transport and ATP hydrolysis is tight and a coupling ratio of 2 between the two reactions is maintained over a wide range of reaction conditions (Hasselbach, 1981). The enzyme also catalyzes ATP synthesis from ADP and Pi by a reversal of the $Ca^{2+}$ pump (de Meis, 1981).

(4) Phosphoenzyme (EP) is formed as an intermediate of the ATPase reaction. The amount of EP can be easily measured by stopping the reaction with trichloroacetic acid (TCA).

Because of these favorable features, considerable progress has been made in the study of the molecular mechanism of coupling in $Ca^{2+}$ transport.

### General properties of the $Ca^{2+}$,$Mg^{2+}$-ATPase

$Ca^{2+}$,$Mg^{2+}ATPase$ and $Mg^{2+}$-ATPase   FSR preparations exhibit two types of ATPase activities, $Ca^{2+}$-independent and -dependent activities (Hasselbach & Makinose, 1961). The former is called the basic ATPase or Mg²⁺-ATPase, and the latter is called the extra ATPase or Ca²⁺,Mg²⁺-ATPase. Since basic ATPase is lost during purification of the Ca²⁺,Mg²⁺-ATPase, it is probable that the Mg²⁺-ATPase originates from subcellular entities other than SR such as transverse tubules and sarcolemma (McFarland & Inesi, 1970; Fernandez, Rosemblatt & Hidalgo, 1980).

*Effects of divalent cations* Full activation of the ATPase reaction requires both Ca²⁺ and Mg²⁺. When the Mg²⁺ concentration is in the millimolar range, ATPase activity increases with increasing Ca²⁺ concentrations and reaches a maximum velocity at about 1 µM (Hasselbach & Makinose, 1963). The Ca²⁺ concentration giving half-maximal ATPase activity is 0.3 to 0.5 µM (Weber, Herz & Reiss, 1966; Yamamoto & Tonomura, 1967). The Hill coefficient for activation by Ca²⁺ is nearly 2 (The & Hasselbach, 1972; Vianna, 1975). This value is consistent with the fact that 2 mol of Ca²⁺ are transported by 1 mol of ATP hydrolyzed. It is known that Sr²⁺ can also be used as a divalent cation instead of Ca²⁺, but its affinity is much lower than that of Ca²⁺ (Weber *et al.*, 1966; Yamada & Tonomura, 1972; Mermier & Hasselbach, 1976).

Mg²⁺ plays at least two roles in the Ca²⁺-dependent ATPase reaction. First, it accelerates the decomposition of phosphorylated intermediates, as will be described later. Second, it forms an equimolar Mg²⁺–ATP that serves as the true substrate. The latter function was suggested by the finding that the ATPase activity shows a constant value independent of the Mg²⁺ concentration when the Mg²⁺ concentration is in excess of that of ATP, and that the dependence of ATPase activity on the Ca²⁺ concentration does not change over a wide range of Mg²⁺ concentrations (Yamamoto & Tonomura, 1967). This effect of Mg²⁺ is also implied in the finding that a straight line is obtained only when the reciprocal of the Ca²⁺-dependent ATPase activity is plotted against the reciprocal of the concentration of the Mg²⁺–ATP complex (Vianna, 1975). Ca²⁺–ATP can also function as a substrate for the formation of phosphorylated intermediates, but its hydrolytic rate is much lower than that of Mg²⁺–ATP (Souza & de Meis, 1976; Yamada & Ikemoto, 1980). Mn²⁺ is, so far, the only known divalent cation that can be used instead of Mg²⁺ (Kalbitzer, Stehlik & Hasselbach, 1978).

*Effects of monovalent cations and pH* Alkali salts accelerate both

$Ca^{2+}$ uptake and ATP hydrolysis by FSR (Shigekawa & Pearl, 1976). This is attributed to the acceleration of decomposition of the phosphorylated intermediate as will be discussed later in this chapter.

The pH dependence of $Ca^{2+},Mg^{2+}$-ATPase activity exhibits a bell-shaped curve with a pH optimum of 6.5–7.5. ATPase activity decreases with an increase of pH in the alkaline range, while the level of the phosphorylated intermediate increases, reaching a maximum at approximately pH 8.5. This is because decomposition of the phosphoenzyme is seriously inhibited in the alkaline pH range (Yamamoto & Tonomura, 1967).

*Substrate specificity* $Ca^{2+},Mg^{2+}$-ATPase shows an ATP dependence in a Michaelis–Menten fashion in the presence of a sufficient amount of $Ca^{2+}$ and $Mg^{2+}$. The $K_m$ of the ATPase for ATP is several μM in a low ATP concentration at neutral pH (Weber *et al.*, 1966; Yamamoto & Tonomura, 1967; Vianna, 1975). The ATPase can also hydrolyze ITP, GTP, CTP, UTP and phosphate compounds such as dinitrophenyl-phosphate, acetylphosphate, *p*-nitrophenylphosphate and carbamyl-phosphate (Hasselbach, 1981). Although the hydrolysis rates of these substrates are much lower than that of ATP, the coupling of 2 mol of $Ca^{2+}$ transported for each mole hydrolyzed is strictly maintained. It was recently shown that furylacryloyl phosphate (FAP) serves not only as an ATP analog but also as a Pi analog in the reverse reaction of ATP hydrolysis (Kurzmack *et al.*, 1981).

In high concentration, ATP acts not only as a substrate but also as a regulator of the enzymatic reaction (Kanazawa *et al.*, 1971). As shown in Fig. 7.4, a Lineweaver–Burk plot of the $Ca^{2+},Mg^{2+}$-ATPase reaction gave two straight lines which crossed at about 0.1 mM ATP. At steady state both the $V_{max}$ and the apparent $K_m$ for the ATPase reaction were greater at higher ATP concentrations than at lower ATP concentrations. More-direct evidence for the dual function of ATP has been presented by Dupont (1977) and Nakamura & Tonomura (1982*b*) who demonstrated that the ATPase possesses two ATP-binding sites with high and low affinities. The mechanism by which a high concentration of ATP is involved in the regulation of this enzymatic reaction will be discussed in detail later in this chapter.

*Chemical modification of $Ca^{2+},Mg^{2+}$-ATPase* Specific residues of arginine (Murphy, 1976), cysteine (Hasselbach & Seraydarian, 1966), lysine (Pick, 1981) and histidine (Tenu *et al.*, 1976) have been found to be essential for $Ca^{2+}$,$Mg^{2+}$-ATPase activity. Arginine residues are known to be functional groups common to energy-transducing ATPases. SH groups have been studied in more detail than any of the other functional groups. The ATPase is inactivated by mercurials or alkylating agents. Inactivation was prevented by ATP through protection of 1 or 2 mol of essential SH groups/mol of ATPase, suggesting that these SH groups exist near the active site. Pick & Racker (1979) reported that, when a particular carboxyl group near the cation-binding site is modified by DCCD, the $Ca^{2+}$ sensitivity of the ATPase is lost. It was suggested that this site is located on the m.w. 20 000 tryptic subfragment of the ATPase polypeptide chain. Recently Mitchinson *et al.* (1982) reported that the binding of 1 mol of FITC to the adenine-binding site led to the complete loss of ATPase activity. The specific lysine residue of the adenine-binding site was found to be located in a position 10 amino acids from the N-terminal of the tryptic fragment with a molecular weight of 45 000. It is still not confirmed where the other functional groups are located.

Fig. 7.4 Lineweaver–Burk plot of the $Ca^{2+}$,$Mg^{2+}$-ATPase reaction. There is a linear relationship between $V_0^{-1}$ and $ATP^{-1}$ at ATP concentrations lower than about $10 \,\mu M$, while at higher concentrations, $V_0^{-1}$ falls below this line.

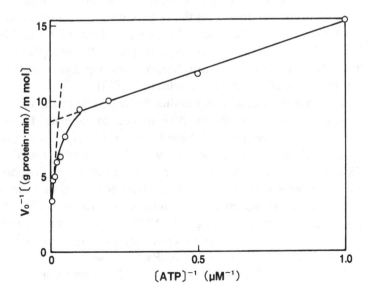

*Phosphorylated intermediate*   The existence of a high-energy, phosphorylated intermediate was proposed by Hasselbach & Makinose (1961) as well as by Ebashi & Lipmann (1962) on the basis of the finding that FSR catalyzes NDP–NTP exchange that exhibits the same dependence on $Ca^{2+}$ concentration as ATP hydrolysis and $Ca^{2+}$ uptake. Later, Yamamoto & Tonomura (1967) and also Makinose (1969) found that an SR protein was phosphorylated when the membrane was incubated with $[\gamma\text{-}^{32}P]ATP$ in the presence of $Ca^{2+}$ and the reaction was quenched by TCA. It was shown that almost 1 mol of EP is formed/mol of ATPase (MacLennan *et al.*, 1971). The dependence of the steady-state EP level on the concentrations of both ATP and $Ca^{2+}$ was in parallel with that of the reaction rate of the $Ca^{2+},Mg^{2+}$-ATPase (Yamamoto & Tonomura, 1968). This finding suggested that EP is an intermediate in the $Ca^{2+},Mg^{2+}$-ATPase reaction. This was later confirmed by kinetic studies of partial reactions of ATPase, as will be discussed in the next section.

The phosphoprotein isolated after quenching with TCA is stable in the acidic pH range but unstable in the alkaline pH range and can be easily hydrolyzed by hydroxylamine to form a hydroxamate (Yamamoto & Tonomura, 1968; Yamamoto, Yoda & Tonomura, 1971). These stability characteristics of the phosphoprotein are similar to those of an acyl-phosphate. These chemical properties are the same as those of the phosphoproteins that were previously found in studies of the $Na^+$, $K^+$-ATPase (Hokin *et al.*, 1965; Nagano *et al.*, 1965; Post, Sen & Rosenthal, 1965). It has been shown that the $\beta$-carboxyl group of an aspartyl residue 153 amino acid residues from the N-terminal of the m.w. 30 000 tryptic fragment of the ATPase was phosphorylated by ATP (see Fig. 7.3). It was also found that the primary structure of the phosphorylated site of the $Na^+$, $K^+$-ATPase is also Ser(or Thr)–Asp–Lys and is similar to that of $Ca^{2+},Mg^{2+}$-ATPase (Bastide *et al.*, 1973).

An important question is whether the aspartyl residue of the active enzyme is phosphorylated by ATP or not. In studying this problem, Webb & Trentham (1981) labelled the $\gamma$-P of ATP stereospecifically with S, $^{16}O$ and $^{17}O$ to obtain ATP-$\gamma$-S. They reacted the ATP-$\gamma$-S with FSR ATPase and studied the stereochemical course of the phosphoryl moiety transfer using NMR analysis. They concluded that the reaction proceeds with retention of the atomic configuration. This conclusion is consistent with the finding that ATP is decomposed via EP as will be described in the next section. Fossel *et al.* (1981) observed that phosphorylation of the $Ca^{2+},Mg^{2+}$-ATPase, as well as of the $Na^+,K^+$-ATPase, produced a resonance of $^{31}P$-NMR occurring at $+17$ ppm and disappearing when hydroxylamine was added. A similar resonance was found after phosphorylation of a model dipeptide L-seryl–L-aspartate. They attributed the resonance of $+17$ ppm to formation of an acylphosphate at an aspartyl residue at the catalytic site of the active enzyme.

## *Elementary steps of the $Ca^{2+}, Mg^{2+}$-ATPase reaction*

In this subsection, we discuss in detail the elementary steps of the $Ca^{2+}$,$Mg^{2+}$-ATPase reaction that are coupled with the translocation of $Ca^{2+}$ through the membrane. For a better understanding of the description below, see Fig. 7.5. This scheme represents the elementary steps of ATP hydrolysis associated with $Ca^{2+}$ transport that have been deduced from the experimental results thus far obtained. It is clear that the use of FSR is essential for the understanding of the coupling mechanism between ATP hydrolysis and $Ca^{2+}$ transport. However, when FSR is used, the vectorial properties of the enzyme reaction must always be distinguished. In addition, it should be noted that the ionic composition of the inner environment of FSR is continuously changing during the reaction, and that the $Ca^{2+}$ gradient or membrane potential may affect the kinetic properties of the ATPase (Yamada *et al.*, 1971;

Fig. 7.5 The mechanism of coupling of ATP hydrolysis with cation transport across the sarcoplasmic reticulum membrane. $E_1$ and $E_2$ are enzyme forms capable of reacting with ATP and Pi, respectively. $Ca_2$–$E_1$–ATP, and $Ca_2$–$E_1$*–ATP represent loose and tight binding of ATP to the Ca–$E_1$ complex, respectively. $Ca_2$–$E_1$P indicates the 'occluded' Ca in the ADP-sensitive EP. $E_2$P and $E_2$Pi are ADP-insensitive EP and an enzyme–Pi complex, respectively.

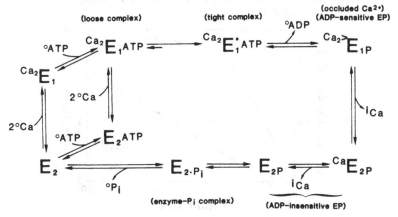

Froehlich & Taylor, 1975; Takisawa & Tonomura, 1978). For this reason, solubilized ATPase or leaky vesicles are used for assays in which anisotropic events do not play a significant role in the reaction.

*EP formation*   The phosphoenzyme (EP) can be obtained as an acid-stable phosphoprotein after the reaction is stopped by TCA. The pre-steady-state reaction can be quantitatively analyzed by assuming that EP is a true intermediate in the ATPase reaction. As shown in Fig. 7.6, the amount of EP increases immediately after addition of ATP and rapidly reaches a steady-state level. On the other hand, the time course of Pi liberation exhibits a lag phase corresponding to the time course of EP formation and then increases linearly. The observed time course of Pi liberation is in good agreement with that calculated from the time course of EP formation, assuming that EP is an intermediate in the reaction with a turnover rate of about 4/sec.

The sequence in which $Ca^{2+}$ and ATP are involved in the reaction of ATP hydrolysis can be determined by measuring the dependence of the initial rate ($v_f$) of EP formation on the concentration of $Ca^{2+}$ and ATP (Kanazawa *et al.*, 1971). As shown in Fig. 7.7, the double reciprocal plot of $v_f$ against ATP concentration gives a straight line for any $Ca^{2+}$ concentration. The value ($V_f$) of $v_f$ at an infinite concentration of ATP decreases with a decrease in $Ca^{2+}$ concentration, while the apparent $K_m$ for ATP remains constant. This indicates that EP is formed

Fig. 7.6 Initial phase of the ATPase reaction. The time course of Pi liberation (open circles) is in good agreement with that calculated from the amount of EP (filled circles), taking the EP turnover rate as 4.2/sec.

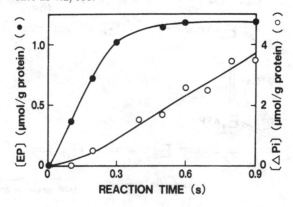

via an enzyme–ATP–Ca²⁺ complex produced by binding ATP and Ca²⁺ to the enzyme in a random sequence. The double-reciprocal plot of $v_f$ against the square of the free Ca²⁺ concentration gives a straight line, indicating that 2 mol of Ca²⁺ are bound/mol of ATPase to form EP. The same result was also obtained by Coffey *et al*. (1975).

Hasselbach & Makinose (1961) and Ebashi & Lipmann (1962) demonstrated that FSR catalyzes a rapid ATP–ADP exchange in the presence of Ca²⁺. They suggested that ATP can be formed by the reverse reaction of EP formation. Direct evidence for this reaction was later provided by Kanazawa *et al*. (1971). As shown in Fig. 7.8, when FSR was reacted with ATP to form EP and then EP formation was terminated by the addition of excess EGTA, EP decomposed by first order kinetics. When ADP and EGTA were added to EP, the phosphoenzyme decayed much more rapidly than in the absence of ADP. The decay of EP was not accompanied by Pi liberation but was accompanied by the formation of ATP in an amount equimolar with the decrease in EP. This finding

Fig. 7.7 Dependence of the initial rate of EP formation on the concentration of ATP at various concentrations of Ca²⁺. Concentrations of Ca²⁺ were changed by adding EGTA in the presence of 0.5 mM CaCl₂ at pH 7.0. EGTA concentrations were 0.4 mM (half-filled circles), 0.45 mM (open circles), 0.51 mM (filled circles), 0.54 mM (open triangles) and 0.57 mM (crosses).

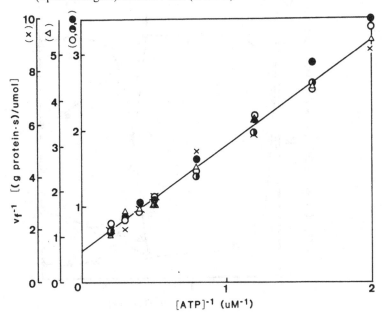

indicated that EP did not contain bound ADP. Indeed, Takisawa & Tonomura (1978) measured the amount of ADP bound during the ATPase reaction coupled with an ATP-regenerating system and showed that it was negligibly small.

*EP decomposition*   In contrast to EP formation, EP decomposition does not require $Ca^{2+}$ but does require $Mg^{2+}$ (Martonosi, 1969; Inesi *et al.*, 1970; Kanazawa *et al.*, 1971; Panet, Pick & Selinger, 1971). The rate constant of EP decomposition can be obtained from the ratio of a steady-state rate of ATP hydrolysis, $v_o$, to [EP] in the steady state, $v_o/EP$, if only one species of EP is present or if several EP species exist but are in rapid equilibrium in the reaction system. Yamada & Tonomura (1972) found that the double reciprocal plot of $v_o/EP$ against $Mg^{2+}$ concentration gave a straight line with a half-effective concen-

Fig. 7.8  Formation of ATP from EP and ADP. At 1.5 s after the initiation of phosphorylation (arrow), EGTA was added with (open triangles, filled triangles) or without (open circles, filled circles) ADP. From Kanazawa *et al.* (1971).

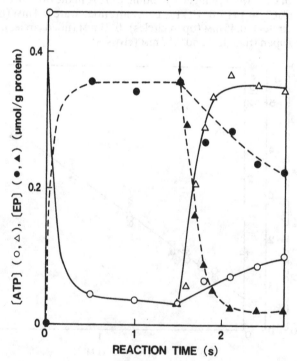

tration of about 50 μM. They suggested that an equimolar EP–$Mg^{2+}$ complex precedes the decomposition of EP.

EP formation can be terminated instantaneously by chelating $Ca^{2+}$ from the medium by adding EGTA. However, even when $Mg^{2+}$ is removed by adding chelators such as EDTA and CDTA, EP decomposition cannot be stopped immediately. Thus, the $Mg^{2+}$ required for EP decomposition is in a state inaccessible to the chelators. Kanazawa *et al.* (1971) suggested that this is because $Mg^{2+}$ functions from the inside of the vesicular membrane and the chelators are impermeable to the membrane. However, the same phenomenon was observed when the membrane was destroyed with detergent. On the basis of this observation, Garrahan *et al.* (1976) and Takakuwa & Kanazawa (1982) suggested that the $Mg^{2+}$ required for EP decomposition is tightly bound to the enzyme.

It should be noted that occlusion of the $Mg^{2+}$-binding site in EP formation has only been estimated kinetically. Therefore, it is still unknown how many moles of $Mg^{2+}$ are occluded per mole of EP formed and whether or not the bound $Mg^{2+}$ is released after EP decomposition.

## ATP binding to the $Ca^{2+}, Mg^{2+}$-ATPase

Since the stoichiometric analysis so far described was carried out at a low ATP concentration, only one kind of E–ATP complex was observed. However, under physiological conditions or in the presence of mM concentrations of ATP, EP is formed via two kinds of E–ATP complexes. Kanazawa *et al.* (1971) found that the rate of EP formation is accelerated by high ATP concentrations, while the reverse reaction, from EP and ADP to ATP, is not accelerated. They suggested that another type of E–ATP complex is formed after the first E–ATP complex and that the rate of formation of the second E–ATP complex is accelerated by a high ATP concentration.

In extremely low concentrations of KCl, ATP bound to the enzyme is replaced only slowly by ATP added to the medium (Nakamura & Tonomura, 1982*b*; Shigekawa & Kanazawa, 1982). E³²P formed under this condition was not chased from the enzyme by the addition of non-radioactive ATP even in excess amounts (Nakamura & Tonomura, 1982*b*). These results can be explained adequately if we assume that: (1) the hydrolysis of ATP proceeds via the same pathway irrespective of ATP and KCl concentrations; (2) a slowly exchangeable E–ATP complex is formed via a rapidly exchangeable E–ATP complex; (3) high concentrations of KCl have a strong tendency to shift the equilibrium between the two E–ATP complexes toward the rapidly exchangeable E–ATP complex; and (4) ATP modulates the equilibrium between these complexes.

*Two conformations of the $Ca^{2+}, Mg^{2+}$-ATPase* As mentioned in the previous section, the kinetic properties of the $Ca^{2+}, Mg^{2+}$-ATPase can be adequately explained by assuming the existence of only one kind of enzyme conformation in the catalytic cycle. On the other hand, more recent studies indicate the existence of two distinct conformations of the $Ca^{2+}, Mg^{2+}$-ATPase, as is the case for the $Na^+, K^+$-ATPase (see Chapter 8). Takisawa & Tonomura (1978) observed that the rate of EP formation initiated by the simultaneous addition of ATP and $Ca^{2+}$ to EGTA-preincubated FSR is much lower than the rate of EP formation initiated by the addition of ATP to FSR preincubated with $Ca^{2+}$ (see Fig. 7.9). The same result was obtained by Sumida *et al.* (1978) and by Guillain *et al.* (1981). These observations suggest that the enzyme can exist in two different conformations, $E_1$ and $E_2$, depending on the presence and absence of $Ca^{2+}$ and that the transition between the two conformations is the rate-determining step in the ATP hydrolytic reaction.

The existence of two distinct forms of the ATPase was also suggested by Dupont & Leigh (1978) and Guillain *et al.* (1981), who found that binding of 2 mol of $Ca^{2+}$ to the high-affinity sites of the enzyme induces a large increase in fluorescence intensity of intrinsic tryptophans (Fig. 7.10). The same conclusion was obtained from experimental observations on the reactivity of SH groups of the enzyme protein (Murphy, 1978; Champeil *et al.*, 1978; Ikemoto, Morgan & Yamada, 1978; Kawakita, Yasuoka & Kaziro, 1980) and from measurements of the ESR spectra of spin-labelled SR (Coan & Inesi, 1977; Coan, Verjovski-Almeida & Inesi, 1979).

The existence of two distinct forms of EP was first demonstrated by Shigekawa & Dougherty (1978), who observed that most EP is incapable of reacting with ADP to form ATP in the presence of extremely low concentrations of KCl, while an increase in KCl concentration results in the formation of EP that readily reacts with ADP. Takisawa & Tonomura (1979) made similar observations of EP reactions with ADP in the presence of high concentrations of KCl and high pH so that EP decomposed very slowly. They found that the ADP-sensitive and

Fig. 7.9 The time courses of EP formation and Pi liberation in the reactions started by adding ATP with $Ca^{2+}$ to SR and by adding ATP to SR preloaded with $Ca^{2+}$. The reactions of EP formation (filled circles, filled triangles) and Pi liberation (open circles, open triangles) were initiated by adding ATP (filled circles, open circles) and ATP + $Ca^{2+}$ (filled triangles, open triangles).

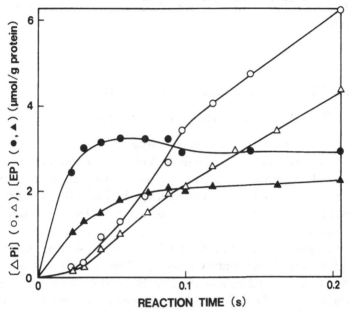

Fig. 7.10 Calcium-concentration dependence of intrinsic fluorescence changes. Open circles, various amounts of $Ca^{2+}$ added to the $Ca^{2+}$,$Mg^{2+}$-ATPase initially suspended in 0.5 mM EGTA ($E_2 \rightarrow E_1$). Open triangles, various amounts of EGTA added to the enzyme initially suspended in 100 μM $Ca^{2+}$ ($E_1 \rightarrow E_2$). From Guillain *et al.* (1981).

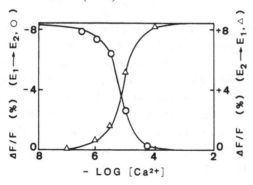

-insensitive forms of EP exist in equilibrium under these conditions and that the equilibrium is shifted in favor of the ADP-insensitive form by $Mg^{2+}$ but is shifted to the ADP-sensitive form by high concentrations of $Ca^{2+}$. Furthermore, the conversion between the two kinds of EP is inhibited by the modification of SH groups in the ATPase protein with $N$-ethylmaleimide (Takisawa & Tonomura, 1979; Kawakita *et al.*, 1980).

## $Ca^{2+},Mg^{2+}$-ATPase and $Ca^{2+}$ transport

### *ATP hydrolysis and $Ca^{2+}$ transport*

As shown in the reaction scheme of Fig. 7.5, the $Ca^{2+},Mg^{2+}$-ATPase catalyzes the hydrolysis of ATP via phosphorylated intermediates, $E_1P$ and $E_2P$, and the enzyme takes two kinds of conformation, $E_1$ and $E_2$, during the reaction cycle. However, the mechanism that involves these reaction intermediates in active $Ca^{2+}$ transport is unknown.

The transport of a solute by a carrier protein in a biomembrane consists of at least three elementary processes (Pardee, 1968*a,b*) in which the solute is recognized by a carrier, translocated by the carrier across the membrane and then released from the carrier. To understand the molecular mechanism that couples ATP hydrolysis to $Ca^{2+}$ transport, it is necessary to clarify the relationship between the elementary processes of cation transport and the elementary steps of the ATPase reaction. In the following subsection, we discuss the coupling mechanism based on a comparison of partial reactions of ATP hydrolysis with the elementary steps of $Ca^{2+}$ transport across the SR membrane.

*Recognition of $Ca^{2+}$*   Since the ATPase itself serves as a carrier protein in the $Ca^{2+}$-transport system, studies on $Ca^{2+}$ binding to this enzyme provide information about the initial step of $Ca^{2+}$ transport. The $Ca^{2+},Mg^{2+}$-ATPase possesses 2 mol of high-affinity $Ca^{2+}$-binding sites-/mol of enzyme on the cytoplasmic side, with a dissociation constant of about 1 μM at room temperature and neutral pH (cf. review by Yamamoto *et al.*, 1979). Further analyses on $Ca^{2+}$ binding to the ATPase as a function of $Ca^{2+}$ concentration showed a cooperative $Ca^{2+}$ binding to the high-affinity binding sites (Inesi *et al.*, 1980).

The specificity of the binding site for Ca²⁺ is approximately 30 000 times that for Mg²⁺ (Yamada & Tonomura, 1972). Specific Ca²⁺ binding to the high-affinity Ca²⁺-binding site was also shown by Kalbitzer *et al.* (1978). Using Mn²⁺-EPR, they observed competition between Mn²⁺ and either Ca²⁺ or Mg²⁺ for this site and estimated that the dissociation constants of Mg²⁺ and Mn²⁺ are, respectively, about 7000 times and 13 000 times that of Ca²⁺.

Figure 7.11 illustrates Ca²⁺ binding to the ATPase as measured by a membrane filtration method (Nakamura & Tonomura, 1982*a*). The figure shows that there are 2 mol of Ca²⁺-binding sites/mol of active site, and that the 2 mol of Ca²⁺ bind cooperatively with a Hill coefficient of 2. The dissociation constant, $K_d$, increases only slightly in the presence of a high concentration of Mg²⁺. In addition, ATP analogs such as AMPPNP do not affect the binding of Ca²⁺ to the site (Yates & Duance, 1976; Watanabe *et al.*, 1981; Nakamura & Tonomura, 1982*a*). These results are consistent with the results of kinetic studies that suggest the following: (1) 2 mol of Ca²⁺ are transported/mol of ATP hydrolysis (see previous section), (2) phosphorylation of the ATPase by ATP requires 2 mol of Ca²⁺ binding/mol of ATPase, (3) Ca²⁺ binding is independent

Fig. 7.11 Calcium binding by the Ca²⁺-ATPase. The amount of Ca²⁺ bound to the enzyme was measured in the presence of 0.2 mM (open squares) or 20 mM MgCl₂ (open circles) at pH 7.0. From Nakamura & Tonomura (1982*a*).

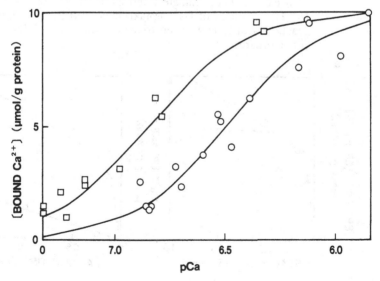

of ATP binding and (4) the $K_d$ for $Ca^{2+}$, measured directly, agrees well with the apparent $K_d$ obtained from the $Ca^{2+}$ dependence of the EP level in the steady state and the $Ca^{2+}$ dependence of the rate of EP formation (Yamamoto & Tonomura, 1967; Makinose, 1969; Martonosi, 1969; Inesi *et al.*, 1970). Accordingly, it can be concluded that $Ca^{2+}$ binding to the high-affinity $Ca^{2+}$-binding site represents the process in which the solute is recognized by a carrier.

*Translocation of $Ca^{2+}$ across the membrane*   By measuring the initial phase of $Ca^{2+}$ uptake, Sumida & Tonomura (1974) first identified the step in which $Ca^{2+}$ is translocated across the membrane. The time course of $Ca^{2+}$ uptake was determined after the addition of EGTA to remove $Ca^{2+}$ outside the SR vesicle. The time course of the $Ca^{2+}$ uptake showed a fast initial phase corresponding to the period when EP was formed. This initial phase was followed by a slow steady phase corresponding to Pi liberation (Fig. 7.12).

On the other hand, when $Ca^{2+}$ uptake was terminated by the simultaneous addition of EGTA and ADP, only the slow phase of $Ca^{2+}$ uptake was observed. Since the EP formed under these conditions is $E_1P$, these experimental results indicate that $Ca^{2+}$ bound to the enzyme becomes

Fig. 7.12 Coupling between ATP hydrolysis and $Ca^{2+}$ translocation. (*a*) Time courses of Pi liberation measured by quenching the reaction with EGTA (filled triangles) and EP formation (open circles). (*b*) Time courses of $Ca^{2+}$ uptake measured by quenching the reaction with EGTA (open triangles) and with EGTA + ADP (open squares).

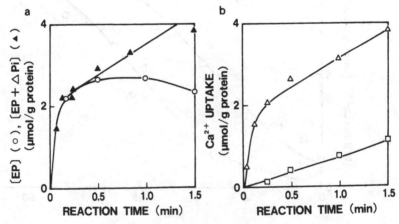

inaccessible to EGTA upon the formation of $E_1P$ and that, when $E_1P$ is converted into E and ATP by adding ADP, the bound $Ca^{2+}$ can be easily removed by EGTA outside the membrane. A similar result was obtained by Dupont (1980). It should be noted, however, that the different investigators reached different stoichiometric conclusions from experiments performed at the same temperature, $0\,°C$. Sumida & Tonomura (1974) reported that 1 mol of $Ca^{2+}$ became inaccessible to EGTA when 1 mol of EP was formed, whereas Dupont (1980) reported that 2 mol of $Ca^{2+}$ became inaccessible to EGTA. The reason for this discrepancy is unknown, but the difference may be related to the observation that there is only 1 mol of high affinity $Ca^{2+}$-binding site/mol of ATPase at $0\,°C$ (Ikemoto, 1974). At room temperature, there are 2 mol of high affinity $Ca^{2+}$-binding sites/mol of ATPase. It was shown unambiguously that 2 mol of $Ca^{2+}$ are transported when 1 mol of EP is formed (Kurzmack & Inesi, 1977; Kurzmack, Verjovski-Almeida & Inesi, 1977; Verjovski-Almeida, Kurzmack & Inesi, 1978).

*Occlusion of Ca$^{2+}$ in EP*   Since $Ca^{2+}$ bound to the enzyme in $E_1P$ was not removed by EGTA in the external medium, the $Ca^{2+}$ should be occluded by the enzyme or exposed to the inner surface of the membrane. Using solubilized ATPase, kinetic measurements indicated that $Ca^{2+}$ is tightly bound to the enzyme in EP. As described earlier in this chapter, the formation of $E_1P$ from $Ca_2$–E–ATP is reversible. This also implies that $Ca^{2+}$ is indispensible for the formation of $Ca_2$–E–ATP from $E_1P$ and ADP. However, even when $Ca^{2+}$ in the outer medium was chelated by EGTA after $E_1P$ was formed, the ADP reactivity of EP decreased only slowly (Takakuwa & Kanazawa, 1979; Takisawa & Tonomura, 1979). This observation indicates that the $Ca^{2+}$ required for the reverse reaction is already bound to the enzyme as $Ca_2$–$E_1P$. By contrast, $Ca^{2+}$ bound to $E_1$ is easily removed by EGTA. Therefore, it is assumed that $Ca^{2+}$ bound to the ATPase is occluded during formation of $E_1P$.

The occlusion of $Ca^{2+}$ in the EP state was shown directly by Takisawa & Makinose (1981) and Nakamura & Tonomura (1982a) using a solubilized ATPase preparation. Figure 7.13 shows a comparison of $Ca_2$–$E_1$ and $Ca_2$–$E_1P$ exchange rates of bound $Ca^{2+}$ for medium $Ca^{2+}$. Before addition of ATP, 2 mol of $Ca^{2+}$ were bound/1 mol of ATPase active site. This bound $Ca^{2+}$ was rapidly exchanged with medium $Ca^{2+}$. When the ATPase was phosphorylated by ATP, 2 mol of bound $Ca^{2+}$/mol of enzyme became slowly exchangeable with $Ca^{2+}$ in the medium. This slow rate of exchange was nearly equal to the turnover rate of EP. In other words, 2 mol of bound $Ca^{2+}$ became exchangeable when EP was converted into the other intermediate. Since most of the EP existed as $E_1P$ in the above two experimental conditions, 2 mol of $Ca^{2+}$ bound to the enzyme were occluded upon formation of 1 mol of $E_1P$.

*Release of $Ca^{2+}$ from the ATPase*  $Ca^{2+}$ occluded in the membrane bound ATPase is released into the SR lumen during the $Ca^{2+}$-transport cycle. The relationship between the release of $Ca^{2+}$ and dephosphorylation of the ATPase was first postulated kinetically by Yamada & Tonomura (1972). They measured the apparent rate constant of EP decomposition, $v_o/EP$, over a wide range of $Mg^{2+}$ and $Ca^{2+}$ concentrations and found that $Mg^{2+}$-dependent EP decomposition was competitively inhibited by $Ca^{2+}$ and that the ratio of affinity of the enzyme for $Ca^{2+}$ to that for $Mg^{2+}$ in this step was 2.5:1. This ratio was far smaller than that in the step of EP formation. As a result, they concluded that the affinity of the ATPase for $Ca^{2+}$ decreases greatly with phosphoryl-

Fig. 7.13 Occlusion of $Ca^{2+}$ in the EP intermediate of the ATPase. The reaction was started by adding 200 μM ATP and the amount of $Ca^{2+}$ bound to the enzyme (open circles) was measured. At 0 min (filled squares), or 1 min (filled circles), unlabeled $CaCl_2 +$ EGTA were added and the time course of the displacement of bound $Ca^{2+}$ was measured.

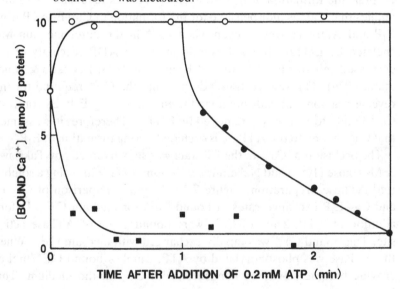

ation of the enzyme. A similar result was also obtained by Ikemoto (1975). However, Ikemoto did not clarify whether $E_1P$ and $E_2P$ formation induces a change in $Ca^{2+}$ affinity. Based on the measurements of the amounts of $E_1P$ and $E_2P$ at various concentration ratios of $Ca^{2+}$ to $Mg^{2+}$, Takisawa & Tonomura (1979) suggested that the affinity of the enzyme for $Ca^{2+}$ decreases greatly when $E_1P$ is converted into $E_2P$.

$Ca^{2+}$ release from the enzyme upon EP formation was observed under conditions that favored the formation of $E_2P$ (Ikemoto, 1975, 1976; Watanabe et al., 1981). It was also found that $Ca^{2+}$ is released from the ATPase at a slower rate than the enzyme is phosphorylated and that there is a lag phase in the rate of $Ca^{2+}$ release but not in the rate of EP formation (Ikemoto, 1976). These observations can be readily explained by assuming that $E_1P$ with a high affinity for $Ca^{2+}$ and $E_2P$ with a low affinity for $Ca^{2+}$ are formed sequentially in a catalytic cycle.

Recently, direct evidence for $Ca^{2+}$ release upon $E_2P$ formation was provided by Nakamura & Tonomura (1982a) and by Takisawa & Makinose (1983). In Fig 7.5, coupling of the elementary steps of $Ca^{2+}$ transport with the elementary steps of the ATPase reaction is schematically represented.

*Heterogeneity of $Ca^{2+}$-binding site*   It has been assumed thus far that the ATPase molecule possesses two high-affinity $Ca^{2+}$-binding sites and that all enzyme molecules are functionally identical. It has also been assumed that 2 mol of $Ca^{2+}$ bound to the high-affinity sites are translocated across the membrane/mol of ATP hydrolysis. However, recent work suggests that only 1 mol $Ca^{2+}$ is bound to the enzyme/mol of active site under some reaction conditions or in some preparations, as described below.

(1) The deoxycholate-solubilized ATPase has about 1 mol of high-affinity $Ca^{2+}$-binding sites/mol of active sites (Takisawa & Makinose, 1981).

(2) 1 mol of $Ca^{2+}$ remains on the enzyme even when $E_2P$ is formed in the presence of a sufficient amount of $Mg^{2+}$ and ATP (Nakamura & Tonomura, 1982a), whereas no $Ca^{2+}$ is bound when $E_2P$ is formed from Pi by the backward reaction. Therefore, it has been indicated that 2 mol of bound $Ca^{2+}$ are released successively/mol of ATPase in the $E_2P$ formation step (Nakamura & Tonomura, 1982a).

(3)  About 50% of the $Ca^{2+}$ bound to FSR dissociates immediately after addition of EGTA, but the remainder dissociates only slowly. Furthermore, the time course of $Ca^{2+}$ uptake after addition of ATP consists of rapid and slow phases. The amounts of $Ca^{2+}$ uptake in the two phases are nearly equal (Ikemoto *et al.*, 1981).

These results suggest that the ATPase molecules exist in the FSR membrane in functionally non-equivalent forms and that they are structurally indistinguishable. Ikemoto *et al.* (1981) suggested that the non-equivalent SR ATPase molecules form dimers in the membrane as functional units. As mentioned earlier in this chapter, structural studies show no heterogeneity in the $Ca^{2+}$-ATPase polypeptide and there is no evidence of any other peptides that modify the chemical structure of the ATPase protein. Therefore, it can be concluded that there is only slight variation in the chemical structure between the ATPase molecules.

## Counter ion of $Ca^{2+}$ transport

The accumulation of $Ca^{2+}$ in the FSR lumen is accompanied by the counter movement of ions. This was first demonstrated by Carvalho & Leo (1967), who found that FSR possesses a fixed cation-binding capacity of about $350\,\mu eq/g$ protein at neutral pH, and that $Mg^{2+}$, $K^+$ and $H^+$ are released from FSR when $Ca^{2+}$ is actively transported through the membrane. In addition, they showed that the total equivalency of these cations released from FSR is equal to that of the $Ca^{2+}$ taken up.

Their experiment, however, did not show whether the FSR membrane is merely permeable to the cations or whether these cations were transported as counter ions.

Using reconstituted vesicles impermeable to inorganic cations such as $K^+$, Zimniak & Racker (1978) found that a membrane potential (inside positive) is generated during $Ca^{2+}$ uptake, and that the rate of $Ca^{2+}$ uptake increases when the membrane potential is discharged using valinomycin and $K^+$. From these findings, they concluded that $Ca^{2+}$-ATPase is an electrogenic pump.

Since the FSR membrane is highly permeable to $K^+$, $H^+$ and $Cl^-$, it was difficult to determine whether or not $Ca^{2+}$-ATPase is electrogenic in the intact FSR. Recently, Beeler (1980) and Meissner (1981) found that when an inside-negative membrane potential is generated with $K^+$

or $H^+$ gradient the rate of $Ca^{2+}$ uptake increases and induced membrane potential disappears with $Ca^{2+}$ uptake. It was concluded, therefore, that the $Ca^{2+}$ pump in FSR is also electrogenic when an inside-negative membrane potential is generated with a $K^+$ or $H^+$ gradient. In their studies, the ratio of the amount of $K^+$ or $H^+$ released to that of $Ca^{2+}$ taken up was about 2:1. Thus, it is not likely that any ions other than $K^+$ or $H^+$ moved during $Ca^{2+}$ uptake.

Meissner and co-workers (McKinley & Meissner, 1978; Meissner & Young, 1980) showed that the FSR membrane has channels specific for $Na^+$, $K^+$ and $H^+$. Accordingly, $K^+$ or $H^+$ can be released through these channels during the $Ca^{2+}$ uptake process. As for the gating behavior of the $K^+$ and $Na^+$ channels, the reader is referred to the excellent paper by Labarca, Coronado & Miller (1980).

Under physiological conditions, $K^+$ or $Na^+$ are considered to be transported passively as counter ions during $Ca^{2+}$ uptake in FSR, since (1) some organic cations that can pass through the $K^+$ and $Na^+$ channels serve as counter ions instead of $K^+$, and (2) in a FSR fraction without $K^+$ or $Na^+$ channels, the amount of monovalent cation released from FSR during $Ca^{2+}$ transport was smaller than the equivalent amount of $Ca^{2+}$ taken up (Meissner, 1981).

On the other hand, the following findings suggest that $H^+$ is an antiport ion of $Ca^{2+}$ in FSR. (1) During the initial phase of $Ca^{2+}$ uptake by FSR, $H^+$ is released rapidly from FSR (Madeira, 1978). (2) The time course of $H^+$ release agrees well with that of $Ca^{2+}$ uptake. Furthermore, the amount of $H^+$ released is almost the same as that of $Ca^{2+}$ accumulated in the initial phase of the reaction (Chiesi & Inesi, 1980). (3) $H^+$ binds to the high-affinity $Ca^{2+}$-binding site of the ATPase in competition with $Ca^{2+}$.

Kinetic analyses formerly indicated that the $Mg^{2+}$ required for EP decomposition might be counter-transported. However, Chiesi & Inesi (1980) measured the movement of $Mn^{2+}$, an analog of $Mg^{2+}$, in $Ca^{2+}$ transport and showed that $Mn^{2+}$ is not counter-transported. Therefore, it is not likely that $Mg^{2+}$ is a counter ion of $Ca^{2+}$ transport.

## Reversal of the $Ca^{2+}$ pump

$Ca^{2+}$ transport coupled to ATP hydrolysis results in the formation of a steep $Ca^{2+}$-concentration gradient across the SR membrane. The $Ca^{2+}$,$Mg^{2+}$-ATPase in SR is capable of utilizing this $Ca^{2+}$ gradient to synthesize ATP from ADP and Pi. FSR provides a simple model

The sarcoplasmic reticulum ATPase

system for studying the molecular mechanism involved in the transformation of energy in this active transport. This section deals with the detailed reaction mechanism of ATP synthesis by the ATPase. We also discuss the energy levels, deduced from thermodynamic studies, of the reaction intermediates in the catalytic cycle.

*Reversal of the $Ca^{2+}$ pump in the presence of a $Ca^{2+}$ gradient*  ATP synthesis by FSR in the presence of a $Ca^{2+}$ gradient is thought to be the reverse reaction of ATP hydrolysis coupled with $Ca^{2+}$ transport. This was illustrated by the following experimental results:

(1) When 1 mol of ATP is formed from ADP and Pi, 2 mol of $Ca^{2+}$ are released from the FSR lumen (Makinose & Hasselbach, 1971; Panet & Selinger, 1972).

(2) When the $Ca^{2+}$ concentration in the medium was reduced below $10^{-8}$ M with EGTA, the $Ca^{2+}$-ATPase was phosphorylated with Pi (see Fig. 7.14). The EP formed in this reaction was an acylphosphate that reacted with ADP to form ATP (Makinose, 1972; Yamada, Sumida & Tonomura, 1972).

(3) As shown in the inset of Fig. 7.14, the double-reciprocal plot of the amount of EP formed against the square of the $Ca^{2+}$-concentration gradient was linear. This suggests that in the reverse reaction, in the presence of a $Ca^{2+}$ gradient, 2 mol of $Ca^{2+}$ are bound to the enzyme inside the membrane to form 1 mol of EP (Yamada et al., 1972; Prager et al., 1979). These results are consistent with the facts that: (a) the stoichiometric ratio of the amount of $Ca^{2+}$ transported to that of ATP hydrolyzed is 2:1; (b) phosphorylated intermediates are formed in the $Ca^{2+}$-transport cycle, and; (c) 2 mol of $Ca^{2+}$ are bound to 1 mol of $E_1P$.

Pi is considered to interact with the ATPase on the exterior of the SR membrane, since phosphorylation of the ATPase by Pi was competitively inhibited by membrane-impermeable ATP added outside the membrane (Yamada et al., 1972). The intravesicular Pi did not serve as a substrate for EP formation or for the ATP–Pi exchange reaction, whereas external Pi served as a substrate for both (de Meis & Carvalho, 1976).

*Phosphorylation of the $Ca^{2+},Mg^{2+}$-ATPase by Pi in the absence of a $Ca^{2+}$ gradient*  The phosphorylation of the ATPase by Pi occurs in the absence of a $Ca^{2+}$ gradient. This has been shown using an ATPase preparation that is incapable of forming a $Ca^{2+}$-concentration gradient (Masuda & de Meis, 1973; Kanazawa, 1975). Phosphorylation by Pi in the presence or absence of a $Ca^{2+}$ gradient is markedly inhibited by external $Ca^{2+}$ in the concentration range (several μM) required for ATP hydrolysis (Kanazawa & Boyer, 1973; Kanazawa, 1975). This indicates that the conformation of the ATPase phosphorylated by Pi is $E_2$, in which the high-affinity $Ca^{2+}$-binding site is not saturated by $Ca^{2+}$ (discussed earlier in this chapter). By monitoring the fluorescence intensity of intrinsic tryptophan, the conversion rate of $E_2$ into $E_1$ was found to be much slower when the ATPase was prephosphorylated by Pi than when the ATPase was not phosphorylated (Lacapère et al., 1981). This finding indicates that $E_2$ is converted into $E_1$ after dephosphorylation of the ATPase and that the enzyme is capable of reacting with Pi only after $E_2$ is formed.

The phosphorylation of the ATPase by Pi in the reverse reaction is $Mg^{2+}$ dependent. Studies of the dependence of the amount of EP on

Fig. 7.14 Phosphorylation of the $Ca^{2+}$-ATPase with Pi in the presence of a $Ca^{2+}$ gradient. Amounts of ATP formed were measured before (open circles) and after (open triangles) addition of ADP ( ↓ ). Inset: the reciprocal of EP was plotted against the square of the $Ca^{2+}$ gradient, $Ca^i/Ca^o$. From Yamada et al. (1972).

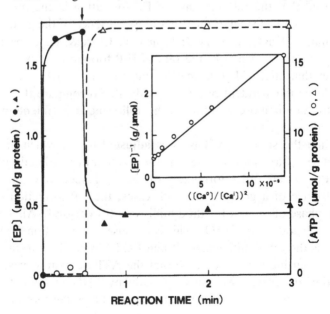

$Mg^{2+}$ and Pi concentrations in an equilibrium state showed that $Mg^{2+}$ and Pi bind to $E_2$ in a random sequence to form $Mg-E_2-Pi$ (Kolassa *et al.*, 1979). The experimental results mentioned above are summarized in the following reaction scheme.

$$
\begin{array}{c}
\qquad\qquad Mg-E_2 \quad Pi \\
Mg \nearrow\qquad\qquad\searrow \\
CaE_1 \overset{C_a}{\underset{Pi}{\longleftrightarrow}} E_2 \qquad\qquad Mg-E_2Pi \longleftrightarrow Mg-E_2P \\
\searrow\qquad\qquad\nearrow \\
\qquad\quad E_2Pi \quad Mg
\end{array}
$$

*Different properties of EP formed from E and Pi in the presence and absence of $Ca^{2+}$ gradient*  EP formed from Pi in the presence of a $Ca^{2+}$ gradient is $E_1P$, which reacts with ADP to synthesize ATP, whereas EP formed in the absence of a $Ca^{2+}$ gradient is $E_2P$, which does not react with ADP. Since it was shown that $Mg^{2+}$ and Pi bind to ATPase in a random sequence, in either the presence or absence of a $Ca^{2+}$ gradient, $E_2P$ must be formed from Pi and $E_2$ in the absence of a $Ca^{2+}$ gradient and then converted into $E_1P$ in the presence of a gradient. However, the sequential formation of $E_1P$ from $E_2P$ was refuted by transient kinetic studies of the backward reaction. It was reported that the rate constant of EP formation in the presence of a gradient is much smaller than that in the absence of the gradient (Beil, v. Chak & Hasselbach, 1977; Rauch, v. Chak & Hasselbach, 1977). If $E_1P$ is formed via $E_2P$, the rate of $E_1P$ formation is expected to be slower than that of $E_2P$ formation, but the rate of EP formation itself should be independent of a $Ca^{2+}$ gradient. To compare the kinetics of EP formation more accurately, the following experiments were performed.

When the state of ATPase was adjusted to $E_1$, which binds 2 mol of external $Ca^{2+}$/mol enzyme, and the reaction was initiated with EGTA and Pi, the rate of EP formation was slow, irrespective of the presence or absence of a gradient. In both cases, the EP formation exhibited a lag phase of several hundred milliseconds (Verjovski-Almeida *et al.*, 1978; Chaloub *et al.*, 1979). This is because the transition from $Ca_2-E_1$ to $E_2$ is the rate-limiting step. When EGTA was added to the medium before starting the reaction, so that the ATPase mainly existed as $E_2$, and then the enzyme was phosphorylated by Pi, the rate of EP formation in the absence of a $Ca^{2+}$ gradient was ten times the rate measured in the presence of a $Ca^{2+}$ gradient. These differences in the kinetics of EP formation strongly suggest that gradient-dependent EP is formed by a route different from that of gradient-independent EP. These experimental results can be reasonably explained by the following scheme (Chaloub *et al.*, 1979).

$$E_2Pi \longleftrightarrow E_2P$$
$$Pi \nearrow \qquad \nwarrow 2^iCa$$
$$Ca_2\text{–}E_1 \longleftrightarrow E_2 \qquad {}^iCa_2\text{–}E_2P \longleftrightarrow Ca_2\text{–}E_1P$$
$$2^iCa \searrow \qquad \nearrow$$
$${}^iCa_2\text{–}E_2 \longleftrightarrow {}^iCa_2\text{–}E_2Pi$$
$$Pi$$

In this scheme, it is assumed that the transition from $E_1$ to $E_2$ is the rate-limiting step of EP formation and that the rate of $E_2P$ formation from $E_2Pi$ through the upper route is sufficiently faster than through the lower route.

## *ATP synthesis from ADP and Pi in the absence of a $Ca^{2+}$ gradient*

As shown in the reaction scheme above, 2 mol of $Ca^{2+}$ bind to $E_2P$ to form $E_1P$, which can react with ADP to synthesize ATP. Knowles & Racker (1975) first demonstrated ATP formation using purified ATPase. The ATPase was phosphorylated with Pi in the presence of $Mg^{2+}$ and EGTA. The subsequent addition of ADP with a large amount of $Ca^{2+}$ produced an amount of ATP equal to that of the decrease in EP. Afterward, de Meis and co-workers (de Meis & Tume, 1977; de Meis, Martins & Alves, 1980) showed that the ATP was synthesized in a single turnover under conditions different from those used by Knowles & Racker. Figure 7.15 illustrates that when a large amount of $Ca^{2+}$ is added to $E_2P$ formed in the reverse reaction at pH 6.0, EP reacts with ADP to form ATP.

The requirement of a high concentration of $Ca^{2+}$ in net ATP synthesis was first suggested by the fact that the rate of exchange between ATP and Pi increases when the $Ca^{2+}$ concentration in the medium is increased by the use of leaky SR membrane (de Meis & Carvalho, 1974; de Meis, Carvalho & Sorenson, 1975; de Meis & Sorenson, 1975). Since net ATP synthesis requires a high concentration of $Ca^{2+}$ in the medium, it was suggested that the $Ca^{2+}$-binding site of $E_2P$ has a low affinity for $Ca^{2+}$ and that $Ca_2\text{–}E_1P$ is formed only when $Ca^{2+}$ has been bound to the

$Ca^{2+}$-binding site that faces the SR lumen. The conversion of $E_2P$ into $E_1P$ can also be brought about by changing pH or the DMSO concentration in the medium to modulate the equilibrium between $E_2$ and $E_1$. For more-detailed studies on ATP synthesis in the absence of a $Ca^{2+}$ gradient, see the excellent monograph by de Meis (1981*b*).

*Energy levels of $E_1P$ and $E_2P$* $E_1P$ and $E_2P$ can be distinguished from each other by their reactivities to ADP. However, after stopping the ATPase reaction with TCA, both phosphates are found to be bound to the enzyme by high-energy acylphosphate bonds. This indicates that the difference in the reactivity to ADP is dependent on the conformation of the enzyme. Accordingly, to clarify the mechanism of energy transduction, it is necessary to determine the energy levels of the intermediates.

There are two methods of measuring thermodynamic parameters: one is through determination of $\Delta H$ directly by calorimetric measurements and the other is through determination of the temperature dependence of the equilibrium constant. Kodama, Kurebayashi & Ogawa (1980) used the former method and obtained a value of about 12 kcal/mol for $\Delta H$ in $E + ATP \rightarrow E_1P + ADP$, and a value of about $-9.6$ kcal/mol for $\Delta H$ in the overall hydrolysis reaction. On the other hand, Kanazawa

Fig. 7.15 ATP formation from ADP and Pi in the absence of a $Ca^{2+}$ gradient. ATP synthesis (open circles) and EP decomposition (filled circles) were measured after adding $Ca^{2+}$ and ADP.

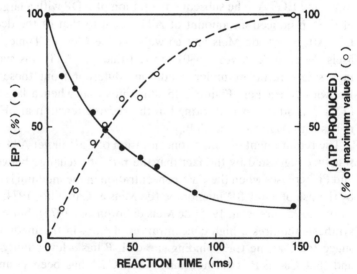

(1975) employed the latter method and obtained values of $+7.9$ e.u. and $+22$ kcal/mol, respectively, for $\Delta S^{\circ\prime}$ and $\Delta H^{\circ\prime}$ in the step $Pi + E_2 \leftrightarrow E_2P$. A similar enthalpy change was also reported by Martin & Tanford (1981).

These are the only thermodynamic parameters that have been obtained so far. Unfortunately, the two experiments mentioned above were performed under different conditions so the results cannot be compared directly. Although the steps $E_1P \leftrightarrow E_2P$ and $E_2 \leftrightarrow E_1$ are thermodynamically important in the $Ca^{2+}$-transport process, the parameters for these steps have not been determined. In the case of myosin ATPase, the thermodynamic studies are more advanced, as described in Chapter 2, and some results have been obtained that are important for clarification of the energy-transducing mechanism. It is expected that the study of this mechanism will also be advanced by understanding the $Ca^{2+}$ pump.

# Molecular mechanism of the active transport of $Ca^{2+}$ in the sarcoplasmic reticulum

This section deals with the molecular mechanism of active $Ca^{2+}$ transport in FSR, based on the descriptions given thus far in this chapter. We will first consider the structural features of the $Ca^{2+}$-pump protein in the membrane as the basis for understanding its function. Some recent studies on the movement or conformational change of the pump protein in the membrane during the catalytic cycle will then be documented. Finally, some molecular models for $Ca^{2+}$ transport in SR will be discussed in detail.

## *$Ca^{2+}$-pump protein in the sarcoplasmic reticulum*

*Asymmetric distribution of ATPase in SR membrane* As described earlier in this chapter, electron micrographs of FSR show that projections with particles (about 4 nm in diameter) attached to stalks (2 nm in length) are arranged on the outer face of the FSR. Many biochemical experimental results, such as immunological reactivity, tryptic fragmentation and chemical modification, demonstrate that these projections represent a part of the ATPase molecule.

Saito, Wang & Fleischer (1978) visualized the asymmetric, trilaminar structure of the SR membrane by staining a thin section of FSR with tannic acid to enhance contrast. The FSR membrane consists of three layers: an electron-dense outer layer about 7 nm wide, and middle and inner layers 2 nm wide. Similar asymmetric structure of the SR membrane was also observed in thin sections of muscle tissues. This asymmetric structure of the membrane has also been deduced from low-angle X-ray scattering of FSR preparations stacked in a lamellar manner (Dupont, Harrison & Hasselbach, 1973; Liu & Worthington, 1974; Herbette *et al.*, 1981). It was shown that the distribution of the electron density perpendicular to the membrane surface exhibits two high-density peaks that are derived from the polar head of lipid and ATPase projections.

*Monomeric and oligomeric ATPase in the SR membrane*    A great deal of experimental evidence supports the suggestion that the $Ca^{2+},Mg^{2+}$-ATPase exists in the membrane as an oligomer. Jilka, Martonosi & Tillack (1975) and Scales & Inesi (1976) found that the density (about $5750/\mu m^2$) of 9-nm particles, revealed in the outer layer of the membrane after freeze-fracture, was about one-fourth that of the 4-nm projections exposed on the surface (about $22\,100/\mu m^2$), and they concluded that the ATPase forms a tetramer as the functional unit for $Ca^{2+}$ transport.

Vanderkooi *et al.* (1977) showed that when two groups of ATPase molecules, labelled separately with two kinds of fluorescent reagents, were mixed in deoxycholate and then reconstituted by removing the detergent, fluorescence energy transfer was induced between the dyes located in the protein molecules. They further demonstrated that energy transfer was hardly affected, even when the lipid/protein weight ratio in the reconstituted membrane was raised. From these findings, they suggested that the ATPase in the reconstituted SR membrane also existed in oligomeric structure.

Using the sedimentation equilibrium method, le Maire *et al.* (1976) confirmed that ATPase fractions with molecular weights of 115 000, 222 000 and 430 000 were obtained from gel filtration of deoxycholate-solubilized ATPase. When the deoxycholate concentration in the gel was increased, only the m.w. 115 000 fraction was obtained, whereas when the gel filtration was carried out in the presence of lipid, there was a strong tendency for these ATPase molecules to self-associate.

Evidence for oligomerization of the $Ca^{2+},Mg^{2+}$-ATPase was provided by a study of rotational movements of the ATPase molecules in the FSR membrane by measuring the anisotropy of flash-induced photo-dichroism of eosin thiocyanate that was covalently attached to the ATPase (Hoffmann, Sarzala & Chapman, 1979). An Arrhenius plot of the correlation time was broken at about 15 °C and at about 35 °C. Hoffmann *et al.*, explained the discontinuities in terms of a temperature-dependent equilibrium existing between the conformationally altered ATPase and its oligomeric forms at this temperature range.

On the other hand, Brady *et al.* (1981) did diffraction analysis of the data obtained by low-angle X-ray scattering using reconstituted vesicle suspensions with different lipid/protein ratios. They proposed a model in which the ATPases have a cylindrical shape (3.5 nm in diameter and 14.2 nm in length) and are arranged asymmetrically in monomeric form in the lipid bilayer (4.1 nm in thickness). Hebdon, Cunningham & Green (1979) showed that the ATPase protein in the membrane was maintained in a monomeric form even after long exposure to the crosslinking reagent, I$^-$- or Cu$^{2+}$-1,10-phenanthroline at low temperature. Recently, Andersen, Møller & Jørgensen (1982) found that when 1 mol of FITC was bound to the ATPase/mol of active site of ATPase, the ATPase activity disappeared. They also found that even when C$_{12}$E$_8$ was added to dissociate the ATPase molecules into monomers, the amount of bound FITC did not change, nor did the ATPase activity increase. From these results, they suggested that the functional unit of Ca$^{2+}$ transport in FSR is a monomeric form. Thus, it cannot be determined clearly by structural studies alone whether the ATPase molecules have an oligomeric or a monomeric form. The clarification of this problem requires a comparison of the elementary steps in the catalytic cycle of the enzyme between the monomeric and oligomeric forms.

*ATPase activity of oligomeric and monomeric ATPases*   Dean & Tanford (1978) showed that when C$_{12}$E$_8$ was added to monomeric ATPase that had been freed of lipid by deoxycholate treatment, capacity for EP formation and for Ca$^{2+}$,Mg$^{2+}$-ATPase activity were restored. Møller *et al.* (1980) obtained a similar result using FSR solubilized with an excess of C$_{12}$E$_8$. These results raise the possibility that the functional unit of Ca$^{2+}$ transport may be a monomer in the intact SR membrane. On the other hand, Yamamoto & Tonomura (1982) showed that when the Ca$^{2+}$,Mg$^{2+}$-ATPase was solubilized as a monomer by a high concentration of C$_{12}$E$_8$, most of the ATP was hydrolyzed via an enzyme–ATP complex that was in equilibrium with EP + ADP. Furthermore, the $Mg^{2+}$ requirements for EP decomposition was completely lost under these conditions. The process of ATP hydrolysis can be represented by the following scheme.

$$E + ATP \longleftrightarrow E\text{–}ATP \rightarrow E^*\text{–}ATP \longleftrightarrow E_1P + ADP$$
$$Mg^{2+}\text{-dependent} \Rightarrow ADP + Pi \quad Pi$$

In this scheme, ATP hydrolysis via $E^*$–ATP is dependent on $Mg^{2+}$, whereas the decomposition of $E_1P$ into E and Pi is completely independent of $Mg^{2+}$, unlike the $Ca^{2+}$,$Mg^{2+}$-ATPase in intact SR. They found that the sensitivity of EP to $Mg^{2+}$ was fully restored when $C_{12}E_8$ was removed from the medium with Bio-Beads. Measurements of fluorescence energy transfer between dyes located on the enzyme and an analysis of the polarization decay of fluorescent dye attached to the enzyme indicate that the monomeric ATPase molecules were associated to form an oligomer in parallel with the restoration of the ATPase reaction coupled to $Ca^{2+}$ transport (Yantorno, Yamamoto & Tonomura, 1983). In fact, several laboratories have been successful in reconstituting $Ca^{2+}$ transport after removal of $C_{12}E_8$ from the mixture of the ATPase monomer and soybean lecithin (Møller et al., 1982; Andoh & Yamamoto, 1985).

## Molecular mechanism of active transport

In the preceding section, the way in which the $Ca^{2+}$,$Mg^{2+}$-ATPase exists in the FSR membrane was discussed mainly in terms of structure. It was suggested that an ATPase oligomer may be the functional unit of $Ca^{2+}$ transport. In this section we will consider the molecular mechanism of the active $Ca^{2+}$ transport from the point of the view of the movement or conformational changes of the ATPase molecule during the elementary steps of the ATPase reaction.

*Conformational changes induced by binding of ligands* As noted previously in this chapter, the binding of $Ca^{2+}$ to the high-affinity site on the $Ca^{2+}$,$Mg^{2+}$-ATPase results in conformational changes in the enzyme protein. The conversion of the enzyme state from $E_2$ to $E_1$, induced by the binding of $Ca^{2+}$, has been indicated by a change in the fluorescence intensity of intrinsic tryptophan (Dupont, 1976; Guillain *et al.*, 1980), the quenching of the fluorescence of a dye attached to the ATPase (Pick, 1981), and the change in the accessibility of the SH reagent (Ikemoto *et al.*, 1978; Murphy, 1978). ATPase conformational changes induced by the binding of ATP have been suggested by the ESR spectra obtained from spin-labelled SR. Coan & Inesi (1977), Coan *et al.* (1979) and Champeil *et al.* (1978) studied the effects of $Ca^{2+}$ and either ATP or AMPPNP on the ESR spectrum. They found that the intramolecular motion of the enzyme is restricted by formation of the E–NTP complex and that enzyme mobility was further reduced by the binding of $Ca^{2+}$ to the E–NTP complex to form a $Ca$–$E_1$–NTP complex.

*Change in ATPase conformation accompanying EP formation*
Observations suggesting that $E_1P$ formation is associated with conformational changes of the ATPase have been made by Miki, Scott & Ikemoto (1981). They showed that when ATPase labelled with ANM ($N$-(1-anilinaphthyl-4)maleimide) reacted with ATP to form $E_1P$, the fluorescence of ANM, which had been partly quenched with acrylamide, rose rapidly to the control level seen in the absence of acrylamide. The fluorescence intensity returned to the original low level when ADP was added to $E_1P$ to form an E–ATP complex. Miki *et al.* suggested that the formation of $E_1P$ was coupled to a reversible movement of the ANM-binding site from the hydrophilic to the hydrophobic region of the enzyme molecule. Dupont (1978) found that when the $Ca^{2+}$-loaded FSR was reacted with Pi in the presence of EGTA and $Mg^{2+}$ to form $E_1P$, a significant increase in intrinsic fluorescence occurred. The intensity of fluorescence increased with an increase in the amount of $Ca^{2+}$ inside the vesicles.

*Changes in the reactivities of functional groups of the ATPase during its catalytic cycle*   Changes in the structure of the ATPase molecule in the enzyme reaction can be detected from changes in the reactivities of functional groups. For example, Tonomura & Morales (1974) measured time courses in which the ESR signal of spin-labelled NEM attached to FSR was quenched by ascorbate added to the external medium. They found that the ESR signal was rapidly quenched by ascorbate when the state of the ATPase was held in either $E_2$ or $E_1$–Ca. They further showed that the reactivity with ascorbate was much lower in the $E_2$–ATP state and disappeared in the $E_1P$ state. Similar conclusions were reached by Yamamoto & Tonomura (1976), who showed that the number of lysine residues susceptible to an impermeant reagent, 2,4,6-trinitrobenzene sulfonate (TNBS) varied with changes in the enzyme state.

## Molecular mechanism of $Ca^{2+}$ transport

*Kinetic and structural background*   As described in the previous section, it has been well established from a number of kinetic studies that there are 2 mol of high-affinity $Ca^{2+}$-binding sites/mol of the enzyme. When these sites are saturated, the enzyme can be phosphorylated by ATP and the $Ca^{2+}$ is occluded by the enzyme. The conversion of the phosphoenzyme from $E_1P$ to $E_2P$ is accompanied by a reduction in the affinity of the enzyme for $Ca^{2+}$, and thus $Ca^{2+}$ is released from the enzyme. However, studies on the structural aspects of the membrane ATPase are still insufficient for a complete understanding of active $Ca^{2+}$ transport.

Although many investigations have provided evidence that changes in the structural properties of the ATPase are induced by binding of ligands such as $Ca^{2+}$ and ATP, it is not yet clear what part of the ATPase molecule is involved or how these changes are brought about. In addition, the mechanism by which lipid interacts with ATPase in the membrane remains unknown although many different proposals have been offered. Furthermore, although it is well known that the ATPase consists of a single peptide with a molecular weight of about 100 000, as described earlier in this chapter, it is still not certain that all of the peptides are identical. As described earlier in this section, the ATPase in monomeric form cannot actively transport $Ca^{2+}$, and many experimental results suggest that the ATPase may exist in dimeric or tetrameric form in the membrane. Therefore, the next section will deal with the functional roles of the oligomeric form in $Ca^{2+}$ transport.

*Monomer–monomer interaction*   To account for active transport of $Ca^{2+}$ through the SR membrane as well as for $Na^+$ and $K^+$ transport across membranes, a flip–flop mechanism has been proposed by several investigators (Froehlich & Taylor, 1975; Verjovski-Almeida *et al.*, 1978; Grosse *et al.*, 1979). In this model, it is assumed that the functional unit of the cation pump consists of two identical ATPase molecules, which catalyze ATP hydrolysis with half of the sites. For example, when phosphorylation occurs at the catalytic site of one ATPase molecule, ATP binds to the catalytic site of the other ATPase molecule. However, in some cases, this model cannot explain active $Ca^{2+}$ transport in FSR. Nakamura & Tanomura (1982*b*) showed that ATP that is bound to the catalytic site of the ATPase in the presence of sufficient amounts of KCl and $MgCl_2$ disappears immediately after the addition of $Ca^{2+}$ to produce $E_1P$. This means that there is little possibility that E–ATP and $E_1P$ exist simultaneously under these conditions. In addition, this model does not explain the fact that, when the concentration ratio of $Ca^{2+}$ to $Mg^{2+}$ in the reaction medium is rather high, $E_1P$ is predominantly formed, while, when this ratio is low, $E_2P$ is mainly formed (Takisawa & Tonomura, 1979).

Watanabe & Ihesi (1982) suggested that the formation of $E_1P$ results in a dissociation of the oligomeric ATPase into a monomeric form. Thus the possibility cannot be ruled out that active $Ca^{2+}$ transport may require an oligomeric structure for the ATPase in the SR membrane. However, at least from a kinetics viewpoint, we can assume that the catalytic functions for $Ca^{2+}$ transport are performed independently by each ATPase monomer. Therefore, to simplify the discussion of transport models, the functional roles of the ATPase–lipid and ATPase–ATPase interactions will henceforth be neglected, but ATPase oligomers will be assumed to form the channel and other structures through which divalent cations are capable of being transported.

*Molecular models for the active transport of $Ca^{2+}$*   One of the most fundamental models for active cation transport through the membrane is the 'circular carrier model', which was proposed by Shaw (1954) to explain the active transport of $Na^+$ and $K^+$ across the erythrocyte membrane, as will be described in Chapter 8. According to this model, $K^+$ and $Na^+$ can only move in the membrane as complexes with specific carriers, X for $K^+$ and Y for $Na^+$. Furthermore, X and Y are interconvertible, with change in cation affinity coupled to the ATPase reaction. This hypothesis provided a ready explanation for the vectorial movement in cation translocation through the membrane and thus became widely accepted as a model on which later models were based. At present, many models have been proposed to account for the active transport of $Ca^{2+}$ across the SR membrane. They can be roughly divided into two major classes, the rotatory models and mobile-pore models, based on the type of motion of the ATPase molecules. According to the rotatory model, the $Ca^{2+}$,$Mg^{2+}$-ATPase protein rotates during the ATPase reaction, with the result that the $Ca^{2+}$-binding site is exposed alternately outside and inside the membrane. As described in the preceding section, Tonomura & Morales (1974) as well as Yamamoto & Tonomura (1976) examined the reactivity of the functional group to an impermeable chemical modifier during changes in the enzymatic states of the reaction cycle. Their observations of continuous changes in reactivity to the reagents are easily explained by the rotatory model. However, as pointed out by Dutton, Rees & Singer (1976), it is thermodynamically unfavourable for a hydrophilic region to pass through a hydrophobic membrane.

The mobile-pore model indicates that several ATPase molecules associate to form a pore for $Ca^{2+}$ transport, and the pore is transformed by an energy-dependent rearrangement of ATPase molecules so that $Ca^{2+}$ passes through it. This model does not require the assumption that ATPase molecules move very much, but it is too simple to explain the continuous change in the enzyme states during the reaction cycle.

Thus, experimental observations do not easily support either the rotatory or the mobile-pore model, and the two models seem insufficient to explain active $Ca^{2+}$ transport. At present, a mobile-gate model or a mobile-channel model can explain the experimental results better than the rotatory or mobile-pore models.

In the mobile-gate model, shown in Fig. 7.16, the existence of a pocket and a pore that are formed by ATPase monomers or oligomers is assumed. The pocket is covered with mobile gates that restrict cation exchange with the outer solution. The pore, which always has a constant diameter, allows hydrated $Ca^{2+}$, but not $Mg^{2+}$, to pass through it. On the outer surface of the mobile gate, 2 mol of $Ca^{2+}$ are bound to 1 mol of ATPase and are translocated to the inner surface of the pocket by rotational movement of the gate, induced by $E_1P$ formation. In this state, EGTA outside the membrane cannot enter the pocket and thus $Ca^{2+}$ enters an occluded state. $Ca^{2+}$ is released into the interior space of the pocket with the conversion of $E_1P$ to $E_2P$. EP decomposition is accelerated by $Mg^{2+}$ in the interior space of the pocket. $Ca^{2+}$, carried by the pocket, is transported through the pore into the interior space of the FSR vesicle, but $Mg^{2+}$ cannot pass through the pore. When $Ca^{2+}$ is transported inside, $Na^+$, $K^+$ and other ions are passively transported from the inside to the outside of the membrane through channels different from those used for active $Ca^{2+}$ transport, thereby maintaining the electroneutrality of the system. This model assumes that cation exchange between the inside and outside of the pocket is greatly limited. The tightness of the coupling between ATP decomposition and $Ca^{2+}$ uptake can be easily explained by this assumption. In addition, the difficulties in interpreting the mechanism of $Mg^{2+}$ function, which were discussed previously, can also be resolved easily.

Figure 7.16 also illustrates the mobile-channel model for active $Ca^{2+}$ transport. This model differs from the mobile-gate model in that it assumes not only that the pocket opens and shuts according to the enzyme state, but also that the channel itself expands. When the pocket is open, 2 mol of $Ca^{2+}$ are bound to 1 mol of ATPase, and, when the pocket shuts with $E_iP$ formation, $Ca^{2+}$ enters the occluded state. In $E_2P$ formation the pocket stays shut, while the channel increases in diameter, and $Ca^{2+}$ released from the ATPase moves inside through the expanded channel.

These transport models are consistent with present experimental results but do not provide details of the physical movement of $Ca^{2+}$ across the membrane. To propose a model for the motion of the molecules in cation transport coupled to ATP hydrolysis, more information concerning the structure of ATPase must be obtained.

Fig. 7.16 Molecular models for active transport of $Ca^{2+}$ through the SR membrane. For explanation, see text. From Yamamoto *et al.* (1979).

**Mobile Gate Model**

**Mobile Channel Model**

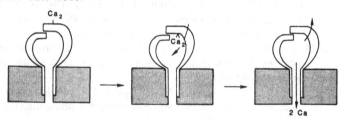

$$^{Ca_2}E_1 ATP \longrightarrow {}^{Ca_2>}E_{1P} \longrightarrow E_{2P} + 2\ Ca$$

# References

Allen, G., Trinnaman, B. J. & Green, N. M. (1980). The primary structure of the calcium ion-transporting adenosine triphosphatase protein of rabbit skeletal sarcoplasmic reticulum: peptide derived from digestion with cyanogen bromide, and the sequences of three long extramembranous segments. *Biochem. J.*, **187**, 591–616.

Andersen, J. P., Møller, J. V. & Jørgensen, P. L. (1982). The functional unit of sarcoplasmic reticulum $Ca^{2+}$-ATPase: active site titration and fluorescence measurements. *J. Biol. Chem.*, **257**, 8300–7.

Andoh, R. & Yamamoto, T. (1985). Restoration of $Ca^{2+}$-transport of sarcoplasmic reticulum ATPase solubilized with octaethyleneglycol mono $n$-dodecyl ether. *J. Biochem.*, **97**, 877–82.

Ashley, C. C., Moisescu, D. G. & Rose, R. M. (1974). Aequorin-light and tension responses from bundles of myofibrils following a sudden change in free calcium. *J. Physiol.*, **241**, 104p–6p.

Bastide, F., Meissner, G., Fleischer, S. & Post, R. L. (1973). Similarity of the active site of phosphorylation of the adenosine triphosphatase for transport of sodium and potassium ions in kidney to that for transport of calcium ions in the sarcoplasmic reticulum of muscle. *J. Biol. Chem.*, **248**, 8385–91.

Beeler, T. J. (1980). $Ca^{2+}$ uptake and membrane potential in sarcoplasmic reticulum vesicles. *J. Biol. Chem.*, **255**, 9156–61.

Beil, F. U., v. Chak, D. & Hasselbach, W. (1977). Phosphorylation from inorganic phosphate and ATP synthesis of sarcoplasmic membranes. *Eur. J. Biochem.*, **81**, 151–64.

Bennet, J. P., McGill, K. A. & Warren, G. B. (1980). The role of lipids in the functioning of a membrane protein: the sarcoplasmic reticulum calcium pump. *Curr. Top. Membr. Transp.*, **14**, 127–64.

Berman, M. C. (1982). Energy coupling and uncoupling of active calcium transport by sarcoplasmic reticulum membranes. *Biochim. Biophys. Acta*, **694**, 95–121.

Brady, G. W., Fein, D. B., Harder, M. E., Spehr, R. & Meissner, G. (1981). A liquid diffraction analysis of sarcoplasmic reticulum. I. Compositional variation. *Biophys. J.*, **34**, 13–34.

Campbell, K. P. & MacLennan, D. H. (1981). Purification and characterization of the 53,000-dalton glycoprotein from the sarcoplasmic reticulum. *J. Biol. Chem.*, **256**, 4626–32.

Carvalho, A. P. & Leo, B. (1967). Effects of ATP on the interaction of $Ca^{2+}$, $Mg^{2+}$, and $K^+$ with fragmented sarcoplasmic reticulum isolated from rabbit skeletal muscle. *J. Gen. Physiol.*, **50**, 1327–52.

Chaloub, R. M., Guimaraes-Motta, H., Verjovski-Almeida, S., de Meis, L. & Inesi, G. (1979). Sequential reaction in Pi utilization for ATP synthesis by sarcoplasmic reticulum. *J. Biol. Chem.*, **254**, 9464–8.

Champeil, P., Buschlen-Boucly, S., Bastide, F. & Gray-Bobo, C. (1978). Sarcoplasmic reticulum ATPase: spin labeling detection of ligand-induced changes in the relative reactivities of certain sulfhydryl groups. *J. Biol. Chem.*, **253**, 1179–86.

Chapman, D., Gomez-Fernandez, J. C. & Goni, F. M. (1979). Intrinsic protein-lipid interactions: physical and biochemical evidence. *FEBS Lett.*, **98**, 211–23.

Chiesi, M. & Inesi, G. (1980). Adenosine 5′-triphosphate dependent fluxes of manganese and hydrogen ions in sarcoplasmic reticulum vesicles. *Biochemistry*, **19**, 2912–18.

Coan, C. R. & Inesi, G. (1977). $Ca^{2+}$-dependent effect of ATP on spin-labeled sarcoplasmic reticulum. *J. Biol. Chem.*, **252**, 3044–9.

Coan, C. R., Verjovski-Almeida, S. & Inesi, G. (1979). $Ca^{2+}$ regulation of conformational states in the transport cycle of spin-labeled sarcoplasmic reticulum ATPase. *J. Biol. Chem.*, **254**, 2968–74.

Coffey, R. L., Lagwinska, E., Oliver, M. & Martonosi, A. N. (1975). The mechanism of ATP hydrolysis by sarcoplasmic reticulum. *Arch. Biochem. Biophys.*, **170**, 37–48.

Dean, W. L. & Tanford, C. (1978). Properties of a delipidated, detergent-activated $Ca^{2+}$-ATPase. *Biochemistry*, **17**, 1683–90.

de Meis, L. (1981). *The Sarcoplasmic Reticulum: Transport and Energy Transduction.* New York: John Wiley & Sons.

de Meis, L. & Boyer, P. D. (1978). Induction by nucleotide triphosphate hydrolysis of a form of sarcoplasmic reticulum ATPase capable of medium phosphate–oxygen exchange in the presence of calcium. *J. Biol. Chem.*, **253**, 1556–9.

de Meis, L. & Carvalho, M. G. C. (1974). Role of the $Ca^{2+}$ concentration gradient in the adenosine 5′-triphosphate-inorganic phosphate exchange catalyzed by sarcoplasmic reticulum. *Biochemistry*, **13**, 5032–8.

de Meis, L. & Carvalho, M. G. C. (1976). On the sidedness of membrane phosphorylation by Pi and ATP synthesis during reversal of the $Ca^{2+}$ pump of sarcoplasmic reticulum vesicles. *J. Biol. Chem.*, **251**, 1413–17.

de Meis, L., Carvalho, M. G. C. & Sorenson, M. M. (1975). ATP–Pi exchange catalyzed by sarcoplasmic reticulum without a $Ca^{2+}$ concentration gradient. In *Concepts of Membranes in Regulation and Excitation*, ed. M. R. Silva & G. S. Kurtz, pp. 7–19. New York: Raven Press.

de Meis, L., Martins, O. B. Z. & Alves, E. W. (1980). Role of water, hydrogen ion, and temperature on the synthesis of adenosine triphosphate by the sarcoplasmic reticulum adenosine triphosphate in the absence of a calcium ion gradient. *Biochemistry*, **19**, 4252–61.

de Meis, L. & Sorenson, M. M. (1975). ATP-Pi exchange and membrane phosphorylation in sarcoplasmic reticulum vesicles: activation by silver in the absence of a $Ca^{2+}$ concentration gradient. *Biochemistry*, **14**, 2739–44.

de Meis, L. & Tume, R. K. (1977). A new mechanism by which an $H^+$ concentration gradient drives the synthesis of adenosine triphosphate, pH jump, and adenosine triphosphate synthesis by the $Ca^{2+}$-dependent adenosine triphosphatase of sarcoplasmic reticulum. *Biochemistry,* **16**, 4455–63.

Dupont, Y. (1976). Fluorescence studies of the sarcoplasmic reticulum calcium pump. *Biochem. Biophys. Res. Commun.,* **71**, 544–50.

Dupont, Y. (1977). Kinetics and regulation of sarcoplasmic reticulum ATPase. *Eur. J. Biochem.,* **72**, 185–90.

Dupont, Y. (1978). Mechanism of the sarcoplasmic reticulum calcium pump. Fluorometric study of the phosphorylated intermediates. *Biochem. Biophys. Res. Commun.,* **82**, 893–900.

Dupont, Y. (1980). Occlusion of divalent cations in the phosphorylated calcium pump of sarcoplasmic reticulum. *Eur. J. Biochem.,* **109**, 231–8.

Dupont, Y., Harrison, S. C. & Hasselbach, W. (1973). Molecular organization in the sarcoplasmic reticulum membrane studied by X-ray diffraction. *Nature,* **244**, 555–8.

Dupont, Y. & Leight, J. B. (1978). Transient kinetics of sarcoplasmic reticulum $Ca^{2+}$ + $Mg^{2+}$ ATPase studied by fluorescence. *Nature,* **273**, 396–8.

Dutton, A., Rees, E. D. & Singer, S. J. (1976). An experiment eliminating the rotating carrier mechanism for the active transport of Ca ion in sarcoplasmic reticulum membranes. *Proc. Natl. Acad. Sci. (USA),* **73**, 1532–6.

Ebashi, S. & Lipmann, F. (1962). Adenosine triphosphate-linked concentration of calcium ions in a particulate fraction of rabbit muscle. *J. Cell Biol.,* **14**, 389–400.

Endo, M. (1964). Entry of a dye into the sarcotubular system of muscle. *Nature,* **202**, 1115–16.

Endo, M. (1977). Calcium release from the sarcoplasmic reticulum. *Physiol. Rev.,* **57**, 71–108.

Endo, M., Tanaka, M. & Ogawa, Y. (1970). Calcium-induced release of calcium from the sarcoplasmic reticulum of skinned skeletal muscle fibres. *Nature,* **228**, 34–6.

Fabiato, A. & Fabiato, F. (1978). Calcium-induced release of calcium from the sarcoplasmic reticulum of skinned cells from adult human, dog, cat, rabbit, rat and frog hearts and from fetal and new-born rat ventricles. *Ann. N.Y. Acad. Sci.,* **307**, 491–522.

Fernandez, J. L., Rosemblatt, M. & Hidalgo, C. (1980). Highly purified sarcoplasmic reticulum vesicles are devoid of $Ca^{2+}$-independent ('basal') ATPase activity. *Biochim. Biophys. Acta,* **599**, 552–68.

Ford, L. E. & Podolsky, R. J. (1972). Intracellular calcium movements in skinned muscle fibres. *J. Physiol.,* **223**, 21–33.

Fossel, E. T., Post, P. L., O'Hara, D. S. & Smith, T. W. (1981). Phosphorus-31 nuclear magnetic resonance of phosphoenzymes of sodium- and potassium-activated and of calcium-activated adenosinetriphosphatase. *Biochemistry,* **20**, 7215–19.

Franzini-Armstrong, C., Landmesser, L. & Pilar, G. (1975). Size and shape of transverse tubule openings in frog twitch muscle fibers. *J. Cell Biol.,* **64**, 493–7.

Froehlich, J. P. & Taylor, E. W. (1975). Transient state kinetic studies of sarco-plasmic reticulum adenosine triphosphatase. *J. Biol. Chem.*, **250**, 2013–21.

Garrahan, P. J., Rega, A. F. & Alonson, G. L. (1976). The interaction of magnesium ions with the calcium pump of sarcoplasmic reticulum. *Biochim. Biophys. Acta*, **448**, 121–32.

Grosse, R., Rapoport, T., Malur, J., Fischer, J. & Repke, K. R. H. (1979). Mathematical modelling of ATP, $K^+$ and $Na^+$ interactions with $(Na^+ + K^+)$-ATPase occurring under equilibrium conditions. *Biochim. Biophys. Acta*, **550**, 500–14.

Guillain, F., Champeil, P., Lacapère, J.-J. & Gingold, M. P. (1981). Stopped flow and rapid quenching measurement of the transient steps induced by calcium binding to sarcoplasmic reticulum adenosine triphospha-tase: competition with $Ca^{2+}$-independent phosphorylation. *J. Biol. Chem.*, **256**, 6140–7.

Guillain, F., Gingold, M. P., Buschlen, S. & Champeil, P. (1979). A direct fluorescence study of the transient steps induced by calcium binding to sarcoplasmic reticulum ATPase. *J. Biol. Chem.*, **225**, 2072–6.

Hardwicke, P. M. D. & Green, N. M. (1974). The effect of delipidation on the adenosine triphosphatase of sarcoplasmic reticulum: electron microscopy and physical properties. *Eur. J. Biochem.*, **42**, 183–93.

Hasselbach, W. (1979). The sarcoplasmic calcium pump: a model of energy transduction in biological membrane. In *Topics in Current Chemistry*, vol. 78, ed. M. J. S. Dewar, K. Hafner, E. Heilbronner, S. Ito, J. M. Lehn, K. Niedenzu, C. W. Ress, K. Schafer, G. Witting & F. L. Boschke, pp. 1–56. Berlin: Springer-Verlag.

Hasselbach, W. (1981). Calcium-activated ATPase of the sarcoplasmic reticulum membranes. In *Membrane Transport*, ed. S. L. Bonting and J. J. H. H. M. de Pont, pp. 183–207. Amsterdam: Elsevier.

Hasselbach, W. & Elfvin, L. G. (1967). Structural and chemical asymmetry of the calcium-transporting membranes of the sarcotubular system as revealed by electron microscopy. *J. Ultrastruct. Res.*, **17**, 598–622.

Hasselbach, W. & Makinose, M. (1961). Die Calciumpumpe der Erschlaffungs-grana' des Muskels und ihre Abhängigkeit von der ATP-Spaltung. *Biochem. Z.*, **333**, 518–28.

Hasselbach, W. & Makinose, M. (1963). Über den Mechanismus des Calcium-transportes durch die Membranen des Sarkoplasmatischen Reticu-lums. *Biochem. Z.*, **339**, 94–111.

Hasselbach, W. & Seraydarian, K. (1966). The role of sulfhydryl groups in calcium transport through the sarcoplasmic membranes of skeletal muscle. *Biochem. Z.*, **345**, 159–72.

Hebdon, G. M., Cunningham, L. W. & Green, N. M. (1979). Cross-linking experiments with the adenosine triphosphatase of sarcoplasmic reti-culum. *Biochem. J.*, **179**, 135–9.

Herbette, L., Scarpa, A., Blasie, J. K., Wang, C. T., Saito, A. & Fleischer, S. (1981). Comparison of the profile structures of isolated and recon-stituted sarcoplasmic reticulum membranes. *Biophys. J.*, **36**, 47–72.

Hesketh, T. R., Smith, G. A., Houslay, M. D., McGill, K. A., Birdsall, N. J. M., Metcalfe, J. C. & Warren, G. B. (1976). Annular lipids determine the ATPase activity of a calcium-transport protein complexed with dipalmitoyl lecithin. *Biochemistry*, **15**, 4145–51.

Hoffmann, W., Sarzala, M. G. & Chapman, D. (1979). Rotational motion and evidence for oligomeric structures of sarcoplasmic reticulum $Ca^{2+}$-activated ATPase. *Proc. Natl. Acad. Sci. (USA)*, **76**, 3860–4.

Hokin, L. E., Sastry, P. S., Galsworthy, P. R. & Yoda, A. (1965). Evidence that a phosphorylated intermediate in a brain transport adenosine triphosphatase is an acylphosphate. *Proc. Natl. Acad. Sci. (USA)*, **54**, 177–84.

Huxley, H. E. (1964). Evidence for continuity between the central elements of the triads and extracellular space in frog sartorius muscle. *Nature*, **202**, 1067–71.

Huxley, A. F. & Taylor, R. E. (1958). Local activation of striated muscle fibres. *J. Physiol.*, **144**, 426–41.

Ikemoto, N. (1974). The calcium-binding sites involved in the regulation of the purified adenosine triphosphatase of the sarcoplasmic reticulum. *J. Biol. Chem.*, **249**, 649–51.

Ikemoto, N. (1975). Transport and inhibitory $Ca^{2+}$-binding sites on the ATPase enzyme isolated from the sarcoplasmic reticulum. *J. Biol. Chem.*, **250**, 7219–24.

Ikemoto, N. (1976). Behavior of the $Ca^{2+}$ transport sites linked with the phosphorylation reaction of ATPase purified from the sarcoplasmic reticulum. *J. Biol. Chem.*, **251**, 7275–7.

Ikemoto, N., Garcia, A. M., Kurobe, Y. & Scott, T. L. (1981). Nonequivalent subunits in the calcium pump of sarcoplasmic reticulum. *J. Biol. Chem.*, **256**, 8593–601.

Ikemoto, N., Morgan, J. F. & Yamada, S. (1978). $Ca^{2+}$-controlled conformational states of the $Ca^{2+}$ transport enzyme of sarcoplasmic reticulum. *J. Biol. Chem.*, **253**, 8027–33.

Inesi, G. (1979). Transport across sarcoplasmic reticulum in skeletal and cardiac muscle. In *Membrane Transport in Biology*, ed. G. Giebisch, D. C. Tosteson and H. H. Ussing, pp. 357–93. Berlin: Springer-Verlag.

Inesi, G., Kurzmack, M., Coan, C. & Lewis, D. E. (1980). Cooperative calcium binding and ATPase activation in sarcoplasmic reticulum vesicles. *J. Biol. Chem.*, **255**, 3025–31.

Inesi, G., Maring, E., Murphy, A. J. & McFarland, B. H. (1970). A study of the phosphorylated intermediate of sarcoplasmic reticulum ATPase. *Arch. Biochem. Biophys.*, **138**, 285–94.

Inesi, G. & Scales, D. (1974). Tryptic cleavage of sarcoplasmic reticulum protein. *Biochemistry*, **13**, 3298–306.

Jilka, R. L., Martonosi, A. N. & Tillack, T. W. (1975). Effect of the purified $[Mg^{2+} + Ca^{2+}]$-activated ATPase of sarcoplasmic reticulum upon the passive $Ca^{2+}$ permeability and ultrastructure of phospholipid vesicles. *J. Biol. Chem.*, **250**, 7511–24.

Jobsis, F. F. & O'Conner, M. J. (1966). Calcium release and reabsorption in the sartorius muscle of the toad. *Biochem. Biophys. Res. Commun.*, **25**, 246–52.

Kalbitzer, H. R., Stehlik, D. & Hasselbach, W. (1978). The binding of calcium and magnesium to sarcoplasmic reticulum vesicles as studied by manganese electron paramagnetic resonance. *Eur. J. Biochem.*, **82**, 245–55.

Kanazawa, T. (1975). Phosphorylation of solubilized sarcoplasmic reticulum by orthophosphate and its thermodynamic characteristics. The dominant role of entropy in the phosphorylation. *J. Biol. Chem.*, **250**, 113–19.

Kanazawa, T. & Boyer, P. D. (1973). Occurrence and characteristics of a rapid exchange of phosphate oxygens catalyzed by sarcoplasmic reticulum vesicles. *J. Biol. Chem.*, **248**, 3163–72.

Kanazawa, T., Yamada, S., Yamamoto, T. & Tonomura, Y. (1971). Reaction mechanism of the $Ca^{2+}$-dependent ATPase of sarcoplasmic reticulum from skeletal muscle. V. Vectorial requirements for calcium and magnesium ions of three partial reactions of ATPase: formation and decomposition of a phosphorylated intermediate and ATP-formation from ADP and the intermediate. *J. Biochem.*, **70**, 95–123.

Kawakita, M., Yasuoka, K. & Kaziro, Y. (1980). Selective modification of functionally distinct sulfhydryl group of sarcoplasmic reticulum $Ca^{2+}$,$Mg^{2+}$-adenosine triphosphatase with N-ethylmaleimide. *J. Biochem.*, **87**, 609–17.

Klip, A., Reithmeier, R. A. F. & MacLennan, D. H. (1980). Alignment of the major tryptic fragments of the adenosine triphosphatase from sarcoplasmic reticulum. *J. Biol. Chem.*, **255**, 6562–8.

Knowles, A. F. & Racker, E. (1975). Formation of adenosine triphosphate from Pi and adenosine diphosphate by purified $Ca^{2+}$-adenosine triphosphatase. *J. Biol. Chem.*, **250**, 1949–51.

Kodama, T., Kurebayashi, N. & Ogawa, Y. (1980). Heat production and proton release during the ATP-driven Ca uptake by fragmented sarcoplasmic reticulum from bullfrog and rabbit skeletal muscle. *J. Biochem.*, **88**, 1259–65.

Kolassa, N., Punzengruber, C., Suko, J. & Makinose, M. (1979). Mechanism of calcium-independent phosphorylation of sarcoplasmic reticulum ATPase by orthophosphate. Evidence of magnesium-phosphoprotein formation. *FEBS Lett.*, **108**, 495–500.

Kurzmack, M. & Inesi, G. (1977). The initial phase of $Ca^{2+}$ uptake and ATPase activity of sarcoplasmic reticulum vesicles. *FEBS Lett.*, **74**, 35–7.

Kurzmack, M., Inesi, G., Tal, N. & Bernhard, S. A. (1981). Transient-state kinetic studies on the mechanism of furylacryloylphosphatase-coupled calcium ion transport with sarcoplasmic reticulum adenosine triphosphatase. *Biochemistry*, **20**, 486–91.

Kurzmack, M., Verjovski-Almeida, S. & Inesi, G. (1977). Detection of an initial burst of $Ca^{2+}$ translocation in sarcoplasmic reticulum. *Biochem. Biophys. Res. Commun.*, **78**, 772–6.

Labarca, P., Coronado, R. & Miller, C. (1980). Thermodynamic and kinetic studies of the gating behavior of a $K^+$-selective channel from the sarcoplasmic reticulum membrane. *J. Gen. Physiol.*, **76**, 397–424.

Lacapère, J.-J., Gingold, M. P., Champeil, P. & Guillain, F. (1981). Sarcoplasmic reticulum ATPase phosphorylation from inorganic phosphate in the absence of a calcium gradient. *J. Biol. Chem.*, **256**, 2302–6.

Lehninger, A. L. (1975). *Biochemistry*, 2nd edn. New York: Worth Publishers.

le Maire, M., Møller, J. V. & Gulik-Krzywicki, T. (1981). Freeze-fracture study of water-soluble, standard proteins and of detergent-solubilized forms of sarcoplasmic reticulum $Ca^{2+}$ ATPases. *Biochim. Biophys. Acta*, **643**, 115–25.

le Maire, M., Møller, J. V. & Tardieuw, A. (1981). Shape and thermodynamic parameters of a $Ca^{2+}$-dependent ATPase: a solution X-ray scattering and sedimentation equilibrium study. *J. Mol. Biol.*, **150**, 273–96.

le Maire, M., Jørgensen, K. E., Roigaard-Peterson, H. & Møller, J. V. (1976). Properties of deoxycholate solubilized sarcoplasmic reticulum $Ca^{2+}$-ATPase. *Biochemistry*, **15**, 5805–12.

Liu, S. C. & Worthington, C. R. (1974). Electron density levels of sarcoplasmic reticulum membranes. *Arch. Biochem. Biophys.*, **163**, 332–42.

McFarland, B. H. & Inesi, G. (1970). Studies of solubilized sarcoplasmic reticulum. *Biochem. Biophys. Res. Commun.*, **41**, 239–43.

McKinley, D. & Meissner, G. (1978). Evidence for a $K^+$, $Na^+$ permeable channel in sarcoplasmic reticulum. *J. Membr. Biol.*, **44**, 159–86.

MacLennan, D. H. (1970). Purification and properties of an adenosine triphosphatase from sarcoplasmic reticulum. *J. Biol. Chem.*, **245**, 4508–18.

MacLennan, D. H. & Holland, P. C. (1975). Calcium transport in sarcoplasmic reticulum. *Annu. Rev. Biophys. Bioeng.*, **4**, 377–404.

MacLennan, D. H., Seeman, P., Iles, G. H. & Yip, C. C. (1971). Membrane formation by the adenosine triphosphatase of sarcoplasmic reticulum. *J. Biol. Chem.*, **246**, 2702–10.

MacLennan, D. H. & Wong, P. T. S. (1971). Isolation of a calcium-sequestering protein from sarcoplasmic reticulum. *Proc. Natl. Acad. Sci. (USA)*, **68**, 1231–5.

MacLennan, D. H., Yip, C. C., Iles, G. H. & Seeman, P. (1973). Isolation of sarcoplasmic reticulum proteins. *Cold Spring Harbor Symp. Quant. Biol.*, **37**, 469–77.

Madeira, V. M. C. (1978). Proton gradient formation during transport of $Ca^{2+}$ by sarcoplasmic reticulum. *Arch. Biochem. Biophys.*, **185**, 316–25.

Makinose, M. (1969). The phosphorylation of the membranal protein of the sarcoplasmic vesicles during active calcium transport. *Eur. J. Biochem.*, **10**, 74–82.

Makinose, M. (1972). Phosphoprotein formation during osmo-chemical energy conversion in the membrane of the sarcoplasmic reticulum. *FEBS Lett.*, **25**, 113–15.

Makinose, M. & Hasselbach, W. (1971). ATP synthesis by the reverse of the sarcoplasmic calcium pump. *FEBS Lett.*, **12**, 271–2.

Martin, D. W. & Tanford, C. (1981). Phosphorylation of calcium adenosinetri-phosphatase by inorganic phosphate: van't Hoff analysis of enthalpy changes. *Biochemistry*, **20**, 4597–602.

Martonosi, A. (1969). Sarcoplasmic reticulum. VII. Properties of a phosphoprotein intermediate implicated in calcium transport. *J. Biol. Chem.* **244**, 613–620.

Martonosi, A. (1975). The mechanism of calcium transport in sarcoplasmic reticulum. In *Calcium Transport in Contraction and Secretion*, ed. E. Carafoli, F. Clementi, W. Drabikowski & A. Margreth, pp. 313–27. Amsterdam: North-Holland.

Martonosi, A., Donley, J. & Halpin, R. A. (1968). Sarcoplasmic reticulum. III. The role of phospholipids in the adenosine triphosphatase activity and $Ca^{2+}$ transport. *J. Biol. Chem.*, **243**, 61–70.

Masuda, H. & de Meis, L. (1973). Phosphorylation of the sarcoplasmic reticulum membrane by orthophosphate. Inhibition by calcium ions. *Biochemistry*, **12**, 4581–5.

Meissner, G. (1975). Isolation and characterization of two types of sarcoplasmic reticulum vesicles. *Biochim. Biophys. Acta*, **389**, 51–68.

Meissner, G. (1981). Calcium transport and monovalent cation and proton fluxes in sarcoplasmic reticulum vesicles. *J. Biol. Chem.*, **256**, 636–43.

Meissner, G., Conner, G. E. & Fleischer, S. (1973). Isolation of sarcoplasmic reticulum by zonal centrifugation and purification of $Ca^{2+}$-pump and $Ca^{2+}$ binding proteins. *Biochim. Biophys. Acta*, **298**, 246–69.

Meissner, G. & Fleischer, S. (1971). Characterization of sarcoplasmic reticulum from skeletal muscle. *Biochim. Biophys. Acta*. **241**, 356–78.

Meissner, G. & Young, R. C. (1980). Proton permeability of sarcoplasmic reticulum vesicles. *J. Biol. Chem.*, **255**, 6814–19.

Mermier, P. & Hasselbach, W. (1976). Comparison between strontium and calcium uptake by the fragmented sarcoplasmic reticulum. *Eur. J. Biochem.*, **69**, 79–86.

Migala, A., Agostini, B. & Hasselbach, W. (1973). Tryptic fragmentation of the calcium transport system in the sarcoplasmic reticulum. *Z. Naturf.*, **28**, 178–82.

Miki, K., Scott, T. L. & Ikemoto, N. (1981). A fluorescence probe study of the phosphorylation reaction of the calcium ATPase of sarcoplasmic reticulum. *J. Biol. Chem.*, **256**, 9382–5.

Miyamoto, H. & Racker, E. (1982). Mechanism of calcium release from skeletal sarcoplasmic reticulum. *J. Membr. Biol.*, **66**, 193–201.

Mitchinson, C., Wilderspin, A. F., Trinnaman, B. J. & Green, N. M. (1982). Identification of a labeled peptide after stoichiometric reaction of fluorescein isothiocyanate with the $Ca^{2+}$-dependent adenosine triphosphatase of sarcoplasmic reticulum. *FEBS Lett.*, **146**, 87–92.

Møller, J. V., Andersen, J. P. & le Maire, M. (1982). The sarcoplasmic reticulum $Ca^{2+}$-ATPase. *Mol. Cell. Biochem.*, **42**, 83–107.

Møller, J. V., Lind, K. E. & Andersen, J. P. (1980). Enzyme kinetics and substrate stabilization of detergent-solubilized and membraneous

($Ca^{2+}$ + $Mg^{2+}$)-activated ATPase from sarcoplasmic reticulum: effect of protein–protein interactions. *J. Biol. Chem.*, **255**, 1912–1920.

Møller, J. W., Mahrous, T. S., Andersen, J. P. & le Maire, M. (1982). Functional significance of quaternary organization of the sarcoplasmic reticulum $Ca^{2+}$-ATPase. *Z. Naturf.*, **37C**, 517–21.

Montecucco, C., Smith, G. X., Warren, G. B. & Metcalfe, J. C. (1977). A spin label assay of the annular lipids around a calcium transport protein. In *Structure and Function of Energy-transducing Membranes*, ed. K. Van Damm & B. F. Van Gelder, pp. 187–192. Amsterdam: Elsevier.

Morii, H. & Tonomura, Y. (1983). The gating behavior of a channel for $Ca^{2+}$-induced $Ca^{2+}$ release in fragmented sarcoplasmic reticulum. *J. Biochem.*, **93**, 1271–85.

Murphy, A. J. (1976). Arginyl residue modification of the sarcoplasmic reticulum ATPase protein. *Biochem. Biophys. Res. Commun.*, **70**, 1048–54.

Murphy, A. J. (1978). Effects of divalent cations and nucleotides on the reactivity of the sulfhydryl groups of sarcoplasmic reticulum membranes: evidence for structural changes occurring during the calcium transport cycle. *J. Biol. Chem.*, **253**, 385–9.

Nagano, K., Kanazawa, T., Mizuno, N., Tashima, Y., Nakao, T. & Nakao, M. (1965). Some acylphosphate-like properties of $^{32}P$-labeled sodium-potassium-activated adenosine triphosphatase. *Biochem. Biophys. Res. Commun.*, **19**, 759–64.

Nakamura, M. & Ohnishi, S. (1975). Organization of lipids in sarcoplasmic reticulum membranes and $Ca^{2+}$-dependent ATPase activity. *J. Biochem.*, **78**, 1039–45.

Nakamura, Y. & Tonomura, Y. (1982a). Changes in affinity for calcium ions with the formation of two kinds of phosphoenzyme in the $Ca^{2+},Mg^{2+}$-dependent ATPase of sarcoplasmic reticulum. *J. Biochem.*, **91**, 449–61.

Nakamura, Y. & Tonomura, Y. (1982b). The binding of ATP to the catalytic and regulatory site of $Ca^{2+},Mg^{2+}$-dependent ATPase of the sarcoplasmic reticulum. *J. Bioenerg. Biomemb.*, **14**, 21–32.

Ogawa, Y. & Ebashi, S. (1976). Ca-releasing action of $\beta,\gamma$-methylene adenosine triphosphate on fragmented sarcoplasmic reticulum. *J. Biochem.*, **80**, 1149–57.

Panet, R., Pick, U. & Selinger, Z. (1971). The role of calcium and magnesium in the adenosine triphosphatase reaction of sarcoplasmic reticulum. *J. Biol. Chem.*, **246**, 7349–56.

Panet, R. & Selinger, Z. (1972). Synthesis of ATP coupled to $Ca^{2+}$ release from sarcoplasmic reticulum vesicles. *Biochim. Biophys. Acta*, **255**, 34–42.

Pardee, A. B. (1968a). Membrane transport proteins. *Science*, **162**, 632–370.

Pardee, A. B. (1968b). Biochemical studies on active transport. *J. Gen. Physiol.*, **52**, 279s–88s.

Peachey, L. D. (1965). The sarcoplasmic reticulum and transverse tubules of the frog's sartorius. *J. Cell Biol.*, **25**, 209–31.

Peachey, L. D. & Schild, R. F. (1968). The distribution of the T-system along the sarcomeres of frog and toad sartorius muscles. *J. Physiol.*, **194**, 249–58.

Pick. U. (1981). Interaction of fluorescein isothiocyanate with nucleotide-binding site of the Ca-ATPase from sarcoplasmic reticulum. *Eur. J. Biochem.*, **121**, 187–95.

Pick. U. & Racker, E. (1979) Inhibition of the ($Ca^{2+}$)ATPase from sarcoplasmic reticulum by dicyclohexylcarbodiimide: evidence for location of the $Ca^{2+}$-binding site in a hydrophobic region. *Biochemistry*, **18**, 108–13.

Porter, K. R. & Palade. G. E. (1957). Studies on the endoplasmic reticulum. III. Its form and distribution in striated muscle cells. *J. Biophys. Biochem. Cytol.*, **3**, 269–99.

Post. R. L., Sen, A. K. & Rosenthal, A. S. (1965). A phosphorylated intermediate in adenosine triphosphate-dependent sodium and potassium transport across kidney membranes. *J. Biol. Chem.*, **240**, 1437–45.

Prager, R., Punzengruber, C., Kolassa, N., Winkler, F. & Suko, J. (1979). Ionized and bound calcium inside isolated sarcoplasmic reticulum of skeletal muscle and its significance in phosphorylation of adenosine triphosphatase by orthophosphate. *Eur. J. Biochem.*, **97**, 239–50.

Rauch, B., v. Chak, D. & Hasselbach, W. (1977). Phosphorylation by inorganic phosphate of sarcoplasmic membranes. *Z. Naturf.*, **32C**, 828–34.

Rayns, D. G., Devine, C. E. & Sutherland, C. L. (1975). Freeze fracture studies of membrane systems in vertebrate muscle. 1. Striated muscle. *J. Ultrastruct. Res.*, **50**, 306–21.

Saito, A., Wang, C. T. & Fleischer, S. (1978). Membrane asymmetry and enhanced ultrastructural detail of sarcoplasmic reticulum revealed with use of tannic acid. *J. Cell Biol.*, **79**, 601–16.

Scales, D. & Inesi, G. (1976). Assembly of ATPase protein in sarcoplasmic reticulum membranes. *Biophys. J.*, **16**, 735–51.

Scott, T. L. & Shamoo, A. E. (1982). Disruption of energy transduction in sarcoplasmic reticulum by trypsin cleavage of ($Ca^{2+} + Mg^{2+}$)-ATPase. *J. Membr. Biol.*, **64**, 137–44.

Shamoo, A. E., Ryan, T. E., Stewart, P. S. & MacLennan, D. H. (1976). Localization of ionophore activity in a 20,000-dalton fragment of the adenosine triphosphatase of sarcoplasmic reticulum. *J. Biol. Chem.*, **251**, 4147–54.

Shaw, T. I. (1954). 'Sodium and potassium movement in red cells'. Unpublished Ph.D. thesis, Cambridge University.

Shigekawa, M. & Dougherty, J. P. (1978). Reaction mechanism of $Ca^{2+}$-dependent ATP hydrolysis by skeletal muscle sarcoplasmic reticulum in the absence of added alkali metal salts. III. Sequential occurrence of ADP-sensitive and ADP-insensitive phosphoenzymes. *J. Biol. Chem.*, **253**, 1458–64.

Shigekawa, M. & Kanazawa, T. (1982). Phosphoenzyme formation from ATP in the ATPase of sarcoplasmic reticulum: effect of KCl or ATP and slow dissociation of ATP from precursor enzyme–ATP complex. *J. Biol. Chem.*, **257**, 7657–65.

Shigekawa, M. & Pearl, L. J. (1976). Activation of calcium transport in skeletal muscle sarcoplasmic reticulum by monovalent cations. *J. Biol. Chem.*, **251**, 6947–52.

Singer, S. J. & Nicolson, G. L. (1972). The fluid mosaic model of the structure of cell membrane. *Science*, **175**, 720–31.

Somlyo, A. V., Gonzalez-Serratos, H., Shuman, H., McClellan, G. & Somlyo, A. P. (1981). Calcium release and ionic changes in the sarcoplasmic reticulum of tetanized muscle: an electron-probe study. *J. Cell Biol.*, **90**, 577–94.

Souza, D. O. G. & de Meis, L. (1976). Calcium and magnesium regulation of phosphorylation by ATP and ITP in sarcoplasmic reticulum vesicles. *J. Biol. Chem.*, **251**, 6355–9.

Stewart, P. S. & MacLennan, D. H. (1974). Surface particles of sarcoplasmic reticulum membranes. Structural features of the adenosine triphosphatase. *J. Biol. Chem.*, **249**, 985–93.

Stewart, P. S., MacLennan, D. H. & Shamoo, A. E. (1976). Isolation and characterization of tryptic fragments of the adenosine triphosphatase of sarcoplasmic reticulum. *J. Biol. Chem.*, **251**, 712–19.

Sumida, M. & Sasaki, S. (1975). Inhibition of $Ca^{2+}$ uptake into fragmented sarcoplasmic reticulum by antibodies against purified $Ca^{2+}$,$Mg^{2+}$-dependent ATPase. *J. Biochem.*, **78**, 757–62.

Sumida, M. & Tonomura, Y. (1974). Reaction mechanism of the $Ca^{2+}$-dependent ATPase of sarcoplasmic reticulum from skeletal muscle. X. Direct evidence for $Ca^{2+}$ translocation coupled with formation of a phosphorylated intermediate. *J. Biochem.*, **75**, 283–97.

Sumida, M., Wang, T., Mandel, F., Froehlich, J. P. & Schwartz, A. (1978) Transient kinetics of $Ca^{2+}$ transport of sarcoplasmic reticulum: a comparison of cardiac and skeletal muscle. *J. Biol. Chem.*, **253**, 8772–7.

Tada, M., Yamamoto, T. & Tonomura, Y. (1978). Molecular mechanism of active calcium transport by sarcoplasmic reticulum. *Physiol. Rev.*, **58**, 1–79.

Takakuwa, Y. & Kanazawa, T. (1979). Slow transition of phosphoenzyme from ADP-sensitive to ADP-insensitive forms in solubilized $Ca^{2+}$,$Mg^{2+}$-ATPase of sarcoplasmic reticulum: evidence for retarded dissociation of $Ca^{2+}$ from the phosphoenzyme. *Biochem. Biophys. Res. Commun.*, **88**, 1209–16.

Takakuwa, Y. & Kanazawa, T. (1982). Reaction mechanism of ($Ca^{2+}$,$Mg^{2+}$)-ATPase of sarcoplasmic reticulum: the role of $Mg^{2+}$ that activates hydrolysis of the phosphoenzyme. *J. Biol. Chem.*, **257**, 426–31.

Takisawa, H. & Makinose, M. (1981). Occluded bound calcium on the phosphorylated sarcoplasmic transport ATPase. *Nature*, **290**, 271–3.

Takisawa, H. & Makinose, M. (1983). Occlusion of calcium in the ADP-sensitive phosphoenzyme of the adenosine triphosphatase of sarcoplasmic reticulum. *J. Biol. Chem.*, **258**, 2986–92.

Takisawa, H. & Tonomura, Y. (1978). Factors affecting the transient phase of the $Ca^{2+}.Mg^{2+}$-dependent ATPase reaction of sarcoplasmic reticulum from skeletal muscle. *J. Biochem.*, **83**, 1275–84.

Takisawa, H. & Tonomura, Y. (1979). ADP-sensitive and insensitive phosphorylated intermediates of solubilized $Ca^{2+}.Mg^{2+}$-dependent ATPase of the sarcoplasmic reticulum from skeletal muscle. *J. Biochem.*, **86**, 425–41.

Tenu, J.-P., Chelis, C., Leger, D. S., Carrette, J. & Chevallier, J. (1976). Mechanism of an active transport of calcium: ethoxy-formylation of sarcoplasmic reticulum vesicles. *J. Biol. Chem.*, **251**, 4322–9.

The, R. & Hasselbach, W. (1972). Properties of the sarcoplasmic ATPase reconstituted by oleate and lysolecithin after lipid depletion. *Eur. J. Biochem.*, **28**, 357–63.

Thorley-Lawson, D. A. & Green, N. M. (1973). Studies on the location and orientation of proteins in the sarcoplasmic reticulum. *Eur. J. Biochem.*, **40**, 403–13.

Thorley-Lawson, D. A. & Green, N. M. (1975). Separation and characterization of tryptic fragments from the adenosine triphosphatase of sarcoplasmic reticulum. *Eur. J. Biochem.*, **59**, 193–200.

Tong, S. W. (1977). The acetylated $NH_2$-terminus of Ca ATPase from rabbit skeletal muscle sarcoplasmic reticulum: a common $NH_2$-terminal acetylated methionyl sequence. *Biochem. Biophys. Res. Commun.*, **74**, 1242–8.

Tong, S. W. (1980). Studies on the structure of the calcium-dependent adenosine triphosphatase from rabbit skeletal muscle sarcoplasmic reticulum. *Arch. Biochem. Biophys.*, **203**, 780–91.

Tonomura, Y. (1972). The $Ca^{2+}$-$Mg^{2+}$-dependent ATPase and the uptake of $Ca^{2+}$ by the fragmented sarcoplasmic reticulum. In *Muscle Proteins, Muscle Contraction and Cation Transport*, ed. Y. Tonomura, pp. 305–56. Tokyo: University of Tokyo Press.

Tonomura, Y. & Morales, M. F. (1974). Change in state of spin labels bound to sarcoplasmic reticulum with change in enzymic state, as deduced from ascorbate-quenching studies. *Proc. Natl. Acad. Sci. (USA)*, **71**, 3687–91.

Vanderkooi, J. M., Ierokomas, A., Nakamura, H. & Martonosi, A. (1977). Fluorescence energy transfer between $Ca^{2+}$ transport ATPase molecules in artificial membranes. *Biochemistry*, **16**, 1262–7.

Verjovski-Almeida, S., Kurzmack, M. & Inesi, G. (1978). Partial reactions in the catalytic and transport cycle of sarcoplasmic reticulum ATPase. *Biochemistry*, **17**, 5006–13.

Vianna, A. L. (1975). Interaction of calcium and magnesium in activating and inhibiting the nucleoside triphosphatase of sarcoplasmic reticulum vesicles. *Biochim. Biophys. Acta*, **410**, 389–406.

Warren, G. B., Houslay, M. D., Metcalfe, J. C. & Birdsall, N. J. M. (1975). Cholesterol is excluded from the phospholipid annulus surrounding an active calcium transport protein. *Nature*, **255**, 684–7.

Warren, G. B., Toon, P. A., Birdsall, N. J. M., Lee, A. G. & Metcalfe, J. C. (1974). Reversible lipid titrations of the activity of pure adenosine triphosphatase-lipid complexes. *Biochemistry*, **13**, 5501–7.

Watanabe, T. & Inesi, G. (1982). Structural effects of substrate utilization on the adenosine triphosphatase chains of sarcoplasmic reticulum. *Biochemistry*, **21**, 3254–9.

Watanabe, T., Lewis, D., Nakamoto, R., Kurzmack, M., Fronticelli, C. & Inesi, G. (1981). Modulation of calcium binding in sarcoplasmic reticulum adenosinetriphosphatase. *Biochemistry*, **20**, 6617–25.

Webb, M. R. & Trentham, D. R. (1981). The stereochemical course of phosphoric residue transfer catalyzed by sarcoplasmic reticulum ATPase. *J. Biol. Chem.*, **256**, 4884–7.

Weber, A., Herz, R. & Reiss, I. (1966). Study of the kinetics of the calcium transport by isolated fragmented sarcoplasmic reticulum. *Biochem. Z.*, **345**, 329–69.

Yamada, S. & Ikemoto, N. (1980). Reaction mechanism of calcium-ATPase of sarcoplasmic reticulum: substrates for phosphorylation reaction and back reaction and further resolution of phosphorylated intermediates. *J. Biol. Chem.*, **255**, 3108–19.

Yamada, S., Sumida, M. & Tonomura, Y. (1972). Reaction mechanism of the $Ca^{2+}$-dependent ATPase of sarcoplasmic reticulum from skeletal muscle. VIII. Molecular mechanism of the conversion of osmotic energy to chemical energy in the sarcoplasmic reticulum. *J. Biochem.*, **72**, 1537–48.

Yamada, S. & Tonomura, Y. (1972). Reaction mechanism of the $Ca^{2+}$-dependent ATPase of sarcoplasmic reticulum from skeletal muscle. VII. Recognition and release of $Ca^{2+}$ ions. *J. Biochem.*, **72**, 417–25.

Yamada, S., Yamamoto, T., Kanazawa, T. & Tonomura, Y. (1971). Reaction mechanism of the $Ca^{2+}$-dependent ATPase of sarcoplasmic reticulum from skeletal muscle. VI. Co-operative transition of ATPase activity during the initial phase. *J. Biochem.*, **70**, 279–91.

Yamamoto, T., Takisawa, H. & Tonomura, Y. (1979). Reaction mechanisms for ATP hydrolysis and synthesis in the sarcoplasmic reticulum. *Curr. Top. Bioenerg.*, **9**, 179–236.

Yamamoto, T. & Tonomura, Y. (1967). Reaction mechanism of the $Ca^{2+}$-dependent ATPase of sarcoplasmic reticulum from skeletal muscle. I. Kinetic studies. *J. Biochem.*, **62**, 558–75.

Yamamoto, T. & Tonomura, Y. (1968). Reaction mechanism of the $Ca^{2+}$-dependent ATPase of sarcoplasmic reticulum from skeletal muscle. II. Intermediate formation of phosphoryl protein. *J. Biochem.*, **64**, 137–45.

Yamamoto, T. & Tonomura, Y. (1976). Chemical modification of the $Ca^{2+}$-dependent ATPase of sarcoplasmic reticulum from skeletal muscle.

II. Use of 2,4,6-trinitrobenzenesulfonate to show functional movements of the ATPase molecule. *J. Biochem.*, **79**, 693–707.

Yamamoto, T. & Tonomura, Y. (1982). Desensitization to $Mg^+$ of the phosphoenzyme in sarcoplasmic reticulum $Ca^{2+}$-ATPase by the addition of a nonionic detergent and the restoration of sensitivity after its removal. *J. Biochem.*, **91**, 477–86.

Yamamoto, T., Yoda, A. & Tonomura, Y. (1971). Reaction mechanism of the $Ca^{2+}$-dependent ATPase of sarcoplasmic reticulum from skeletal muscle. IV. Hydroxamate formation from a phosphorylated intermediate and 2-hydroxy-5-nitrobenzyl hydroxylamine. *J. Biochem.*, **69**, 807–9.

Yantorno, R. E., Yamamoto, T. & Tonomura, Y. (1983). Energy transfer between fluorescent dyes attached to $Ca^{2+}$,$Mg^{2+}$-ATPase in the sarcoplasmic reticulum. *J. Biochem.*, **94**, 1137–45.

Yates, K. W. & Duance, V. C. (1976). The binding of nucleotides and bivalent cations to the calcium- and magnesium-ion-dependent adenosine triphosphatase from rabbit muscle sarcoplasmic reticulum. *Biochem. J.*, **159**, 719–28.

Zimniak, P. & Racker, E. (1978). Electrogenicity of $Ca^{2+}$ transport catalyzed by the $Ca^{2+}$-ATPase from sarcoplasmic reticulum. *J. Biol. Chem.*, **253**, 4631–7.

# 8

# $Na^+,K^+$-ATPase
# in the plasma membrane

## Plasma membrane and active transport of $Na^+$ and $K^+$

### Active transport of $Na^+$ and $K^+$

Cytoplasms in almost all animal cells contain a high concentration of $K^+$ and a low concentration of $Na^+$, while the tissue fluid and serum that surround cells contain a low concentration of $K^+$ and a high concentration of $Na^+$. The concentration gradients of these cations across the plasma membrane disappear when energy metabolism of the cell is blocked. With these findings it has been suggested that $Na^+$ and $K^+$ are continuously transported across the plasma membrane against an electrochemical gradient by using metabolic energy.

It has been shown that metabolic inhibitors such as cyanide and 2,4-dinitrophenol inhibit the $Na^+$ efflux from nerve and muscle cells and that oxidative metabolism serves as the energy source for this $Na^+$ efflux in these cells. In erythrocytes, the active transport of $Na^+$ and $K^+$ ceases when an inhibitor of glycolysis is added or when glucose is removed. This indicates that glycolysis serves as the energy source for this transport in erythrocytes. These observations led to a suggestion that ATP is the source of energy for the active transport of these cations. Direct evidence for this was first reported in the squid giant axon by Hodgkin & Keynes (1955). Later, more conclusive evidence was reported by Brinley & Mullins (1967, 1968), who showed that the rate of $Na^+$ efflux decreases markedly when the giant axon is perfused without any supply of metabolic energy. The rate of the efflux was restored when ATP was perfused into the axon, but it was not restored when high-energy phosphate compounds other than ATP, such as acetyl phosphate and phosphoenol pyruvate, were added. In addition, it was shown in erythrocytes and squid giant axons that, coupled to this $Na^+$ efflux, $K^+$ is actively transported through the membrane in the opposite direction (Hodgkin, 1951; Glynn,

1956; Post & Jolly, 1957). This coupled transport occurs with a stoichiometry of 3 mol of $Na^+$ and 2 mol of $K^+$/mol of ATP hydrolyzed (Sen & Post, 1964; Whittam & Agar, 1965; Garrahän & Glynn, 1967). With these findings, the presence of the ATP-dependent active transport of $Na^+$ and $K^+$ (the $Na^+$-pump) in the plasma membrane became generally accepted.

In vertebrates, the amount of energy used for the active transport of $Na^+$ and $K^+$ is considerable. In man, for example, the energy used for kidney function is 6–7% of the total basal energy metabolism, and that used by the nervous system is another 20%; in both cases the majority of this energy is used for the transport of $Na^+$ and $K^+$. Indeed, the active transport of these cations plays many roles in physiologically important processes: it keeps the intracellular ion environment constant, provides the energy required for nerve excitation and supports the secondary active transport of sugars and amino acids.

## $Na^+, K^+$-ATPase

The biochemical study of the $Na^+$-pump began with the discovery of a $Na^+, K^+$-ATPase in the plasma membrane (Skou, 1957). Skou found an ATPase in the microsomal fraction of crab nerve tissues that exhibits high activity only when $Na^+$, $K^+$ and $Mg^{2+}$ coexist in the reaction mixture. Furthermore, this ATPase was shown to be completely inhibited by the cardiac glycoside, ouabain, a specific inhibitor of the active transport of $Na^+$ and $K^+$ (Post *et al.*, 1960; Skou, 1960).

The following observations provide rational evidence that the $Na^+, K^+$-ATPase itself is indeed the $Na^+$ pump (Post *et al.*, 1960; Bonting & Caravaggio, 1963; Glynn & Karlish, 1975): (1) the activities of the $Na^+, K^+$-ATPase and the active $Na^+$ and $K^+$ transport are both found in the membrane fraction, (2) in membranes from various tissues, the higher the transport activity, the higher the $Na^+, K^+$-ATPase activity, (3) both activities show the same concentration dependence and specificities for the substrates and effectors (nucleotides and monovalent cations), (4) cardiac glycosides inhibit both activities with the same concentration dependence and (5) the phospholipid vesicles reconstituted from purified $Na^+, K^+$-ATPase are capable of actively transporting $Na^+$ and $K^+$ (Hilden & Hokin, 1975). The transport ratio of $Na^+$ to $K^+$ upon hydrolysis of ATP in the reconstituted vesicles is about 3:2, which is identical with that in the intact cells. In the following sections, the coupling mechanism between the active $Na^+$ and $K^+$ transport and the

ATPase reaction in the plasma membrane will be discussed in detail, but other topics will be treated only briefly. The reader is referred to the excellent reviews and monographs by Askari, 1974; Glynn & Karlish, 1975*a*; Post, 1979; Skou & Norby, 1979; Jørgensen, 1982; Hoffman & Forbush, 1983.

## Structure of Na+,K+-ATPase

### *Purification*

The Na+,K+-ATPase has been purified from various tissues, such as the kidney and brain of mammals, rectal gland of sharks, electric organ of electric eels and salt gland of ducks (Jørgensen, 1982). The Na+,K+-ATPase is an integral membrane protein embedded deeply in the plasma membrane. Nakao and co-workers (Nakao *et al.*, 1963) succeeded in purifying the ATPase by treatment of brain microsomes with 2 M sodium iodide. This sample was nearly free of ouabain-insensitive ATPase and had an activity of 3–5 μmol/mg·min at 37 °C. Later, a number of purification methods using detergents were developed. These methods can be classified into two types: 'negative purification', in which most of the peripheral proteins are dissociated from the plasma membrane and only the membrane fraction containing the Na+,K+-ATPase is collected, and 'positive purification', in which the Na+,K+-ATPase is solubilized and separated from insoluble materials. A typical example of 'negative purification' is the method developed by Jørgensen (1974) that uses sodium dodecylsulfate (SDS). In this method, microsomes of the outer medulla of kidney are treated with SDS in the presence of ATP and then centrifuged in a sucrose density gradient. An ATPase preparation with specific activity of 32–37 μmol/mg·min at 37 °C is obtained by this method. Because of its simplicity and yield of high-activity preparation, this method is now widely used. One of the 'positive purification' methods was developed by Esmann *et al.* (1980). Microsomes from the outer medulla of kidney are solubilized with a non-ionic detergent, dodecyloctaethyleneglycol monoether, $C_{12}E_8$, and then fractionated by a molecular sieving technique.

### *Protein components and chemical structure*

The purified Na+,K+-ATPase contains three types of polypeptide chains, $\alpha$, $\beta$ and $\gamma$. Its lipid content is 0.8 g/g of protein.

The $\alpha$-subunit, with a molecular weight of about 100 000, is called the catalytic subunit (Kyte, 1971). It can be phosphorylated by ATP to form a phosphoenzyme (EP, to be discussed later in this chapter). Like the $Ca^{2+}.Mg^{2+}$-ATPase of SR, the $\beta$-carboxyl group of its aspartic residue is phosphorylated (Post & Kume, 1973). The amino acid sequence in the vicinity of the aspartic residue, the phosphorylation site, has been reported to be –Thr (or Ser)–Asp–Lys–. Chemical modification experiments show that cysteine, tyrosine and arginine, in addition to the aspartic residue, are also essential for the enzyme activity. The same essential groups are reported for the other energy-transducing ATPases, such as myosin-ATPase and the $Ca^{2+}.Mg^{2+}$-ATPase of SR. In contrast to the $Ca^{2+}.Mg^{2+}$-ATPase of SR, the primary structure of the $\alpha$-subunit of the $Na^{+}.K^{+}$-ATPase remains virtually unknown, but a high content of hydrophobic amino acids is characteristic of the $\alpha$-subunit (Jørgensen, 1982).

A protein polymorphism is present in the catalytic subunit: in addition to the $\alpha$-type seen in the kidney and many other tissues, the $\alpha^{+}$-type with a molecular weight 2000 higher than that of $\alpha$-type has been isolated from rat brain axonemma (Sweadner, 1979). The ATPase containing this $\alpha^{+}$-subunit has lower affinities for $K^{+}$ and ouabain than the ATPase containing the $\alpha$-subunit (Urayama & Nakao, 1979). In addition, the $\alpha$-subunit has functional heterogeneity that may be due to its heterogeneous components which were reported recently (Yamaguchi & Post, 1983).

The $\beta$-subunit is a glycoprotein with a molecular weight of about 50 000, composed of a protein portion of about 40 000 and a sugar chain of about 10 000. The sugar chain contains galactosamine, sialic acid, fucose, mannose, galactose and glucose. This glycoprotein constitutes the integral subunit of $Na^{+}.K^{+}$-ATPase, as is widely accepted from the following facts: (1) the glycoprotein coexists with the $\alpha$-subunit in all highly purified preparations, (2) antibodies to this glycoprotein inhibit the activity of $Na^{+}.K^{+}$-ATPase and the binding of ouabain to the enzyme, (3) studies on the biosynthesis of $Na^{+}.K^{+}$-ATPase reveal that the rates of synthesis and degradation are the same for both the $\alpha$-subunit and the glycoprotein and (4) the $Na^{+}.K^{+}$-ATPase containing both $\alpha$- and $\beta$-subunits can be purified by affinity chromatography using resins to which concanavalin A has been attached. However, the physiological role of this $\beta$-subunit in cation transport still remains unknown.

The $\gamma$-subunit is a proteolipid with a molecular weight of 11 000. It exists at a ratio of about 1 mol/mol of $\alpha$-subunit in all of the $Na^{+}.K^{+}$-

ATPases isolated from sharks, birds and mammals, and its amino acid composition is very similar among these sources (Reeves, Collins & Schwartz, 1980; Hardwicke & Freytag, 1981). Since some of the cardiac glycoside affinity labels can bind to the γ-subunit (Forbush, Kaplan & Hoffman, 1978; Rogers & Lazdunski, 1979), it may be one of the components of the Na⁺,K⁺-ATPase. However, its removal from membrane preparations does not affect the ATPase activity (Hardwicke & Freytag, 1981).

## Structural features of the ATPase in the membrane

Electron micrographs of negatively stained ATPase preparations reveal the presence of structures (3–5 nm in size) protruding from both sides of the membrane (Vogel *et al.*, 1977). It has been claimed that the cytoplasmic-side projection is composed of the stalk knob derived from the α-subunit, and the extracellular-side projection is composed of the ill-defined structure derived from β-subunit. Cardiac glycosides bind to the α-subunit from the extracellular side (Ruoho & Kyte, 1974), whereas the phosphorylation of the α-subunit by ATP and the binding of anti-α-subunit antibodies occur on the cytoplasmic side of the membrane. Both the cardiac glycosides and anti-α-subunit antibodies bind simultaneously to the membrane (Kyte, 1975). These observations suggest that the α-subunit protrudes from both sides of the membrane asymmetrically.

Jørgensen (1982) studied the arrangement of the α-subunit in the hydrophobic region of the lipid bilayer and proposed the model shown in Fig. 8.1. In this model he took account of the following: from which side of the membrane trypsin attacks the polypeptide chain; to which tryptic subfragment the photoaffinity label of ouabain binds and to which subfragment the lipid-soluble, photoreactive probes such as iodonaphthylazide and adamantane diazirine bind.

There are only a few studies on the tertiary structure of Na⁺,K⁺-ATPase in the membrane. It has been speculated, from analysis of the infrared absorption spectra of purified Na⁺,K⁺-ATPase, that about 20% of the polypeptide chains are in α-helix and about 25% in β-sheet structures (Brazhnikov, Chetverin & Chirgadze, 1978). Recently, it was reported that purified membrane preparations of Na⁺,K⁺-ATPase form two-dimensional crystals in the presence of high concentrations of vanadate and $MgCl_2$ (Jørgensen, 1982). In electron micrographs these crystals are arranged in a non-orthogonal lattice of 41 Å × 52 Å. Further studies along these lines are expected to gain insights into the tertiary structure of $Na^+,K^+$-ATPase.

## Oligomeric structure of the ATPase

The $Na^+,K^+$-ATPase is composed mainly of $\alpha$- and $\beta$-subunits. However, how many of each subunit form the functional unit of the $Na^+$-pump has not yet been established.

The amount of each of the subunits was determined by analyzing the amino acid compositions of protein bands that were separated on SDS-gel electrophoresis (Craig & Kyte, 1980; Peterson & Hokin, 1981). As a result, it was found that the molar ratio of $\alpha$-subunit to $\beta$-subunit in the purified $Na^+,K^+$-ATPase is 1:1.

The next problem that needs to be solved is how many 1:1 complexes of $\alpha$- and $\beta$-subunits combine to form a functional unit of the $Na^+,K^+$-ATPase in the membrane. The $Na^+,K^+$-ATPase can be solubilized by a detergent without loss of its activity. Taking advantage of this fact, Brotherus, Møller & Jørgensen (1981) solubilized the purified $Na^+,K^+$-ATPase with $C_{12}E_{18}$ and observed that the activity unit of the enzyme has a molecular weight of $179\,000 \pm 9000$ under the conditions used. It was also reported that the maximal amounts of ATP, TNP-ATP (fluorescent ATP analogue) and ouabain that bind to the purified $Na^+,K^+$-

Fig. 8.1 Model for the arrangement of the $\alpha$-subunit in the membrane-bound $Na^+,K^+$-ATPase. P shows the position of the aspartyl phosphate formed from ATP. The upward pointing arrows mark the sites of primary tryptic cleavage in the presence of NaCl or KCl. Modified from Jørgensen (1982).

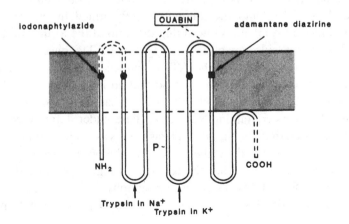

ATPase, as well as the amount of phosphate incorporated into the enzyme, are each almost 1 mol/mol of the m.w. 160 000–170 000 protein (Moczydlowski & Fortes, 1981*a*, *b*; Peters *et al.*, 1981; Matsui *et al.*, 1983). These results suggest that the 1 : 1 complex of the $\alpha$- and $\beta$-subunits is sufficient to be the minimal unit of the ATPase activity. However, it is possible that the functional unit of ATPase activity found in non-ionic detergent does not have the capacity for cation transport in the membrane. Thus, studies on the functional unit of Na⁺,K⁺-ATPase possessing cation transport activity in the membrane are important.

Results suggesting that the Na⁺,K⁺-ATPase functions as a unit of $\alpha_2\beta_2$ or $\alpha_4\beta_4$ were obtained using the radiation inactivation method, the crosslinking method or the ligand-binding method. Using the radiation inactivation method, based on the target theory, Nakao *et al.* (1967) estimated that the size of the functional unit of the Na⁺,K⁺-ATPase in the membrane is 500 000. Later, Kepner & Macey (1968) modified this method and obtained a value of 190 000–300 000.

From the results of experiments using the crosslinking reagent, cupric phenanthroline, Kyte (1975) and Giotta (1976) suggested that the Na⁺,K⁺-ATPase exists as a dimer of $\alpha$-subunits in the membrane. On the other hand, Askari & Huang (1980), using a similar method but under different conditions, suggested that the Na⁺,K⁺-ATPase exists as a tetramer of $\alpha$-subunits in the membrane.

In analyzing the results of crosslinking experiments, however, it must be generally noted that the formation of crosslinks depends not only on the number of subunits that constitute the functional unit but also, possibly, on such factors as the movement of the ATPase molecules and the reactivity of functional groups.

Kudoh *et al.*, (1979) measured the amount of ouabain bound to the Na⁺,K⁺-ATPase and the amount of EP formed in the presence of Na⁺,Mg² and ATP and showed that the ratio of the former to the latter is 2:1 over a wide range of Na⁺, Mg²⁺ and ATP concentrations. This observation suggests that the ATPase molecule in the membrane is at least a dimer of $\alpha$-subunits and also that an interaction exists between the subunits. Hansen *et al.* (1979) made similar measurements on microsomal fractions containing Na⁺,K⁺-ATPase under different conditions. He suggested that the Na⁺,K⁺-ATPase exists as a tetramer of $\alpha$-subunits in the membrane and, furthermore, that the occurrence of interaction between $\alpha$-subunits within the tetramer is highly possible. In support of these observations, electron micrographs of freeze-fractured preparations of purified Na⁺,K⁺-ATPase showed the presence of the particles in the membrane with sizes corresponding to either $\alpha_2\beta_2$ or $\alpha_4\beta_4$ (Haase & Koepsell, 1979; Maunsbach, Skriver & Jørgensen, 1979).

### Role of phospholipid in the interaction between the subunits of the ATPase

As described above, an interaction between $\alpha$-subunits may take place during the catalytic function of Na+,K+-ATPase. Studies on the role of phospholipid in the function of Na+,K+-ATPase have provided interesting insights into the interaction between the subunits. Ottolenghi (1979) analyzed the pattern of reactivation of the delipidated Na+,K+-ATPase by titrating it with phospholipid. He found that a partial reaction, the K+-PNPPase activity (a phosphatase activity using $p$-nitrophenyl phosphate as the substrate; to be discussed below), is reactivated in a hyperbolic manner in the lower concentration range of phospholipid, while the Na+,K+-ATPase activity reappears in a sigmoidal manner in its higher concentration range. From these results he proposed the following model: the Na+,K+-ATPase is composed of two subunits, and K+-PNPPase activity is recovered when the phospholipid-binding site on either of the two subunits is saturated, whereas the recovery of Na+,K+-ATPase activity occurs only when the phospholipid-binding sites on both of the subunits are saturated. According to this model, the functional unit of Na+,K+-ATPase is a dimer ($\alpha_2\beta_2$) and the phospholipids are required for both the ATPase activity and the subunit interaction within this dimer. In connection with this model, there have been interesting reports that the K+-PNPPase activity is brought about by a smaller structural unit ($\alpha_1\beta_1$) than the dimer ($\alpha_2\beta_2$) required for the activity of Na+,K+-ATPase (Winter & Moss, 1979).

# Reaction mechanism of Na+,K+-ATPase

Since Na+,K+-ATPase serves not only as an energy transducer for the Na+-pump but also as a carrier of Na+ and K+, the elucidation of the reaction mechanism of ATP hydrolysis is important for the understanding of the molecular mechanism of this pump. In this section, general properties of the steady-state Na+,K+-ATPase reaction will be discussed and the elementary steps of the ATPase reaction will be described.

## Na+, K+-ATPase in the plasma membrane

### Steady-state reaction

*Optimal conditions*   The Na+,K+-ATPase shows the maximal activity in the presence of 100–150 mM NaCl, 5–10 mM KCl, 5 mM MgCl₂ and 3 mM ATP at pH 7.2–7.6. The highest activity reported so far is about 50 μmol/mg·min at 37 °C (Jørgensen, 1982). A trace of Na+,K+-independent, ouabain-insensitive and $Mg^{2+}$-dependent ATPase activity is detected even in the purified Na+,K+-ATPase preparation. Similar to the $Ca^{2+}$,$Mg^{2+}$-ATPase of SR, the Arrhenius plot of Na+,K+-ATPase activity bends at about 20 °C. The enthalpy and entropy of activation in the range of higher temperatures are smaller than those in the range of lower temperatures.

*Specificities for substrates and cations*   The Na+,K+-ATPase shows a high specificity for the substrate nucleotide, and the relative activities with ATP, dATP, ITP, CTP, GTP and UTP are reported to be 100:49:2.4:2.3:0.6:0.6, respectively.

   The requirement for Na+ is quite strict. Only several per cent of the activity in the presence of Na+ is observed when Li+ is substituted for Na+. On the contrary, K+ can be replaced by several kinds of monovalent cations, with the effectiveness being Tl+ > K+ ≈ Rb+ > NH₄+ > Cs+ > Li+.

   $Mg^{2+}$ is essential for the hydrolysis of ATP. $Mn^{2+}$, $Co^{2+}$, $Fe^{2+}$, $Ca^{2+}$ or $Ni^{2+}$ can be substituted for $Mg^{2+}$. However, since the enzyme activity decreases by an order of magnitude or more when these cations are used as an activator, they act as inhibitors in the presence of $Mg^{2+}$. On the other hand, $Zn^{2+}$, $Cu^{2+}$, $Ba^{2+}$, $Sr^{2+}$, $Be^{2+}$ and $Pb^{2+}$ cannot replace $Mg^{2+}$ and they act only as inhibitors. Among these cations, $Ca^{2+}$ is of particular interest, since $Ca^{2+}$ alters the properties of EP (Fukushima & Post, 1978). $Pb^{2+}$ can induce the formation of EP even in the absence of Na+ (Siegel & Fogt, 1977).

*Activating effect of ATP*   ATP participates in the reaction of Na+,K+-ATPase not only as a substrate but also as a regulatory factor (Kanazawa, Saito & Tonomura, 1967, 1970; Neufeld & Levy, 1969). Using brain microsomes, these investigators found that the Lineweaver–Burk plot of the ATPase bends sharply downward at around 10 μM ATP as the concentration of ATP increases (Fig. 8.2). This results in $V_{max}$ and $K_m$

values far greater than those in the lower concentration range of ATP. They also found that this bend disappears when the concentration of K+ is lowered. From these two different Michaelis constants, Kanazawa *et al.* (1967, 1970) inferred the existence of two ATP-binding sites, the 'high-affinity site' ($K_m = 0.3–3.6\,\mu\text{M}$) and the 'low-affinity site' ($K_m = 0.3–0.5\,\text{mM}$). They proposed that the former is the catalytic site and the latter is the regulatory site. Similar results have also been observed in erythrocyte, kidney and other materials (Glynn & Karlish, 1975; Robinson, 1976; Sakamoto & Tonomura, 1980). Moreover, the stimulation of the activity by a high concentration of ATP has also been reported for the $Ca^{2+}$,$Mg^{2+}$-ATPase of SR, as has been described in Chapter 7. The mechanism involved in this activation will be discussed in detail later in this chapter.

*Partial reactions*   The Na+,K+-ATPase catalyzes the Na+-dependent ATP–ADP exchange (Fahn, Koval & Albers, 1966*b*; Fahn *et al.*, 1966*a*) and the K+-dependent [$^{18}$O]Pi–[$^{18}$O]$H_2O$ exchange (Dahms & Boyer, 1974) as partial reactions of the Na+,K+-ATPase.

Fig. 8.2   Lineweaver–Burk plot of the steady-state rate of ATP hydrolysis by Na+,K+-ATPase. 1 mM $MgCl_2$, 140 mM NaCl and 0.6 mM KCl were used.

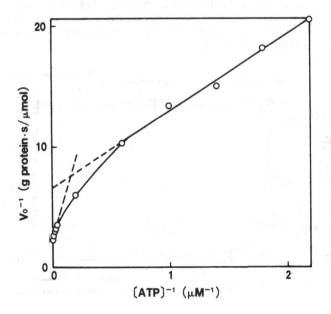

The Na$^+$,K$^+$-ATPase exhibits a phosphatase activity (*p*-nitrophenyl phosphate (PNPP) is usually used as a substrate) in the presence of K$^+$ and Mg$^{2+}$. Na$^+$ and ATP inhibit this phosphatase activity when either Na$^+$ or ATP is present but promote it when coexisting in the presence of a relatively low concentration of K$^+$ (Nagai & Yoshida, 1966).

## *Intermediates and elementary reactions*

The kinetic properties of various reaction intermediates, in the absence of K$^+$ or in the presence of low concentrations of K$^+$, have been studied by Post, Albers and other investigators. From such studies, the reaction scheme shown in Fig. 8.3 has been proposed. Although this scheme requires modifications when both Na$^+$ and K$^+$ are present at optimal concentrations, it can explain many elementary reactions. Thus, the kinetics of Na$^+$,K$^+$-ATPase are discussed here according to this scheme.

*Formation of EP*   Post, Sen & Rosenthal (1965) found that the Na$^+$,K$^+$-ATPase is phosphorylated by ATP in the presence of Na$^+$ and Mg$^{2+}$, whereas the dephosphorylation is strongly promoted by K$^+$. The

Fig. 8.3 Reaction scheme I for Na$^+$,K$^+$-ATPase. The subscripts, 1 and 2, designate the different conformations of the enzyme. $^{Na}E_1ATP$ and $E_2ATP$ are the enzyme–substrate complexes with loosely bound ATP, whereas $^{Na}E^*ATP$ represents the complex with tightly bound ATP.

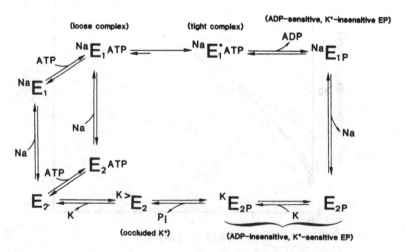

phosphorylation is achieved by the formation of an acylphosphate bond between the γ-phosphate group of ATP and the β-carboxyl group of a specific aspartic residue of the α-subunit (Nagano *et al.*, 1965; Kyte, 1971; Post & Kume, 1973).

The identity of this EP as the reaction intermediate was finally proven by kinetic analysis of the formation and decomposition of EP (Fig. 8.4) (Kanazawa *et al.*, 1970). When the reaction is initiated by the addition of ATP to the enzyme in the presence of $Na^+$, $Mg^{2+}$ and a small amount of $K^+$, the formation of EP occurs without a lag phase and approaches steady state in 1 s. The liberation of Pi from ATP, on the other hand, occurs in two phases, an initial lag phase and the subsequent steady phase. The initial phase of Pi liberation agrees well with the curve calculated by assuming that the specific turnover rate of EP in the initial phase is the same as that in the steady state. Furthermore, Kanazawa *et al.* (1970) observed that the ATP-concentration dependence of the

Fig. 8.4 Initial rates of Pi liberation and EP formation. The curve drawn through the open circles is the time course of Pi liberation calculated from the rate of EP formation, assuming the turnover rate of EP to be 4.4/sec. Reactants: 1 μM ATP, 1 mM $MgCl_2$ and: (filled circles, open circles), 140 mM NaCl + 0.6 mM KCl; (filled triangles), 14 mM KCl; (open triangles), no NaCl + no KCl.

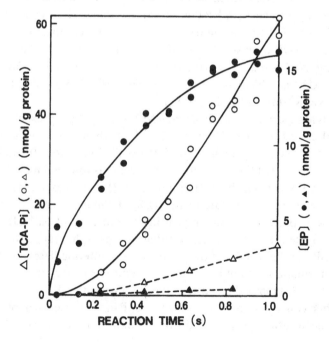

rate of EP formation is not affected by drastic changes in the concentration of $Na^+$ (e.g., from 140 mM to 0.5 mM), and they suggested that $Na^+$ and ATP bind to the enzyme to form the ternary complex in random sequence. As described in Chapter 7, Fossel *et al.* (1981) showed that this EP is not an artificial product formed by adding TCA or a detergent to terminate the reaction but is a true intermediate formed *in situ* during the ATPase reaction. They demonstrated that (1) a peak appears at +17 ppm in the $^{31}$P-NMR signal after the addition of $[\gamma\text{-}^{31}P]ATP$ to the $Na^+,K^+$-ATPase in the presence of $Na^+$ and $Mg^{2+}$, (2) the peak is observed at the same position in the NMR signal of a phosphoaspartate, a model of the phosphorylated site of the enzyme and (3) this peak disappears upon addition of $K^+$.

$E_1P$ *and* $E_2P$  The presence of two types of EP intermediates during the ATPase reaction was first suggested by Fahn *et al.* (1966*a, b*) from the analysis of the ATP–ADP exchange reaction. Later, Post *et al.* (1969) provided conclusive evidence for this by analyzing the kinetics of the decay of EP that was formed in the presence of $Na^+$ and $Mg^{2+}$ but not in the presence of $K^+$. They showed that a rapid decay of EP occurs when $K^+$ is added to this enzyme after chase with unlabeled ATP (Fig. 8.5). This decay of EP does not occur when ADP is added in place of $K^+$. The EP formed under such conditions is now called the $K^+$-sensitive, ADP-insensitive EP, or $E_2P$. On the other hand, the EP formed with the NEM-treated enzyme is not affected by $K^+$ but decays rapidly when ADP is added to the enzyme. This EP is now called the ADP-sensitive, $K^+$-insensitive EP, or $E_1P$.

EP is almost exclusively $E_2P$ when $Na^+$ and $Mg^{2+}$ are present at physiological concentrations. However, it is $E_1P$ when (1) the concentration of $Na^+$ is extremely high, e.g., 500 mM (Tobin *et al.*, 1973; Hara & Nakao, 1981), (2) the concentration of $Mg^{2+}$ is low (Fahn *et al.*, 1966*b*; Stahl, 1968; Yamaguchi & Tonomura, 1977), (3) oligomycin is present, (4) the enzyme is pretreated with NEM (Fahn *et al.*, 1966*a*) or (5) the enzyme is pretreated with trypsin in the presence of $Na^+$ (Jørgensen, 1975). Since the same aspartic residues are phosphorylated in both $E_1P$ and $E_2P$, it is believed that these two types of intermediates are different in conformation but not in chemical structure (Post, 1979).

Post *et al.* (1969) showed that the ADP-sensitive $E_1P$ is formed first and then converted to the $K^+$-sensitive $E_2P$ in the presence of physiological concentrations of $Na^+$ and $Mg^{2+}$. Fukushima & Nakao (1981) also showed that the conversion of $E_1P$ to $E_2P$ occurs in the presence of $Na^+$ and $Ca^{2+}$.

*Decomposition of $E_2P$*   As described above, $K^+$ promotes the decomposition of $E_2P$. Post, Hegyvary & Kume (1972) showed that the dissociation of $K^+$ from the enzyme occurs slowly after Pi has been released from the enzyme. On the basis of these findings, they proposed that the $K^+$ remains tightly bound to the enzyme even after the decomposition of $E_2P$, i.e., the $K^+$ is 'occluded', shown by $^K{>}E$. They also observed that this $K^+$-dissociation step is promoted by a high concentration of ATP. Their suggestion has since been supported by the direct measurement of the rate of $Rb^+$ dissociation from the enzyme (Glynn & Richards, 1980) and by comparing the time courses of $Rb^+$ binding and EP decomposition (Yamaguchi & Tonomura, 1980*c*).

Fig. 8.5 Sensitivity of EP to splitting in the presence of ADP or $K^+$. (*A*) The native enzyme was first phosphorylated by the addition of [³²P]ATP. At zero time the radioactive ATP was chased with a 100-fold excess of unlabeled ATP and additions were made after 5 s, as indicated. (*B*) The same experiment as in (*A*) was performed using NEM-treated enzyme. From Post *et al.*, 1969.

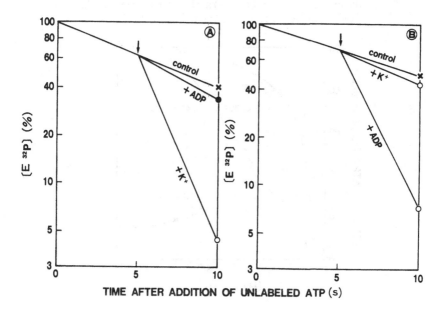

*Reverse reaction* According to the ATPase reaction scheme shown in Fig. 8.3, ADP must dissociate from the enzyme first, followed by Pi. Taniguchi & Post (1975) provided evidence for this sequence by synthesizing ATP by the reverse reaction. They showed that the $K^+$-sensitive $E_2P$ is formed when $Mg^{2+}$ and a small amount of $Na^+$ are added to the enzyme in the presence of Pi and that the addition of a large amount of $Na^+$ causes the conversion of $E_2P$ to $E_1P$. The addition of ADP to this $E_1P$ causes the formation of ATP. Figure 8.6 shows the dependences of the amount of $E_2P$ remaining and the amount of ATP synthesized on the concentration of $Na^+$. They concluded that, because the conversion of $E_2P$ to $E_1P$ requires a high concentration of $Na^+$, the affinity of $E_2P$ for $Na^+$ is much lower than that of $E_1P$. This conclusion was later supported by a direct demonstration that the binding affinity of $Na^+$ for the ATPase is much higher in $E_1P$ than in $E_2P$ (to be discussed later in this chapter; Yamaguchi & Tonomura, 1980*c*).

Fig. 8.6 Effects of concentration of $Na^+$ on synthesis of ATP from $E_2P$. After formation of $E_2P$ from E and Pi, ADP and CDTA (1,2-cyclohexylenedinitrilotetraacetic acid) with various quantities of $Na^+$ were added to start the synthesis of ATP. After 2 s (crosses, open circles) or 80 s (filled circles) the reaction was stopped and the amounts of $E_2P$ (crosses) and ATP (open circles, filled circles) were measured. From Taniguchi & Post, 1975.

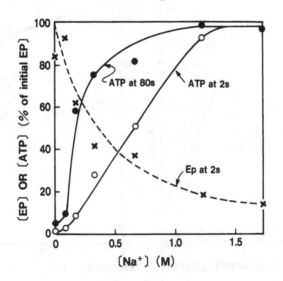

*Two routes in the ATPase reaction*  As described above, many experimental results can be explained by the reaction scheme shown in Fig. 8.3. However, it must be noted that most of the experiments described so far have been carried out under nonphysiological conditions in which $K^+$ is not present. Thus, to elucidate the reaction mechanism of $Na^+,K^+$-ATPase, it is necessary to clarify the pathway by which ATP is hydrolyzed under physiological conditions.

From the results of experiments concerning the formation and decomposition of EP and the binding of ADP to the enzyme, we have proposed the reaction scheme shown in Fig. 8.7. In this scheme the reaction is assumed to proceed through an inner pathway ($E_2 \rightarrow {}^{Na}E_1ATP \rightarrow {}^{Na}E_{1P}^{ADP} \rightarrow {}^{K}E_2^{ADP} \rightarrow {}^{K>}E_2 \rightarrow E_2$) under physiological conditions in which $Na^+$, $K^+$ and $Mg^{2+}$ are present, but through an outer pathway ($E_2 \rightarrow {}^{Na}E_1ATP \rightarrow {}^{Na}E_{1P}^{ADP} \rightarrow {}^{Na}E_{1P} \rightarrow E_{2P} \rightarrow {}^{K>}E_2 \rightarrow E_2$) in the absence of $K^+$. The outer pathway is in agreement with the reaction scheme (I) in Fig. 8.3. Kanazawa *et al.* (1970) determined the rate constant for the decomposition of EP during ATPase reaction as the ratio of the velocity of Pi release in the steady state, $v_0$, to the concentration of EP, [EP], $v_0/[EP]$. They also determined the apparent first-order rate constant, $k_d$, for the decay of EP. In this experiment, $E^{32}P$ formation was started by the addition of $[^{32}P]ATP$ in the presence of $Mg^{2+}$, $Na^+$ and $K^+$, and then stopped at various times either by a chase with a high concentration of unlabeled ATP or by chelation of $Mg^{2+}$ with EDTA to measure the

Fig. 8.7 Reaction scheme II for $Na^+,K^+$-ATPase. ${}^{Na}E_{1P}^{ADP}$ is ADP-insensitive, $K^+$-insensitive EP with bound ADP. ${}^{K}E_2^{ADP}$ is the dephosphoenzyme with bound ADP. The other designations are the same as in Fig. 8.3.

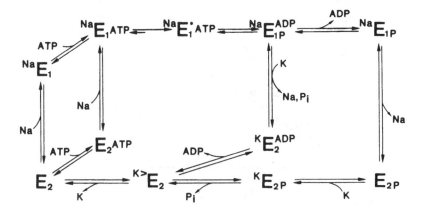

time course of the decay of $E^{32}P$. Figure 8.8 shows the comparison between $v_0/[EP]$ and $k_d$ at various concentrations of KCl. It can be seen that the ratio of $v_0/[EP]$ to $k_d$ increases as the concentration of $K^+$ increases. This result can be explained by assuming the presence of an enzyme–ATP complex that is in equilibrium with EP:

$$E_1{}^*ATP \underset{}{\overset{K}{\rightleftharpoons}} E_{1P}^{ADP} \overset{k}{\longrightarrow} E + Pi + ADP$$

where $K$ is the equilibrium constant between $E_1{}^*ATP$ and $E_{1P}^{ADP}$ and $k$ is the rate constant for the release of Pi and ADP from $E_{1P}^{ADP}$. Here, $k = v_0/[EP]$ and $k_d = k/(1 + K)$. Thus, the observed increase in the ratio of $k/k_d$ means that $K^+$ shifts the equilibrium toward $E_1{}^*ATP$. The following experimental results support this mechanism: (1) The time course of Pi release following the addition of EDTA coincides with the time course of EP decay. However, the amount of Pi released is larger than

Fig. 8.8 Dependence of EP decomposition on KCl concentration. $k_d$ is the rate constant of EP decay (see text). $v_0/[EP]$ is the ratio of the rate of Pi release in the steady state ($v_0$) to the concentration of EP([EP]). Concentrations of ATP: (open circles, filled circles), 1.1 μM; (open squares, filled squares), 31 μM; (open triangles, filled triangles), 301 μM.

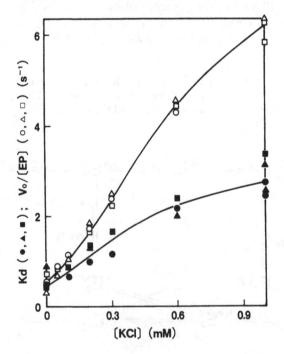

that of EP decayed, with the ratio of the former to the latter being the same as the ratio of $v_0/[EP]$ to $k_d$; (2) when K$^+$ is added along with EDTA, a portion of EP decays rapidly and the rest decays slowly in the first-order reaction. This rapid decay of EP can be quantitatively explained in terms of the K$^+$-induced increase in $K$. Fukushima & Tonomura (1973) provided more-direct evidence by showing that the rapid decrease in the amount of EP caused by the simultaneous addition of K$^+$ and EDTA accompanies the formation of corresponding amounts of ATP.

Blostein (1975) also reported that the amount of Pi released following the termination of EP formation by the addition of EDTA is much larger than the amount of EP decayed. Klodos & Nørby (1979) reported that the value of $k_d$ is smaller than that of $v_0/[EP]$ in the presence of not only K$^+$ but also Li$^+$. These results also support the above mechanism.

Hexum, Samson & Himes (1970) and Robinson (1976) showed that Na$^+$,K$^+$-ATPase is competitively inhibited by ADP and noncompetitively inhibited by phosphate in the presence of a sufficient concentration of K$^+$. These results suggest that Pi is released first from $E_{1P}^{ADP}$ in the presence of both Na$^+$ and K$^+$. On the other hand, from kinetic analysis of the ADP inhibition, Sakamoto & Tonomura (1980) reported that the inhibition is competitive over a wide range of K$^+$ concentrations and is noncompetitive only in the absence of K$^+$. These findings indicate that the reaction of Na$^+$,K$^+$-ATPase proceeds via two pathways differing in the order of release of ADP and Pi. In one pathway, Pi is released first. This is the main pathway under physiological conditions in which a high concentration of K$^+$ is present. In the other pathway, ADP is released first. This pathway operates only in the absence of K$^+$.

Direct evidence for the above two-route mechanism was provided by Yamaguchi & Tonomura (1978). They measured the amount of ADP bound to the enzyme in the steady state in the presence of a sufficient amount of ATP-regenerating system (Fig. 8.9). In the absence of K$^+$, the amount of EP is larger than that of bound ADP. When the concentration of K$^+$ increases, however, only the amount of EP decreases markedly, becoming smaller than the amount of bound ADP. In addition, the ATPase activity remains proportional to the amount of bound ADP over a wide range of K$^+$ concentrations. These results strongly suggest that, under physiological conditions, i.e., in the presence of optimal amounts of both Na$^+$ and K$^+$, the main pathway is the route in which $E_{1P}^{ADP}$ is formed first and Pi is then released to form $E_2^{ADP}$.

However, the order in which ADP and Pi dissociate from $E_{1P}^{ADP}$ may be dependent not only on the reaction conditions but also on the structure of substrate. When the fluorescent ATP analogue formycin triphosphate (FTP) is used as the substrate, FDP (formycin diphosphate) dissociates first from the $E_{1P}^{FDP}$ complex, followed by the formation of $E_{1P}$ even in the presence of $K^+$ (Karlish, Yates & Glynn, 1976).

*Role of lipid in the elementary reactions*   It is generally accepted that phospholipids such as phosphatidyl serine and phosphatidyl inositol play an essential role in the function of $Na^+,K^+$-ATPase. The role of phospholipid in the elementary steps of the ATPase reaction was first studied in $Na^+,K^+$-ATPase from beef brain by Taniguchi & Tonomura (1971). They treated the enzyme preparation with phospholipase A and examined the effects of delipidation on the formation and decomposition of EP. It was found that the rate of EP formation is reduced by as

Fig. 8.9 Dependences on the KCl concentration of bound ADP, EP and ATPase activity ($v_0$) in the steady state. Creatine kinase and creatine phosphate were used as an ATP-regenerating system.

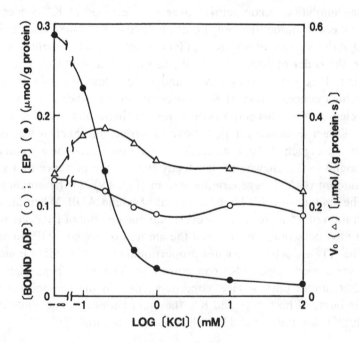

much as 40% by the phospholipase treatment, while the amount of EP in the steady state remains constant. The original rate of EP formation can be restored by the re-addition of phospholipid. It was also found that treatment of the enzyme with phospholipase A resulted in a reduction in the sensitivity of EP to K$^+$. Furthermore, the rate of EP decomposition is also reduced by as much as 93% of the control. Thus, it has become clear that phospholipid is required for both the phosphorylation step and dephosphorylation step and is particularly essential to the EP decomposition step.

Hegyvary *et al.* (1980) reported that the Na$^+$,K$^+$-ATPase prepared from cardiac muscle exhibits the K$^+$-PNPPase activity but not the Na$^+$,K$^+$-ATPase activity when 80–90% of the total phospholipid has been removed from the ATPase. They also reported that, although this delipidated preparation is capable of forming only E$_1$P from ATP, it is still capable of forming E$_2$P from Pi. These results suggest that the inactivation of the ATPase by delipidation is due to the prevention of the conversion of E$_1$P to E$_2$P. This is supported by the finding that the synthesis of ATP from E$_2$P is inhibited by the delipidation of ATPase preparations (Taniguchi & Post, 1975).

# Molecular mechanism of Na$^+$,K$^+$-ATPase

## *Cation exchanges catalyzed by Na$^+$,K$^+$-ATPase*

In the study of the Ca$^{2+}$,Mg$^{2+}$-ATPase of SR it has been possible to obtain direct information on the mechanism of coupling between the elementary steps of the ATPase reaction and Ca$^{2+}$ translocation across the membrane, since the partial reactions of ATP hydrolysis and the transport of Ca$^{2+}$ can be simultaneously measured with FSR. In the case of the Na$^+$-pump, on the other hand, studies using reconstituted vesicles have only recently begun and most of these studies are still concerned at present with the overall transport of Na$^+$ and K$^+$. 'Ghost' cells prepared from erythrocytes by reversible hemolysis provide a useful system for studying partial reactions of cation transport and ATPase reaction, since these cells are capable of controlling intracellular ion composition. Glynn and co-workers (see reviews by Glynn & Karlish, 1975*a*, *b*) have greatly contributed to this line of research. They classified the behavior of the Na$^+$-pump into five modes as shown in Fig. 8.10.

*Normal transport*   As described earlier in this chapter, efflux of 3 mol of the intracellular Na+ and concomitant influx of 2 mol of the extracellular K+ coupled to the hydrolysis of 1 mol of ATP take place under physiological conditions. No other ions are transported in a manner directly coupled to the transport of these two cations. Thus, this transport is electrogenic (Dixon & Hokin, 1980).

*Reversed transport*   When the ratio of [ADP] × [Pi] to [ATP] is increased and the concentration gradients of Na+ and K+ across the membrane are made far steeper than those under physiological conditions, the direction of cation transport is reversed and ATP is formed from ADP and Pi. It must be noted here that the synthesis of ATP by the reverse reaction of Na+,K+-ATPase was achieved even without the concentration gradient of Na+ and K+ across the membrane by the use of leaky vesicles (see section above on the reverse reaction).

*Na+–Na+ exchange*   When Na+ is present both inside and outside the cell and K+ is absent outside, an exchange reaction takes place

Fig. 8.10 The five modes of behavior of the Na+-pump. From Glynn & Karlish, 1975*b*.

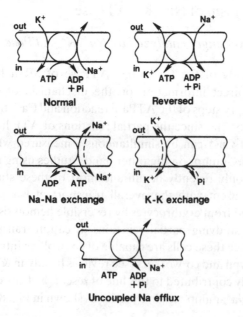

between the extracellular Na$^+$ and intracellular Na$^+$. The transport ratio is about 1:1, and the affinity for Na$^+$ is much higher inside than outside the cell. For this reaction to take place, Mg$^{2+}$, ADP and ATP must be present inside the cell. This reaction is accompanied by the ATP–ADP exchange in which the $\gamma$-phosphate of ATP is transferred to ADP but not accompanied by the net hydrolysis of ATP (Cavieres & Glynn, 1976). However, since oligomycin inhibits the Na$^+$–Na$^+$ exchange but not the ATP–ADP exchange, it is assumed that the release of ADP from the enzyme takes place prior to the release of Na$^+$ (see Fig. 8.3):

$$^iNaE*ATP \xrightleftharpoons{\quad\text{ADP}\quad} \,^{Na}E_1P \xrightleftharpoons[\text{oligomycin}]{\qquad\qquad} E_2P + {}^ONa$$

*K$^+$–K$^+$ exchange*  When K$^+$ is present both inside and outside the cell and Na$^+$ is absent inside, the K$^+$–K$^+$ exchange takes place with a transport ratio of about 1:1. The affinity of the enzyme for K$^+$ is low at the inner surface and high at the outer surface of the cell membrane. This pattern is opposite to that of the affinity of the enzyme for Na$^+$. This exchange reaction requires intracellular Mg$^{2+}$ and is stimulated by intracellular Pi. This K$^+$–K$^+$ exchange is accompanied by the K$^+$-dependent Pi–H$_2$O exchange (Dahms & Boyer, 1974), as shown in the scheme given below. Both exchange reactions are stimulated by a high concentration of intracellular ATP. The unhydrolyzable ATP analogs, adenylyl ($\beta,\gamma$-methylene)-diphosphonate (AMPPCP) and adenylyl-imido-diphosphate (AMPPNP), can replace ATP in this stimulatory effect. This ATP is thought to correspond to the ATP that stimulates the release of K$^+$ from the enzyme with occluded K$^+$($^{K}{>}E_2$) (see Fig. 8.3):

$$^iK^+ + E_1 \xrightleftharpoons[\text{ATP stimulation}]{\qquad\qquad} {}^{K}{>}E_2 \xrightleftharpoons[\substack{Pi \quad H_2O}]{\substack{Pi \quad H_2O}} KE_2P$$

*Uncoupled Na$^+$ efflux*  When Na$^+$ is present inside the cell and neither Na$^+$ nor K$^+$ is present outside, efflux of Na$^+$ takes place without the inward movement of cations. This efflux is accompanied by the hydrolysis of ATP, and is thought to be coupled to the 'Na$^+$-ATPase activity' that is found in the absence of K$^+$ and presence of Na$^+$.

From these five modes of cation movement observed by Glynn and co-workers, it has been concluded that (1) the Na$^+$-pump is reversible,

(2) the outward movement of $Na^+$ is coupled to the formation of EP and (3) the inward movement of $K^+$ is coupled to the decomposition of EP.

## Affinity changes in the ligand-binding sites

*Bindings of $Na^+$ and $K^+$*　From the study using a sedimentation technique, Kaniike *et al.* (1976) reported that 3 mol of $Na^+$/mol of ouabain-binding sites bind to the ATPase with the apparent dissociation constant ($K_d$) of 0.2 mM. Using a similar technique, Matsui *et al.* (1977) reported that 2 mol of $K^+$/mol of ouabain-binding sites bind to the enzyme with a $K_d$ of 0.05 mM and suggested the presence of co-operativity between these $K^+$ bindings. The ratio of the amount of bound $Na^+$ to that of bound $K^+$ agrees well with the transport ratio of $Na^+$ to $K^+$.

The binding of $Na^+$ and $K^+$ to the ATPase has also been studied using an ion-selective electrode (Hastings, 1977; Hara & Nakao, 1979) and ion-exchange resin (Beauge & Glynn, 1979). Furthermore, measurement of cation binding during the ATPase reaction using a membrane filtration technique was carried out by Yamaguchi & Tonomura (1979, 1980a, b, c). As shown in Fig. 8.11, 3 mol of $Na^+$-binding sites are present-/mol of active center, and $Na^+$ binds co-operatively (Hill coefficient about 2.5–3.0) to these sites with $K_d$ of 0.2–0.3 mM. On the other hand, the binding sites for $K^+$ are present at the ratio of 2 mol/mol of active center, and $K^+$, $Rb^+$ and $Na^+$ bind co-operatively (Hill coefficient of 1.5–2.0) to these sites with $k_d$ of 0.044, 0.024 and 2.8 mM, respectively.

Knowing whether $Na^+$ and $K^+$ bind consecutively or simultaneously to the ATPase is important for the elucidation of the molecular mechanism of the active transport of $Na^+$ and $K^+$ (Garahan & Garray, 1976). Various reaction intermediates to which $Na^+$ and $K^+$ are simultaneously bound have been suggested from kinetic studies in the steady state (Gache, Rossi & Lazdunski, 1977; Wang, Lindenmayer & Schwartz, 1977; Grosse *et al.*, 1979). Yamaguchi & Tonomura (1980a) showed that 2.5 mol of $Na^+$ and 2 mol of $Rb^+$/mol of active center can bind simultaneously to the $Na^+$-binding sites and $K^+$-binding sites, respectively. Matsui and Homareda (1982), on the other hand, reported that $Na^+$ and $K^+$ do not bind simultaneously to the cation-binding sites. However, since all of these measurements were carried out in the absence of $Mg^{2+}$ and ATP, it is still unknown whether the cation-binding sites are occupied simultaneously or consecutively by $Na^+$ and $K^+$ during the ATPase reaction.

*Affinity changes in the Na⁺ and K⁺ binding sites* As described earlier in this chapter, it has been suggested that the affinities of the cation-binding sites for Na⁺ and K⁺ change during ATPase reaction. To obtain direct evidence for this suggestion, Yamaguchi & Tonomura (1979, 1980c) determined the changes in the amounts of bound Na⁺ and K⁺ during EP formation. As shown in Fig. 8.12, they found that the formation of $E_2P$ is accompanied by a pronounced decrease in the affinity of the Na⁺-binding site for Na⁺ and, contrary, a pronounced

Fig. 8.11 Dependences of the amounts of bound Na⁺ and Rb⁺ on the concentrations of free Na⁺ and Rb⁺. (*A*) The Na⁺ bindings in the presence (filled circles, filled triangles) and absence (open circles, open triangles) of 30 mM KCl were measured with (open triangles, filled triangles) and without (open circles, filled circles) ATP. (*B*) The Rb⁺ bindings in the presence (filled circles) and absence (open circles) of 100 mM KCl without ATP.

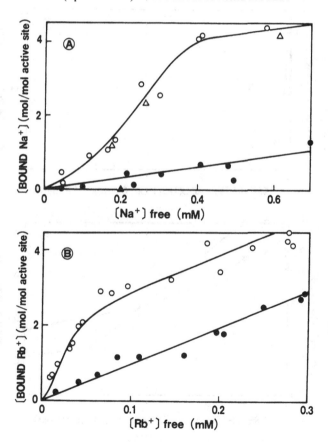

increase in the affinity of the K⁺-binding site for K⁺. They also found that, during the $E_2P$ formation, the affinity of the K⁺-binding site for Rb⁺ increases 28-fold and the Hill coefficient remains unchanged. On the other hand, they showed with the NEM-treated enzyme that the affinities of the cation-binding sites for Na⁺ and Rb⁺ do not change during the formation of the ADP-sensitive $E_1P$, and they concluded that the conversion of the ADP-sensitive $E_1P$ to the K⁺-sensitive $E_2P$ is accompanied by the release of Na⁺ from the Na⁺-binding site. This conversion is also accompanied by an increase in the affinity of the K⁺-binding site for K⁺. This conclusion is consistent with the observation

Fig. 8.12 (*A*) K⁺ binding to the enzyme and (*B*) Na⁺ release from the enzyme upon EP formation. The Na⁺-dependent EP (filled circles) represents the amounts of EP formed in the presence of NaCl.

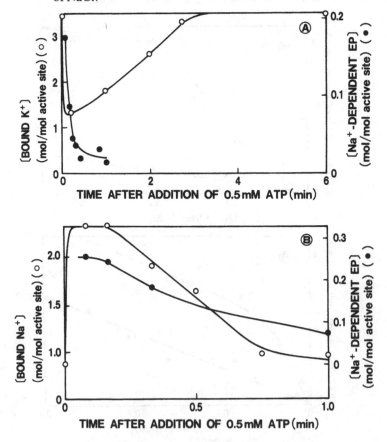

that the equilibrium between $E_1P$ and $E_2P$ is shifted toward $E_1P$ by an extremely high concentration of $Na^+$ (Tobin *et al.*, 1973; Taniguchi & Post, 1975; Hara & Nakao, 1981). It was also observed that $Rb^+$ remains bound for some time after the decomposition of EP. This shows that the enzyme still has a high affinity for $K^+$ even after the decomposition of $E_2P$, as has been suggested from kinetic studies by Post *et al.* (1972) (discussed earlier in this chapter).

*Binding of ATP*  Using a flow dialysis technique, Hegyvary & Post (1971) and Nørby & Jensen (1971) measured the binding of ATP to the $Na^+$,$K^+$-ATPase in the absence of divalent cations. They found that 1 mol of ATP/mol of active center binds to the enzyme with a $K_d$ of $0.12$–$0.22\ \mu M$ in the absence of $K^+$, that the $K_d$ increases as the concentration of $K^+$ increases ($K_d = 0.7\ \mu M$ at 3 mM $K^+$) and that $Na^+$ antagonizes this effect of $K^+$. These findings show that 1 mol of the high-affinity ATP-binding site is present/mol of active center and that $K^+$ reduces the affinity of the binding site for ATP.

Furthermore, Hegyvary & Post (1971) observed that more than 1 mol of ATP/mol of active center binds to the enzyme when ATP is present at a high concentration. They concluded that the enzyme contains the low-affinity ATP-binding site in addition to the high-affinity ATP-binding site. Later, using an improved membrane filtration technique, Yamaguchi & Tonomura (1980*b*) examined the binding of ATP to the enzyme. They found, per mol of active center, at least 2 mol of low-affinity ATP-binding sites ($k_d \geqq 0.2$ mM) as well as 1 mol of the high-affinity ATP-binding site ($K_d = 0.5\ \mu M$). These low-affinity and high-affinity ATP-binding sites correspond well with the regulatory ($K_m = 0.3$–$0.5$ mM) and catalytic ($K_m = 0.3$–$3.6\ \mu M$) ATP-binding sites estimated from kinetic studies of the $Na^+$,$K^+$-ATPase (see earlier section in this chapter on activating effect of ATP).

Beauge & Glynn (1979, 1980) developed a rapid method to measure the binding of ligands using a resin column and showed that the release of $Rb^+$ from the enzyme with occluded $Rb^+$ ($^{Rb>}E_2$) is accelerated by a high concentration of ATP. Yamaguchi & Tonomura (1980*b*) also showed that $Na^+$ binding to $^{K>}E_2$ is slow in the absence of ATP but is markedly accelerated by the binding of ATP to the low-affinity ATP-binding sites. These findings indicate that the transition from $^{K>}E_2$ to $E_1$ is accelerated by the binding of ATP to the low-affinity ATP-binding sites. The ATP-induced activation of $Na^+$,$K^+$-ATPase described previously can be attributed to this acceleration of the transition from $^{K>}E_2$ to $E_1$.

However, as suggested by Moczydlowski & Fortes (1981$a,b$), it should be noted that the activation of the ATPase at a high concentration of ATP can be explained by the mechanism in which the binding of ATP to a catalytic site with low affinity for ATP in the $E_2$ form accelerates the transition from $^{K>}E_2$ to $E_1$.

*Binding of $Mg^{2+}$*   No one has yet succeeded in directly measuring the binding of $Mg^{2+}$ to the Na+,K+-ATPase. However, O'Connor & Grisham (1979), taking advantage of the fact that $Mn^{2+}$ can replace $Mg^{2+}$ in catalyzing the Na+,K+-ATPase reaction, studied the properties of the $Mn^{2+}$-binding site by measuring changes in ESR signal. They reported that $Mn^{2+}$ binds to the Na+,K+-ATPase competitively with $Mg^{2+}$ and that the enzyme contains 1 mol of high-affinity $Mn^{2+}$-binding sites ($K_d = 0.21$ μM)/mol of ouabain-binding sites. In addition, the value of this $K_d$ was found to be about the same as the $Mn^{2+}$ concentration that gives half-maximum activation of the Na+,K+-ATPase. These results suggest that this high-affinity $Mn^{2+}$-binding site is identical with the $Mg^{2+}$-binding site required for the Na+,K+-ATPase reaction.

As has already been described in the preceding section, the binding of ATP to the Na+,K+-ATPase occurs without $Mg^{2+}$ but the formation

Fig. 8.13 Distances between ligand-binding sites in Na+,K+-ATPase. From O'Connor & Grisham, 1980.

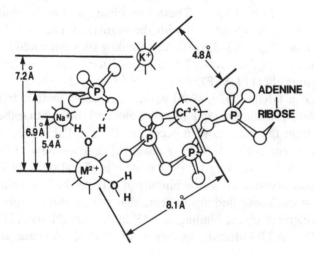

Table 8.1 *Two conformations of Na$^+$,K$^+$-ATPase*

| Characteristic | E$_1$ conformation | E$_2$ conformation |
|---|---|---|
| Favored monovalent cation | Na$^+$ | K$^+$ |
| Number of cation-binding sites for translocation | 3 | 2 |
| Favored carrier sidedness | cytoplasmic | external |
| Favored cation exchange | Na$^+$–Na$^+$ | K$^+$–K$^+$ |
| ATP affinity | high | low |
| Favored molecular exchange | ATP–ADP | Pi–H$_2$O |
| Favored EP | E$_1$P | E$_2$P |

of EP requires Mg$^{2+}$ (Post *et al.*, 1969; Kanazawa *et al.*, 1970). Once E$_1$P and E$_2$P are formed, the Mg$^{2+}$ bound to the enzyme becomes difficult to remove by treatment with a chelator. Fukushima & Nakao (1980) showed that E$_1$P and E$_2$P can be formed even when Ca$^{2+}$ is used in place of Mg$^{2+}$ and that, unlike Mg$^{2+}$, Ca$^{2+}$ can easily be removed from E$_1$P and E$_2$P by a chelator. In addition, Ca$^{2+}$ was found to be removed by a chelator more easily from E$_1$P than from E$_2$P. These results suggest that the affinity of the enzyme for Mg$^{2+}$ is increased upon conversion of E$_1$P to E$_2$P.

O'Connor & Grisham (1980) determined the distances between these ligand-binding sites with ESR and NMR techniques using $^7$Li$^+$, $^{205}$Rb$^+$, Mn$^{2+}$, $^{31}$P and CrATP from Na$^+$, K$^+$, Mg$^{2+}$, Pi and MgATP, respectively. The results they obtained are summarized in Fig. 8.13. All the findings described so far show that the Na$^+$,K$^+$-ATPase molecule contains specific binding sites for each of Na$^+$, K$^+$, Mg$^{2+}$ and ATP.

## Conformational changes of the ATPase

As described previously, kinetic studies and ligand-binding studies suggest that the Na$^+$,K$^+$-ATPase reaction proceeds through two conformational states, E$_1$ and E$_2$, in a cycle of ATP hydrolysis. The results are summarized in Table 8.1. In this section, it will be shown that the conformational changes do, in fact, take place during ATPase reaction.

*Changes in susceptibility to trypsin*   Jørgensen (1975) studied the inactivation of the ATPase during tryptic digestion and the changes in the patterns of SDS-PAGE. He found that the time course of inactivation is monophasic and fast in the presence of $K^+$, whereas it is biphasic and slow in the presence of $Na^+$. Patterns of SDS-PAGE revealed that in the presence of $K^+$ the $\alpha$-subunit is dissociated into two fragments with molecular weights of 58 000 and 41 000 (see Fig. 8.1) In the presence of $Na^+$, on the other hand, the gel patterns failed to show any change that corresponds to the first phase of the biphasic inactivation, but they revealed that, corresponding to the second phase, the $\alpha$-subunit is dissociated into a 77 000 fragment and a small fragment. These results can be accounted for by the difference in the conformation of the ATPase in the membrane; the comformation in the presence of $K^+$ corresponds to the $E_2$ form and the conformation in the presence of $Na^+$ corresponds to the $E_1$ form (see Fig. 8.3 and Table 8.1).

When a high concentration of ATP is added in the presence of $K^+$, the pattern of the time course of tryptic inactivation changes to biphasic, a pattern similar to that seen in the presence of $Na^+$ (Jørgensen & Petersen, 1979). This observation agrees well with the results of kinetic and ligand-binding studies (see previous sections of this chapter) that showed a high concentration of ATP accelerates the conversion of $^{K>}E_2$ to $E_1$. With these additional experiments they showed that the conformation of $E_2P$ is similar to that of $E_2$.

*Changes in the fluorescence of tryptophan*    Karlish & Yates (1978) found that the fluorescence intensity of the intrinsic tryptophan residues of purified $Na^+,K^+$-ATPase is 2–3% higher in the presence of $K^+(E_2)$ than in the presence of $Na^+(E_1)$. They also found that the fluorescence in the presence of $Na^+$ increases to the level in $E_2$ when the ATPase reaction is initiated by the addition of ATP in the presence of $Na^+$, $K^+$ and $Mg^{2+}$, and it decreases again to the original low level in $E_1$ upon disappearance of ATP. These findings demonstrate the occurrence of change from $E_1$ conformation to $E_2$ conformation during ATP hydrolysis.

Jørgensen & Petersen (1979) and Jorgen & Karlish (1980) showed by the determination of tryptophan fluorescence that the conformations of $E_1$ and $E_2$ are similar to those of $E_1P$ and $E_2P$, respectively. Furthermore, the measurement of the time course of the change in this intrinsic fluorescence using the stopped-flow technique revealed that the conversion of $^{K>}E_2$ to $E_1$ induced by the addition of $Na^+$ is markedly accelerated by a high concentration of ATP. This effect of ATP was also observed with the unhydrolyzable ATP analogs, AMPPCP and AMPPNP (Beauge & Glynn, 1980).

*Studies on the conformational changes using fluorescent probes*   It has recently become possible to obtain more detailed information about the conformational change of the $Na^+,K^+$-ATPase by using fluorescent probes such as $S$-mercuric-$N$-dansyl cystein (Harris & Stahl, 1977), fluorescein isothiocyanate (FITC) (Karlish, 1980; Hedgyvary & Jørgensen, 1981) and eosin (Skou & Esmann, 1981). In particular, FITC binds covalently to the purified $Na^+,K^+$-ATPase at the ratio of 1 mol/mol of active center and inhibits the ATPase activity. This inhibition is protected by ATP. The FITC-labelled enzyme shows a 20% increase in fluorescence when $K^+$ is added and, therefore, provides a useful tool for detecting various conformations.

Taniguchi, Suzuki & Iida (1982, 1983) have recently found that $N$-[$P$-(2-benzimidazolyl)phenyl]maleimide (BIPM) can be used as a fluorescent probe that does not strongly inhibit the ATPase activity. They observed the largest difference in the fluorescent intensity (10%) during the conversion of the enzyme state from $E_1P$ to $E_2P$. They also demonstrated the transient decrease and the subsequent increase in the fluorescence in the initial phase of the ATPase reaction. These changes corresponded well to the conversions of the enzyme state from $E_1$ to $E_1P$ and from $E_1P$ to $E_2P$.

## Molecular models of the Na+-pump

*Kinetic model*   It is possible to design a probable kinetic model of the $Na^+$-pump by combining the studies on the reaction mechanism, cation transport, ligand binding and conformational changes of $Na^+,K^+$-ATPase. This model is schematically shown in Fig. 8.14. It assumes that the $Na^+,K^+$-ATPase molecule contains three $Na^+$-binding sites on the inner (cytoplasmic side) surface and two $K^+$-binding sites on the outer surface of the membrane. When $Na^+$ and ATP are added in the absence of $Mg^{2+}$, 3 mol of $Na^+$ bind to these $Na^+$-binding sites on the inner surface of the membrane and form $^{Na3}E^{ATP}$. When $Mg^{2+}$ is added, the enzyme is phosphorylated and forms $^{Na3}E_1P$ or $^{Na3}E_{1P}^{ADP}$. Since the $Na^+$-dependent ATP–ADP exchange reaction that induces $Na^+$–$Na^+$ exchange proceeds through $E_1P$, the $Na^+$-binding sites are already translocated to the outer surface of the membrane in the $E_1P$ state. However, since the affinities of the cation-binding sites still remain unchanged, the release of $Na^+$ does not yet occur. When $K^+$ is not present outside the cell, $E_1P$ is converted to $E_2P$ in which the affinities of the cation-binding sites are now changed. At this stage, the affinity

Fig. 8.14 Kinetic models for the $Na^+$-pump. (*a*) A model based on the inner pathway of reaction scheme II in Fig. 8.7 (see text). (*b*) A model based on the outer pathway of scheme II in Fig. 8.7.

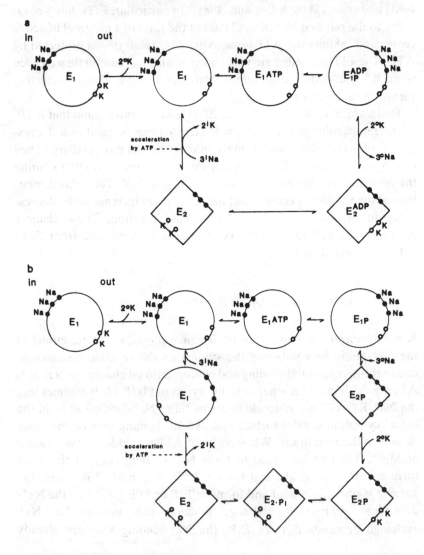

of the Na$^+$-binding site for Na$^+$ is markedly lowered and, as a result, Na$^+$ is released from the cell. When K$^+$ is present outside the cell, on the other hand, it binds to E$_2$P and forms $^{K2}$E$_2$P, which in turn undergoes rapid dephosphorylation and forms $^{K2>}$E$_2$. Alternatively, E$_{1P}^{ADP}$ undergoes rapid dephosphorylation and forms $^{K2}$E$_2^{ADP}$, which in turn releases ADP and forms $^{K2>}$E$_2$. The affinity for K$^+$ is high in the $^{K2>}$E$_2$ state and thus the release of K$^+$ does not occur. $^{K2>}$E$_2$ is slowly converted back to E$_1$, releasing K$^+$ into the cytoplasm. This step is accelerated when a high concentration of ATP is present. In this way, the active transport of Na$^+$ and K$^+$ occurs as the result of cyclic changes in the conformation of ATPase and the affinities of its cation-binding sites for Na$^+$ and K$^+$ during ATP hydrolysis.

*Flip–flop model*  Most of the experimental results on the reaction mechanism of Na$^+$,K$^+$-ATPase can be explained by a scheme that assumes that the $\alpha_1\beta_1$ complexes function independently. However, as has been described earlier in this chapter, data have also been accumulated that are difficult to explain without assuming that the ATPase functions only through interaction between $\alpha$-subunits.

Taking this interaction between $\alpha_1\beta_1$ complexes into consideration, Schön, Dittrich & Repke (1974) and Stein *et al.* (1973) proposed the flip–flop model and the half-of-the-sites-reactive model, respectively. In principle, these models both argue that, to minimize the loss of energy, two homogeneous subunits (two $\alpha_1\beta_1$ complexes) are coupled to each other by shifting the phase of ATPase reaction by a half cycle each. For example, two ATPase molecules form a pair in such combinations as E$_1$ATP–E$_2$P and E$_2$P–E$_1$ATP. According to these models, the transport of Na$^+$ and the transport of K$^+$ occur consecutively on individual subunits, but occur simultaneously as a whole.

As described earlier, it has recently been shown that the binding of ATP, ouabain and other ligands to the enzyme and the formation of EP all occur at the ratio of 1 mol/mol of the $\alpha_1\beta_1$ complex. Many investigators claim that these results disprove the flip–flop model and the half-of-the-sites-reactive model. However, it must be noted that all of the above measurements were carried out under non-physiological conditions in which the ATPase does not undergo turnovers. Therefore, these results do not truly reflect the behavior of the enzyme under physiological conditions in which it does undergo turnovers. Thus, these results do not necessarily deny the above two models.

*Molecular model*   As described earlier, the kinetic analysis of the $Na^+$-pump has already advanced considerably. In contrast, the progress in the structural analysis of the $Na^+$-pump has been slow and, as a result, it is almost impossible at present to discuss the molecular mechanism of the active transport of $Na^+$ on the basis of the structure of the ATPase molecule in the membrane. In analogy with the molecular models of the $Ca^{2+}$-pump of SR described in Chapter 7, the models so far proposed for the $Na^+$-pump can be classified into three types:

(1)  Rotational model. This model argues that the cation-binding sites are translocated as a result of an extensive rotation of the ATPase molecule as a whole. A model of this type was first proposed by Opit & Charnock (1965). Thermodynamically, however, the possibility that an extensive movement occurs throughout the ATPase molecule in the membrane is small. In fact, Kyte (1974) has shown that antibodies to $Na^+,K^+$-ATPase that are bound to the cytoplasmic side of the α-subunit do not inhibit the activity of the enzyme, and he suggested that the entire molecule of $Na^+,K^+$-ATPase does not undergo extensive movement when the cation pump functions.

(2)  Channel model. The early version of the channel model did not consider the morphological changes of the channel (Skou, 1963). However, after being proposed by Singer & Nicolson (1972), the mobile-channel model has become one of the most probable models. The simplest form of this model can be seen in the models proposed by Jardetzky (1966) and Lowe (1968). Being considered at present are more-complex models that take into consideration the fact that the $Na^+$-pump is an oligomer. A typical example of these models is that proposed by Robinson & Flashner (1979).

(3)  Mobile-gate and fixed-channel model. As has been described in detail for the $Ca^{2+}$-pump of SR (see Chapter 7), this model argues that the pump consists of a mobile gate and a fixed channel and that a conformational change in the gate protein causes cation translocation and affinity change. It has already been described that a variety of experimental data consistent with the mobile-gate and fixed-channel model is being accumulated for the $Ca^{2+}$-pump. In the case of the $Na^+$-pump, the mobile-channel model and the mobile-gate and fixed-

channel model are the two most plausible models. However, discussion of a practical model of the $Na^+$-pump awaits clarification of various problems. For example, although it is being established that the conformation of $Na^+,K^+$-ATPase changes during the ATPase reaction, little is known about the concrete nature of this conformational change. Furthermore, although it is thought that two or four $\alpha_1\beta_1$ complexes aggregate to form a $Na^+$-pump, there is virtually no information about changes in the interaction between these $\alpha_1\beta_1$ complexes during the ATPase reaction.

As has already been described for muscle contraction (Chapter 3), studies by X-ray diffraction, fluorescent probes and EPR have provided various lines of information on the movement of crossbridges. The study of membrane ATPase by X-ray diffraction has only recently begun on the $Ca^{2+},Mg^{2+}$-ATPase of SR. We hope that, in the future, studies of the movement of the $Na^+$-pump at the molecular level using X-ray diffraction, fluorescent probes and EPR will be fruitful in the elucidation of the molecular mechanism of the $Na^+$-pump.

## References

Askari, A., ed. (1974). Properties and functions of $(Na^+ + K^+)$-activated adenosinetriphosphatase. *Ann. N.Y. Acad. Sci.*, **242**, 1–174.

Askari, A. & Huang, W. (1980). $Na^+,K^+$-ATPase: half-of-the-subunits crosslinking reactivity suggests an oligomeric structure containing a minimum of four catalytic subunits. *Biochem. Biophys. Res. Commun.*, **93**, 448–53.

Beauge, L. A. & Glynn, I. M. (1979). Occlusion of K ions in the unphosphorylated sodium pump. *Nature*, **280**, 510–12.

Beauge, L. A. & Glynn, I. M. (1980). The equilibrium between different conformations of the unphosphorylated sodium pump: effects of ATP and of potassium ions, and their relevance of potassium transport. *J. Physiol.*, **299**, 367–83.

Blostein, R. (1975). $Na^+$-ATPase of the mammalian erythrocyte membrane, reversibility of phosphorylation at 0°C. *J. Biol. Chem.*, **250**, 6118–24.

Bonting, S. L. & Caravaggio, L. L. (1963). Studies on Na:K-activated ATPase. V. Correlation of enzyme activity with cation flux in six tissues. *Arch. Biochem. Biophys.*, **101**, 37–46.

Brazhnikov, E. V., Chetverin, A. B. & Chirgadze, Yu. N. (1978). Secondary structure of $Na^+,K^+$-dependent adenosine triphosphatase. *FEBS Lett.*, **93**, 125–8.

Brinley, F. J. & Mullins, L. J. (1967). Sodium extrusion by internally dialyzed squid axons. *J. Gen. Physiol.*, **50**, 2303–31.

Brinley, F. J. & Mullins, L. J. (1968). Sodium fluxes in internally dialyzed squid axons. *J. Gen Physiol.*, **52**, 181–211.

Brotherus, J., Møller, J. V. & Jørgensen, P. L. (1981). Soluble and active renal Na,K-ATPase with maximum protein molecular mass 170 000 ± 9000 daltons; formation of larger units by secondary aggregation. *Biochem. Biophys. Res. Commun.*, **100**, 146–54.

Cavieres, J. D. & Glynn, I. M. (1976). $Na^+$–$Na^+$ exchange by the $Na^+$ pump requires ATP as well as ADP. *J. Physiol.*, **192**, 189–216.

Craig, W. S. & Kyte, J. (1980). Stoichiometry and molecular weight of the minimum asymmetric unit of canine renal sodium and potassium ion-activated adenosine triphosphatase. *J. Biol. Chem.*, **255**, 6262–9.

Dahms, A. S. & Boyer, P. D. (1974). Oxygen exchanges catalyzed by and the mechanism of acyl phosphate formation in transport ATPase: properties and functions of $(Na^+,K^+)$-activated adenosine triphosphatase. *Ann. N.Y. Acad. Sci.*, **242**, 133–8.

Dixon, J. F. & Hokin, L. E. (1980). The reconstituted (Na,K)-ATPase is electrogenic. *J. Biol. Chem.*, **255**, 10681–6.

Esmann, M., Christiansen, C., Karlsson, K. A., Hansson, G. C. & Skou, J. C. (1980). Hydrodynamic properties of solubilized $(Na^++K^+)$-ATPase from rectal glands of *squalus Acanthias*. *Biochim. Biophys. Acta*, **603**, 1–12.

Fahn, S., Hurley, M. R., Koval, G. J. & Albers, R. W. (1966a). Sodium-potassium-activated adenosine triphosphatase of electrophorus electric organ. II. Effects of *N*-ethylmaleimide and other sulfhydryl reagents. *J. Biol. Chem.*, **241**, 1890–5.

Fahn, S., Koval, G. J. & Albers, R. W. (1966b). Sodium-potassium-activated adenosine triphosphatase of electrophorus electric organ. I. An associated sodium-activated transphosphorylation. *J. Biol. Chem.*, **241**, 1882–9.

Forbush, B., Kaplan, J. H. & Hoffman, J. F. (1978). Characterization of a new photoaffinity derivative of ouabain: labeling of the large polypeptide and a proteolipid component of the Na, K-ATPase. *Biochemistry*, **17**, 3667–76.

Fossel, E. T., Post, R. L., O'Hara, D. S. & Smith, T. W. (1981). Phosphorus-31 nuclear magnetic resonance of phosphoenzymes of sodium and potassium activated and of calcium activated adenosine triphosphatase. *Biochemistry*, **20**, 7215–19.

Fukushima, Y. & Nakao, M. (1980). Changes in affinity of $Na^+$ and $K^+$-transport ATPase for divalent cations during its reaction sequence. *J. Biol. Chem.*, **255**, 7813–19.

Fukushima, Y. & Nakao, M. (1981). Transient state in the phosphorylation of sodium and potassium transport adenosine triphosphatase by adenosine triphosphate. *J. Biol. Chem.*, **256**, 9136–43.

Fukushima, Y. & Post, R. L. (1978). Binding of divalent cation to phosphoenzyme of sodium and potassium transport adenosine triphosphatase. *J. Biol. Chem.*, **253**, 6853–62.

Fukushima, Y. & Tonomura, Y. (1973). Two kinds of high energy phosphory-lated intermediate, with and without bound ADP, in the reaction of $Na^+$,$K^+$-dependent ATPase. *J. Biochem.*, **74**, 135–42.

Gache, C., Rossi, B. & Lazdunski, M. (1977). Mechanistic analysis of the $(Na^+,K^+)$ ATPase using new pseudosubstrates. *Biochemistry*, **16**, 2957–65.

Garrahan, P. J. & Garry, R. P. (1976). The distinction between sequential and simultaneous models for sodium and potassium transport. *Curr. Top. Memb. Transp.*, **8**, 29–97.

Garrahan, P. J. & Glynn, I. M. (1967). The stoichiometry of the sodium pump. *J. Physiol.*, **192**, 217–35.

Giotta, G. J. (1976). Quaternary structure of $(Na^+ + K^+)$-dependent adenosine triphosphatase. *J. Biol. Chem.*, **251**, 1247–52.

Glynn, I. M. (1956). Sodium and potassium movements in human red cells. *J. Physiol.*, **134**, 278.

Glynn, I. M. & Karlish, S. J. D. (1975a). The sodium pump. *Annu. Rev. Physiol.*, **37**, 13–55.

Glynn, I. M. & Karlish, S. J. D. (1975b). Different approaches to the mechanism of the sodium pump. In *Energy Transformation in Biological Systems* (CIBA Symposium 31), ed. G. E. W. Wolstenholme & D. W. Fitzsi-mons, pp. 205–23. Amsterdam: Associated Scientific Publishers.

Glynn, I. M. & Richards, D. E. (1980). Factors affecting the release of occluded rubidium ions from the sodium pump. *J. Physiol.*, **308**, 58.

Grosse, R., Rapoport, T., Malur, J., Fischer, J. & Repke, K. R. H. (1979). Mathematical modelling of ATP, $K^+$ and $Na^+$ interactions with $(Na^+ + K^+)$-ATPase occuring under equilibrium conditions. *Biochim. Biophys. Acta*, **550**, 500–14.

Haase, W. & Koepsell, H. (1979). Substructure of membrane-bound $Na^+$-$K^+$-ATPase protein. *Pflügers Arch.*, **381**, 127–35.

Hansen, O., Jensen, J., Nørby, J. G. & Ottolenghi, P. (1979). A new proposal regarding the subunit composition of $(Na^+ + K^+)$ATPase. *Nature*, **280**, 410–12.

Hara, Y. & Nakao, M. (1979). Detection of sodium binding to $Na^+$,$K^+$-ATPase with a sodium sensitive electrode. In *Cation Flux Across Biomem-branes*, ed. Y. Mukohata & L. Packer, pp. 21–8. New York: Aca-demic Press.

Hara, Y. & Nakao, M. (1981). Sodium ion discharge from pig kidney $Na^+$,$K^+$-ATPase; $Na^+$-dependency of $E_1P \rightleftharpoons E_2P$ equilibrium in the absence of KCl. *J. Biochem.*, **90**, 923–31.

Hardwicke, P. M. D. & Freytag, J. W. (1981). A proteolipid associated with Na, K-ATPase is not essential for ATPase activity. *Biochem. Bio-phys. Res. Commun.*, **102**, 250–7.

Harris, W. E. & Stahl, W. L. (1977). Conformational changes of purified $(Na^+ + K^+)$-ATPase detected by sulfhydryl fluorescence probe. *Bio-chim. Biophys. Acta*, **485**, 203–14.

Hastings, D. F. (1977). Differential titration of potassium binding to mem-brane proteins using ion selective electrodes. *Anal. Biochem.*, **83**, 416–32.

Hegyvary, C., Chigurupati, R., Kang, K. & Mahoney, D. (1980). Reversible alterations in the kinetics of cardiac sodium and potassium activated adenosine triphosphatase after partial removal of membrane lipids. *J. Biol. Chem.*, **255**, 3068–74.

Hegyvary, C. & Jørgensen, P. L. (1981). Conformational changes of renal sodium plus potassium ion transport adenosine triphosphatase labeled with fluorescein. *J. Biol. Chem.*, **256**, 6296–303.

Hegyvary, C. & Post, R. L. (1971). Binding of adenosine triphosphate to sodium and potassium ion-stimulated adenosine triphosphatase. *J. Biol. Chem.*, **246**, 5234–40.

Hexum, T., Samson, F. F. & Himes, R. H. (1970). Kinetic studies of membrane (Na$^+$ + K$^+$ + Mg$^{2+}$)-ATPase. *Biochim. Biophys. Acta*, **212**, 322–31.

Hilden, S. & Hokin, L. E. (1975). Active potassium transport coupled to active sodium transport in vesicles reconstituted from purified sodium and potassium ion-activated adenosine triphosphatase from the rectal glands of squalus acanthias. *J. Biol. Chem.*, **250**, 6296–303.

Hodgkin, A. L. (1951). The ionic basis of electrical activity in nerve and muscle. *Biol. Rev.*, **26**, 339–409.

Hodgkin, A. L. & Keynes, R. D. (1955). Active transport of cation in giant axons from sepia and loligo. *J. Physiol.*, **128**, 28–60.

Hoffman, J. F. & Forbush, B., eds (1983). *Structure, Mechanism and Function of the Na/K Pump*, Current Topics in Membranes and Transport, vol. 19. New York: Academic Press.

Jardetzky, O. (1966). Simple allosteric model for membrane pumps. *Nature*, **211**, 969–70.

Jørgensen, P. L. (1974). Purification and characterization of (Na$^+$ + K$^+$)-ATPase. III. Purification from the outer medulla of mammalian kidney after selective removal of membrane components by sodium dodecyl sulphate. *Biochim. Biophys. Acta*, **356**, 36–52.

Jørgensen, P. L. (1975). Purification and characterization of (Na$^+$,K$^+$)-ATPase. V. Conformational changes in the enzyme. Transitions between the Na-form and the K-form studied with tryptic digestion as a tool. *Biochim. Biophys. Acta*, **401**, 399–415.

Jørgensen, P. L. (1982). Mechanism of the Na$^+$,K$^+$ pump: protein structure and conformations of the pure (Na$^+$ + K$^+$)-ATPase. *Biochim. Biophys. Acta*, **694**, 27–68.

Jørgensen, P. L. & Karlish, S. J. D. (1980). Defective conformational response in a selectively trypsinized (Na$^+$ + K$^+$)-ATPase studied with tryptophan fluorescence. *Biochim. Biophys. Acta*, **597**, 305–17.

Jørgensen, P. L. & Petersen, J. (1979). Protein conformations of the phosphorylated intermediates of purified Na,K-ATPase studied with tryptic digestion and intrinsic fluorescence as tools. In *Na,K-ATPase: Structure and Kinetics*, ed. J. C. Skou & J. G. Nørby, pp. 143–55. New York: Academic Press.

Kanazawa, T., Saito, M. & Tonomura, Y. (1967). Properties of a phosphorylated protein as a reaction intermediate of Na$^+$-K$^+$ sensitive ATPase. *J. Biochem.*, **61**, 555–66.

Kanazawa, T., Saito, M. & Tonomura, Y. (1970). Formation and decomposition of a phosphorylated intermediate in the reaction of Na$^+$-K$^+$ dependent ATPase. *J. Biochem.*, **67**, 693–711.

Kaniike, K., Lindenmayer, G. E., Wallick, E. T., Lane, L. K. & Schwartz, A. (1976). Specific sodium-22 binding to a purified sodium + potassium adenosine triphosphatases. Inhibition by ouabain. *J. Biol. Chem.*, **251**, 4794–5.

Karlish, S. J. D. (1980). Characterization of conformational changes in (Na,K) ATPase labelled with fluorescein at the active site. *J. Bioenerg. Biomemb.*, **12**, 111–36.

Karlish, S. J. D. & Yates, D. W. (1978). Tryptophan fluorescence of (Na$^+$ + K$^+$)-ATPase as a tool for study of the enzyme mechanism. *Biochim. Biophys. Acta*, **527**, 115–30.

Karlish, S. J. D., Yates, D. W. & Glynn, I. M. (1976). Transient kinetics of (Na$^+$ + K$^+$)ATPase studied with a fluorescent substrate. *Nature*, **263**, 251–3.

Kepner, G. R. & Macey, R. I. (1968). Membrane enzyme systems, molecular size determinations by radiation inactivation. *Biochim. Biophys. Acta*, **163**, 188–203.

Klodos, I. & Nørby, J. G. (1979). Effect of K$^+$ and Li$^+$ on intermediary steps in the Na,K-ATPase reaction. In *Na,K-ATPase: Structure and Kinetics*, ed. J. C. Skou & J. G. Nørby, pp. 331–42. New York: Academic Press.

Kudoh, F., Nakamura, S., Yamaguchi, M. & Tonomura, Y. (1979). Binding of ouabain to Na$^+$,K$^+$-dependent ATPase during the ATPase reaction: evidence for a dimer structure of the ATPase. *J. Biochem.*, **86**, 1023–8.

Kyte, J. (1971). Phosphorylation of a purified (Na$^+$ + K$^+$) adenosine triphosphatase. *Biochem. Biophys. Res. Commun.*, **43**, 1259–65.

Kyte, J. (1974). The reactions of sodium and potassium ion-activated adenosine triphosphatase with specific antibodies. *J. Biol. Chem.*, **249**, 3652–60.

Kyte, J. (1975). Structual studies of sodium and potassium ion-activated adenosine triphosphatase. *J. Biol. Chem.*, **250**, 7443–9.

Lowe, A. G. (1968). Enzyme mechanism for the active transport of sodium and potassium ions in animal cells. *Nature*, **219**, 934–6.

Matsui, H., Hayashi, Y., Homareda, H. & Kimimura, M. (1977). Ouabain-sensitive 42K binding to Na$^+$,K$^+$-ATPase purified from canine kidney outer medulla. *Biochem. Biophys. Res. Commun.*, **75**, 373–80.

Matsui, H., Hayashi, T., Homareda, H. & Taguchi, M. (1983). Stoichiometrical binding of ligands to less than 160 kilodaltons of Na, ATPase. *Curr. Top. Memb. Transp.*, **19**, 145–8.

Matsui, H. & Homareda, H. (1982). Interaction of sodium and potassium ions with Na$^+$,K$^+$-ATPase. I. Ouabain-sensitive alternative binding of three Na$^+$ or two K$^+$ to the enzyme. *J. Biochem.*, **92**, 193–217.

Maunsbach, A. B., Skriver, E. & Jørgensen, P. L. (1979). Ultrastructure of purified Na, K-ATPase membranes. In *Na,K-ATPase: Structure and*

*Kinetics*, ed. J. C. Skou & J. G. Nørby, pp. 2–13. New York: Academic Press.

Moczydlowski, E. G. & Fortes, P. A. G. (1981*a*). Characterization of 2′,3′-O-(2,4,6-trinitrocyclohexadienylidine)-adenosine 5′-triphosphate as a fluorescent probe of the ATP site of sodium and potassium transport adenosine triphosphatase. Determination of nucleotide binding stoichiometry and ion-induced changes in affinity for ATP. *J. Biol. Chem.*, **256**, 2346–56.

Moczydlowski, E. G. & Fortes, P. A. G. (1981*b*). Inhibition of sodium and potassium adenosine triphosphatase by 2′,3′-O-(2,4,6-trinitrocyclohexadienylidene)adenine nucleotides. Implications for the structure and mechanism of the Na:K pump. *J. Biol. Chem.*, **256**, 2357–66.

Nagai, K. & Yoshida, H. (1966). Biphasic effects of nucleotides on potassium-dependent phosphatase. *Biochim. Biophys. Acta*, **128**, 410–12.

Nagano, K., Kanazawa, T., Mizuno, N., Tashima, Y., Nakao, T. & Nakao, M. (1965). Some acyl phosphate-like properties of ³²P-labelled sodium-potassium activated adenosine triphosphatase. *Biochem. Biophys. Res. Commun.*, **19**, 759–64.

Nakao, M., Nagano, K., Nakao, T., Mizuno, N., Tashima, Y., Fujita, M., Maeda, H. & Matsudaira, H. (1967). Molecular weight of Na,K-ATPase approximated by the radiation inactivation method. *Biochem. Biophys. Res. Commun.*, **29**, 588–92.

Nakao, T., Nagano, K., Adachi, K. & Nakao, M. (1963). Separation of two adenosine triphosphatases from erythrocyte membrane. *Biochem. Biophys. Res. Commun.*, **13**, 444–8.

Neufeld, A. H. & Levy, H. M. (1969). A second ouabain-sensitive sodium-dependent adenosine triphosphatase in brain microsomes. *J. Biol. Chem.*, **244**, 6493–7.

Nørby, J. G. & Jensen, J. (1971). Binding of ATP to brain microsomal ATPase. *Biochim. Biophys. Acta*, **233**, 104–16.

O'Connor, S. E. & Grisham, C. M. (1979). Manganese electron paramagnetic resonance studies of sheep kidney (Na⁺ + K⁺)-ATPase. Interactions of substrates and activators at a single Mn²⁺-binding site. *Biochemistry*, **18**, 2315–23.

O'Connor, S. E. & Grisham, C. M. (1980). Distance determinations at the active site of kidney (Na⁺ + K⁺)-ATPase by Mn(II) ion electron paramagnetic resonance. *FEBS Lett.*, **118**, 303–7.

Opit, L. J. & Charnock, J. S. (1965). A molecular model for a sodium pump. *Nature*, **208**, 471–4.

Ottolenghi, P. (1979). The relipidation of delipidated Na,K-ATPase. An analysis of complex formation with dioleoyl-phosphatidylcholine and with dioleoylphosphatidylethanolamine. *Eur. J. Biochem.*, **99**, 113–33.

Peters, W. H. M., Swarts, H. G. P., de Pont, J. J. H. H. M., Schuurmans Stekhoven, F. M. A. H. & Bonting, S. L. (1981). (Na⁺ + K⁺)ATPase has one functioning phosphorylation site per subunit. *Nature*, **290**, 338–9.

Peterson, G. L. & Hokin, L. E. (1981). Molecular weight and stoichiometry of the sodium and potassium activated adenosine triphosphatase subunits. *J. Biol. Chem.*, **256**, 3751–61.

Post, R. L. (1979). A perspective on sodium and potassium ion transport adenosine triphosphatase. In *Cation Flux Across Biomembranes*, ed. Y. Mukohata & L. Packer, pp. 3–19. New York: Academic Press.

Post, R. L., Hegyvary, C. & Kume, S. (1972). Activation by adenosine triphosphate in the phosphorylation kinetics of sodium and potassium ion transport adenosine triphosphatase. *J. Biol. Chem.*, **247**, 6530–40.

Post, R. L. & Jolly, P. C. (1957). The linkage of sodium, potassium, and ammonium active transport across the human erythrocyte membrane. *Biochim. Biophys. Acta*, **25**, 118–28.

Post, R. L. & Kume, S. (1973). Evidence for an aspartyl phosphate residue at the active site of sodium and potassium ion transport adenosine triphosphatase. *J. Biol. Chem.*, **248**, 6993–7000.

Post, R. L., Kume, S., Tobin, T., Orcutt, B. & Sen, A. K. (1969). Flexibility of an active center in sodium-plus-potassium adenosine triphosphatase. *J. Gen. Physiol.*, **54**, 306S–25S.

Post, R. L., Merritt, C. R., Kinsolving, C. R. & Albright, C. D. (1960). Membrane adenosine triphosphatase as a participant in the active transport of sodium and potassium in the human erythrocyte. *J. Biol. Chem.*, **235**, 1796–1802.

Post, R. L., Sen, A. K. & Rosenthal, A. S. (1965). A phosphorylated intermediate in adenosine triphosphate-dependent sodium and potassium transport across kidney membranes. *J. Biol. Chem.*, 1437–45.

Reeves, A. S., Collins, J. H. & Schwartz, A. (1980). Isolation and characterization of (Na,K)-ATPase proteolipid. *Biochem. Biophys. Res. Commun.*, **95**, 1591–8.

Robinson, J. D. (1976). Substrate sites of the $(Na^+ + K^+)$-dependent ATPase. *Biochim. Biophys. Acta*, **429**, 1006–19.

Robinson, J. D. & Flashner, M. S. (1979). The $(Na^+,K^+)$-activated ATPase: enzymatic and transport properties. *Biochem. Biophys. Acta*, **549**, 145–76.

Rogers, T. & Lazdunski, M. (1979). Photoaffinity labelling of the digitalis receptor in the (sodium + potassium)-activated adenosinetriphosphatase. *Biochemistry*, **18**, 135–40.

Ruoho, A. & Kyte, J. (1974). Photoaffinity labeling of the ouabain-binding site on $(Na^+ + K^+)$adenosinetriphosphatase. *Proc. Natl. Acad. Sci. (USA)*, **71**, 2352–6.

Sakamoto, J. & Tonomura, Y. (1980). Order of release of ADP and Pi from phosphoenzyme with bound ADP of $Ca^{2+}$-dependent ATPase from sarcoplasmic reticulum and of $Na^+,K^+$-dependent ATPase studied by ADP-inhibition patterns. *J. Biochem.*, **87**, 1721–7.

Schon, R., Dittrich, F. & Repke, K. R. H. (1974). Thermodynamic evaluation of flip-flop mechanism for transport and ATP synthesis function of (Na,K)-ATPase. *Acta Biol. Med. Germ.*, **33**, k9–k16.

Sen, A. K. & Post, R. L. (1964). Stoichiometry and localization of ATP dependent Na⁺,K⁺ transport in the erythrocyte. *J. Biol. Chem.*, **239**, 345–52.

Siegel, G. J. & Fogt, S. M. (1977). Inhibition by lead ion of electrophorus electroplax (Na⁺ + K⁺)-adenosine triphosphatase and K⁺-*p*-nitrophenylphosphatase. *J. Biol. Chem.*, **252**, 5201–5.

Singer, S. J. & Nicolson, G. L. (1972). The fluid mosaic model of the structure of all membranes. *Science*, **175**, 720–31.

Skou, J. C. (1957). The influence of some cations on an adenosine triphosphatase from peripheral nerves. *Biochim. Biophys. Acta*, **23**, 394–403.

Skou, J. C. (1960). Further investigations on a Mg²⁺ + Na⁺-activated adenosine-triphosphatase, possibly related to the active, linked transport of Na⁺ and K⁺ across the nerve membrane. *Biochim. Biophys. Acta*, **42**, 6–23.

Skou, J. C. (1963). Enzymatic aspects of active transport of Na⁺ and K⁺ across the cell membrane. In *Drugs and Membranes*, vol. 4, ed. C. A. M. Hogben, pp. 41–69. London: Pergamon Press.

Skou, J. C. & Esmann, M. (1981). Eosin, a fluorescent probe of ATP binding to the (Na⁺ + K⁺)-ATPase. *Biochem. Biophys. Acta*, **647**, 232–40.

Skou, J. C. & Nørby, J. G., eds. (1979). *Na, K-ATPase: Structure and Kinetics*. New York: Academic Press.

Stahl, W. L. (1968). Sodium stimulated ¹⁴C adenosine diphosphate adenosine triphosphate exchange activity in brain microsomes. *J. Neurochem.*, **15**, 511–18.

Stein, W. D., Lieb, W. R., Karlish, S. J. D. & Eilam, Y. (1973). A model for active transport of sodium and potassium ions as mediated by a tetrameric enzyme. *Proc. Natl. Acad. Sci. (USA)*, **70**, 275–8.

Sweadner, K. J. (1979). Two molecular forms of (Na + K)-stimulated ATPase in brain. Separation and difference in affinity for strophanthidin. *J. Biol. Chem.*, **254**, 6060–7.

Taniguchi, K. & Post, R. L. (1975). Synthesis of adenosine triphosphate and exchange between inorganic phosphate and adenosine triphosphate in sodium and potassium ion transport adenosine triphosphatase. *J. Biol. Chem.*, **250**, 3010–18.

Taniguchi, K., Suzuki, K. & Iida, S. (1982). Conformational change accompanying transition of ADP-sensitive phosphoenzyme to potassium-sensitive phosphoenzyme of (Na⁺,K⁺)-ATPase modified with *N*-*p*-(2-benzimidazolyl)-phenyl maleimide. *J. Biol. Chem.*, **257**, 10659–67.

Taniguchi, K., Suzuki, K. & Iida, S. (1983). Stopped flow measurement of conformational change induced by phosphorylation in (Na⁺,K⁺)-ATPase modified with *N*-*p*-(2-benzimidazolyl)phenyl maleimide. *J. Biol. Chem.*, **258**, 6927–31.

Taniguchi, K. & Tonomura, Y. (1971). Inactivation of Na⁺-K⁺-dependent ATPase by phospholipase treatment and its reactivation by phospholipids. *J. Biochem.*, **69**, 543–57.

Tobin, T., Akera, T., Baskin, S. I. & Brody, T. M. (1973). Calcium ion and sodium- and potassium-dependent adenosine triphosphatase: its mechanism of inhibition and identification of the $E_iP$ intermediate. *Mol. Pharmacol.*, **9**, 336–49.

Urayama, O. & Nakao, M. (1979). Organ specificity of rat Na,K-ATPase. *J. Biochem.*, **86**, 1371–81.

Vogel, F., Meyer, H. W., Grosse, R. & Repke, K. R. H. (1977). Electron microscopic visualization of the arrangement of the two protein components of (Na$^+$ + K$^+$)-ATPase. *Biochim. Biophys. Acta*, **470**, 497–502.

Wang, T., Lindenmayer, G. E. & Schwartz, A. (1977). Steady-state kinetic study of magnesium and ATP effects on ligand affinity and catalytic activity of sheep kidney sodium, potassium adenosine triphosphatase. *Biochim. Biophys. Acta*, **484**, 140–60.

Whittam, R. & Agar, M. E. (1965). The connection between active cation transport and metabolism in erythrocyte. *Biochem. J.*, **97**, 214–27.

Winter, C. G. & Moss, A. J., Jr. (1979). Ultracentrifugal analysis of the enzymatically active fragments produced by digitonin action on Na,K-ATPase. In *Na, K-ATPase: Structure and Kinetics*, ed. J. C. Skou & J. G. Nørby, pp. 25–33. New York: Academic Press.

Yamaguchi, M. & Post, R. L. (1983). Isoelectric focussing of the catalytic subunit of (Na,K)-ATPase from pig kidney. *J. Biol. Chem.*, **258**, 5260–8.

Yamaguchi, M. & Tonomura, Y. (1977). Kinetic studies on the ADP-ATP exchange reaction catalyzed by Na$^+$,K$^+$-dependent ATPase. *J. Biochem.*, **81**, 249–60.

Yamaguchi, M. & Tonomura, Y. (1978). Binding of adenosine diphosphate to reaction intermediates in the Na$^+$,K$^+$dependent ATPase from porcine kidney. *J. Biochem.*, **83**, 977–87.

Yamaguchi, M. & Tonomura, Y. (1979). Simultaneous binding of three Na$^+$ and two K$^+$ ions to Na$^+$,K$^+$-dependent ATPase and changes in its affinities for the ions induced by the formation of a phosphorylated intermediate. *J. Biochem.*, **86**, 509–23.

Yamaguchi, M. & Tonomura, Y. (1980*a*). Binding of monovalent cations to Na$^+$,K$^+$-dependent ATPase purified from porcine kidney. I. Simultaneous binding of three sodium and two potassium or rubidium ions to the enzyme. *J. Biochem.*, **88**, 1365–75.

Yamaguchi, M. & Tonomura, Y. (1980*b*). Binding of monovalent cations to Na$^+$,K$^+$-dependent ATPase purified from porcine kidney. II. Acceleration of transition from a K$^+$-bound form to a Na$^+$-bound form by binding of ATP to a regulatory site of the enzyme. *J. Biochem.*, **88**, 1377–85.

Yamaguchi, M. & Tonomura, Y. (1980*c*). Binding of monovalent cations to Na$^+$,K$^+$-dependent ATPase purified from porcine kidney. III. Marked changes in affinities for monovalent cations induced by formation of an ADP-insensitive but not an ADP-sensitive phosphoenzyme. *J. Biochem.*, **88**, 1387–97.

# 9    Epilogue

Thus far we have discussed the molecular basis of the structure–function relationship in energy-transducing ATPases, the key enzymes of the contractile system and the active-transport system of cations and proton. In this final chapter we will briefly summarize the similarities and differences of these various types of ATPases and consider them from an evolutionary point of view. The evolutionary relationships among various energy-transducing systems are presented in Fig. 1.2.

## Contractile ATPase

There are two types of motile-cell systems that utilize ATP: the myosin–actin system and the dynein–tubulin system. The coupling mechanism between the ATPase reaction and the mechanical work, as well as the regulatory mechanism of each system, was discussed in Chapters 2–4 (the myosin–actin system) and Chapter 5 (the dynein–tubulin system). These two contractile ATPases share close similarities. Myosin and dynein each form an enzyme–P–ADP complex as a key reaction intermediate during their catalytic cycle. The ATPase activity of myosin or dynein is greatly accelerated by actin or tubulin, respectively, and mechanical work is generated by the sliding of the protein filaments past each other. A reaction cycle of muscle contraction is composed of the following three elementary steps. (1) Formation of crossbridges by binding of the myosin heads to the actin filaments. (2) Sliding of actin filaments past myosin filaments, caused by rotational movement of the crossbridge. This movement is strictly coupled to the hydrolysis of ATP by the myosin head through the E–P–ADP complex. (3) Dissociation of myosin heads from actin filaments. Similarly, the dynein–

tubulin system involves sliding of the microtubule doublets past each other to generate mechanical work. The sliding between the paired doublets occurs when the dynein molecule, which is bound to one of the doublets, reacts with ATP and undergoes myosin-like movements.

As described in Chapters 3 and 4, the manner in which actomyosin ATPase is controlled by $Ca^{2+}$ exhibits considerable diversity, and various control mechanisms exist in different types of muscle cells and nonmuscle cells. However, it should be noted that in all cases, $Ca^{2+}$ directly regulates the ATPase reaction in the myosin–actin system, while in the dynein–tubulin system it regulates the direction of movement of the cilia and flagella (Chapter 5). The evolutionary relationship between the myosin–actin and dynein–tubulin systems is still unknown. It was reported that myosin and actin had been isolated and purified from *Escherichia coli*, as mentioned in Chapter 5, but this report has not yet been confirmed. Therefore, it is not known whether or not myosin and/or dynein exists in prokaryotes. Despite these uncertainties, the myosin–actin and dynein–tubulin systems seem to have provided an important background for evolution of eukaryotes from prokaryotes, as suggested by Watson (1976).

## Transport ATPase

We have discussed the coupling mechanism between the ATPase reaction and the transport of ions across the membrane in the $Ca^{2+}$,$Mg^{2+}$-ATPase of sarcoplasmic reticulum (Chapter 7) and the $Na^+$,$K^+$-ATPase of plasma membranes of animal cells (Chapter 8). Many other transport ATPases have been isolated from various membranes, such as the $Ca^{2+}$,$Mg^{2+}$-ATPase of erythrocytes (Niggli, Penniston & Carafoli, 1979; Vincenzi & Hinds, 1981; Schatzmann, 1982) and the $H^+$,$K^+$-ATPase from gastric cells (Ganser & Forte, 1973; Lee *et al.*, 1979; Stewart, Wallmark & Sachs, 1981; Sachs, Faller & Rabon, 1982). All of these transporting ATPases are similar to each other in that they all contain a single peptide with a molecular weight of about 100 000 as the catalytic subunit and require lipids for their enzymatic reaction. The ATPase proteins are arranged asymmetrically in the membrane, probably as an oligomer, to form the functional unit for cation transport. In addition, they all catalyze ATP hydrolysis through a phosphoenzyme (EP) reaction intermediate. In all the cases studied, the $\gamma$-phosphate of ATP is incorporated as an acylphosphate into the Asp residue of the enzyme. Furthermore, these ATPases can be assumed to form sequentially the two conformations, $E_1$ and $E_2$, during the reaction cycle and their cation-binding affinities change greatly depending on which state they are in. Additional characteristics common to ATPases are that they are controlled by ATP itself and also synthesize ATP from ADP and Pi by the reverse action of the cation pump.

The presence of the cation-transporting ATPase in prokaryotes has long been doubted despite various investigations. This is because the transport of substances including cations occurs as an antiport of $H^+$ in prokaryotes. But recently an ATP-dependent transport of $Na^+$ was reported in *Streptococcus faecalis* (Heefner & Harold, 1982). *S. faecalis* has been known to be capable of glycolysis-dependent $Na^+$-extrusion even in the presence of the protonophore. Using everted vesicles of *S. faecalis*, Heefner & Harold observed that the uptake of $Na^+$ is dependent on ATP and is not inhibited by a $H^+$ conductor. They also demonstrated that $Na^+$ stimulates the ATPase activity. The presence of a $Ca^{2+}$-pump, on the other hand, has not yet been demonstrated in prokaryotes. In the discussion of the $Na^+,K^+$-ATPase of the plasma membrane of eukaryotes in Chapter 8, it was shown that divalent cations such as $Ca^{2+}$ and $Mg^{2+}$ play an important role in the regulation of EP formation and decomposition, and that the affinity of the enzyme for $Ca^{2+}$ dramatically changes upon formation of EP. Since this affinity change of the ATPase for divalent cations during the catalytic cycle is also observed in the $Ca^{2+}$-pump system of eukaryote membranes (Chapter 7), the $Na^+$-pump may have emerged first in prokaryotes and then evolved into the $Ca^{2+}$-pump in eukaryotes.

## Proton pump

The $H^+$-pump is extremely important in bioenergetics. Two types are known to exist, each exhibiting characteristics similar to contractile ATPase or cation-transporting ATPase. As described in Chapter 6, the reaction mechanisms of the $F_1$-ATPase of the bacterial plasma membrane or the mitochondrial membrane are analogous to those of contractile ATPase in many respects, including the formation of the E–P–ADP intermediate. In addition, the subunit structure of $F_1$-ATPase is complex, like that of myosin ATPase (Chapter 2). However, $F_1$-

ATPase also possesses characteristics closely related to those of the transport ATPases in that it catalyzes the ATP hydrolysis-mediated transport of $H^+$ only when it forms a complex with $F_0$, present in the lipid membrane. This complex can synthesize ATP from ADP and Pi by utilizing the electrochemical potential of the proton gradient established in the membrane.

Recently, a new type of $H^+$-ATPase was found in fungal and bacterial plasma membranes (Goffeau & Slayman, 1981). Like the $F_1$–$F_0$-type ATPase, this proton-translocating ATPase generates an electrochemical gradient of $H^+$ in the membranes by hydrolyzing ATP. This electrochemical proton gradient is used for incorporating nutrients and ions. Such proton-translocating ATPases were found in the plasma membranes of *Schizosaccharomyces pombe* (Dufour & Goffeau, 1978; Amory, Foury & Goffeau, 1980; Villalobo, 1982), *Saccharomyces cerevisiae* (Willsky, 1979; Foury, Amory & Goffeau, 1981; Malpartida & Serrano, 1981), *Neurospora crassa* (Bowman & Slayman, 1977; Dame & Scarborough, 1981), *E. coli* (Epstein & Laimins, 1980) and *Mycobacterium* (Yoshimura & Brodie, 1981). Biochemical studies, using ATPase preparations partially purified from the plasma membrane of these organisms and reconstituted membrane systems, revealed that all of the $H^+$-ATPases mentioned above are composed of a single peptide with a molecular weight of about 100 000, have an optimum pH of about 5.5 and are inhibited by Dio 9, NaF, vanadate and SH reagents. These ATPases were also reported to form EP as a reaction intermediate. The plasma membrane ATPases of *Neurospora* (Dame & Scarborough, 1981) and *Schizosaccharomyces* (Amory & Goffeau, 1982) were shown to have phosphorylated $\beta$-carboxyl groups of the Asp residues. These biochemical characteristics are common to the cation-transporting ATPases but differ from those of the $F_1$–$F_0$-type $H^+$-ATPases.

Thus, two types of $H^+$-pump exist in the membrane: the contractile ATPase type and the cation-transporting ATPase type. In addition, as Mitchell has successfully demonstrated in his chemiosmotic theory (Mitchell, 1961), the $H^+$-pump built on the $F_1$–$F_0$ ATPase not only synthesizes ATP but also functions as a $H^+$-transport-coupled transport system of cations and other substances, thus playing important roles in living cells. Therefore, it is reasonable to assume that a primitive $H^+$-pumping system appeared at a very early stage of evolution. The proposed primitive $H^+$-pump, unlike the highly evolved contractile ATPase and cation-transporting ATPase, may have been able to use various high-energy phosphate compounds other than ATP as substrates.

It may have had a relatively simple reaction mechanism, lacking the steps in which the reaction rate is affected by different controls, such as acceleration of the formation of a reaction intermediate by actin or tubulin as seen in the contractile ATPase, or acceleration of $E_1 \rightarrow E_2$ transfer by ATP as seen in the cation-transporting ATPase.

Of these two proton-pump types, the $F_1$–$F_0$ ATPase may be evolutionally closer to the primitive $H^+$-pump because the mitochondria and chloroplasts are protected from the evolutionary process by the intracellular environment. These organelles exist inside the cell as independent entities, which perhaps allows the $F_1$–$F_0$ ATPase to retain the characteristics of the primitive enzyme (John & Whatley, 1975). During the earliest period of evolution, $F_1$–$F_0$ ATPase could conceivably have functioned as an ATP-driven $H^+$-pump, helping the cells to maintain a proper level of osmotic pressure by pumping out the $Na^+$ that flowed in, and also incorporating nutrients from outside. At a later stage of evolution, as the cell's $H^+$-transporting ability increased due to the emergence of another type of $H^+$-pump system, the electron-transfer chain, the $F_1$–$F_0$-type ATPase would have catalyzed ATP synthesis by utilizing the downhill movement of $H^+$ across the membrane. With regard to this, it should be noted that the $H^+$-ATPase of chromaffin granules is structurally similar to the cation-transporting ATPase but retains the functional characteristics of the $F_1$–$F_0$-type ATPase (Apps & Schatz, 1979; Roisin, Scherman & Henry, 1980). The similarity of the myosin–actin system and the dynein–tubulin system to $F_1$-ATPase suggests that they may have evolved from $F_1$–$F_0$-ATPase, while the other evolutionary pathway for the primitive proton-pump was toward the cation-transporting ATPase systems that produce EP, such as the proton-ATPase of fungi and the $H^+,K^+$-ATPase of higher animals. The ATP-driven $Na^+$-pump of *Streptococcus* may also be of this class, evolving into the $Na^+,K^+$-ATPase that now plays important roles in eukaryotes. The $Na^+$-pump may also have evolved further into the $Ca^{2+}$-pump of SR and erythrocytes, because the $Na^+,K^+$-ATPase exhibits the characteristics of affinity change not only for $Na^+$ and $K^+$, but also for $Ca^{2+}$, in the reaction cycle of ATP hydrolysis.

## References

Amory, A., Foury, F. & Goffeau, A. (1980). The purified plasma membrane ATPase of the yeast *Schizosaccharomyces pombe* forms a phosphorylated intermediate. *J. Biol. Chem.*, **255**, 9353–7.

Amory, A. & Goffeau, A. (1982). Characterization of the $\beta$-aspartyl phosphate intermediate formed by the $H^+$-translocating ATPase from the yeast *Schizosaccharomyces pombe. J. Biol. Chem.*, **257**, 4723–30.

Apps, D. K. & Schatz, G. (1979). An adenosine triphosphatase isolated from chromaffin-granule membranes is closely similar to F1-adenosine triphosphatase of mitochondria. *Eur. J. Biochem.*, **100**, 411–19.

Blitz, A. L., Fine, R. E. & Tosell, P. A. (1977). Evidence that coated vesicles isolated from brain are calcium-sequestering organelles resembling sarcoplasmic reticulum. *J. Cell Biol.*, **75**, 135–47.

Bowman, B. J. & Slayman, C. W. (1977). Characterization of plasma membrane adenosine triphosphatase of *Neurospora crassa. J. Biol. Chem.*, **252**, 3357–63.

Dame, J. B. & Scarborough, G. A. (1980). Identification of hydrolytic moiety of the *Neurospora* plasma membrane $H^+$-ATPase and demonstration of a phosphorylenzyme intermediate in its catalytic mechanism. *Biochemistry*, **19**, 2931–7.

Dame, J. B. & Scarborough, G. A. (1981). Identification of the phosphorylated intermediate of the *Neurospora* plasma membrane $H^+$-ATPase as $\beta$-aspartyl phosphate. *J. Biol. Chem.*, **256**, 10724–30.

Dufour, J. P. & Goffeau, A. (1978). Solubilization by lysolecithin and purification of the plasma membrane ATPase of the yeast *Schizosaccharomyces pombe. J. Biol. Chem.*, **253**, 7026–32.

Epstein, W. & Laimins, L. (1980). Potassium transport in *Escherichia coli*: diverse systems with common control by osmotic forces. *Trends Biochem. Sci.*, **5**, 21–3.

Foury, F., Amory, A. & Goffeau, A. (1981). Large-scale purification and phosphorylation of a detergent-treated adenosine triphosphatase complex from plasma membrane of *Saccharomyces cerevisiae. Eur. J. Biochem.*, **119**, 395–400.

Ganser, A. L. & Forte, J. G. (1973). $K^+$-stimulated ATPase in purified microsome of bullfrog oxyntic cells. *Biochim. Biophys. Acta*, **307**, 169–180.

Goffeau, A. & Slayman, C. (1981). The proton-translocating ATPase of the fungal plasma membrane. *Biochim. Biophys. Acta*, **639**, 197–227.

Heefner, D. L. & Harold, F. M. (1982). ATP-driven sodium pump in *Streptococcus faecalis. Proc. Natl. Acad. Sci (USA)*, **79**, 2798–802.

John, P. & Whatley, F. R. (1975). *Paracoccus denitrificans* and the evolutionary origin of the mitochondrion. *Nature*, **254**, 495–8.

Lee, H. C., Breitbart, H., Berman, M. & Forte, J. G. (1979). Potassium-stimulated ATPase activity and hydrogen transport in gastric microsomal vesicles. *Biochim. Biophys. Acta*, **553**, 107–31.

Malpartida, F. & Serrano, R. (1981). Phosphorylated intermediate of the ATPase from the plasma membrane of yeast. *Eur. J. Biochem.*, **116**, 413–17.

Mitchell, P. (1961). Coupling of phosphorylation to electron and hydrogen transfer by a chemi-osmotic type of mechanism. *Nature*, **191**, 144–8.

Niggli, V., Penniston, J. T. & Carafoli, E. (1979). Purification of the ($Ca^{2+}$-$Mg^{2+}$)-ATPase from human erythrocyte membranes using calmodulin affinity column. *J. Biol. Chem.*, **254**, 9955–8.

Robinson, J. D. (1978). Calcium-stimulated phosphorylation of a brain (Ca + Mg)-ATPase preparation. *FEBS Lett.*, **87**, 261–4.

Roisin, M. P., Scherman, D. & Henry, J. P. (1980). Synthesis of ATP by an artificially imposed electrochemical proton gradient in chromaffin granule ghosts. *FEBS Lett.*, **115**, 143–7.

Sachs, G., Faller, L. D. & Rabon, E. (1982). Proton/hydrogen transport in gastric and intestinal epithelia. *J. Memb. Biol.*, **64**, 123–35.

Schatzmann, H. J. (1982). The plasma membrane calcium pump of erythrocytes and other animal cells. In *Membrane Transport of Calcium*, ed. E. Carafoli, pp. 41–108. New York: Academic Press.

Schneider, C., Mottla, C. & Romeo, D. (1979). Calcium ion-dependent adenosine triphosphatase activity and plasma-membrane phosphorylation in the human neutrophil. *Biochem. J.*, **182**, 655–60.

Stewart, B., Wallmark, B. & Sachs, G. (1981). The interaction of $H^+$ and $K^+$ with the partial reactions of gastric ($H^+ + K^+$)-ATPase. *J. Biol. Chem.*, **256**, 2682–90.

Villalobo, A. (1982). Potassium transport coupled to ATP hydrolysis in reconstituted proteoliposomes of yeast plasma membrane ATPase. *J. Biol. Chem.*, **257**, 1824–8.

Vincenzi, F. F. & Hinds, Th. R. (1981). Calmodulin and plasma membrane calcium transport. In *Calcium and Cell Function*, ed. W. Y. Cheung, vol. 1, pp. 127–65. New York: Academic Press.

Watson, J. D. (1976). *Molecular Biology of the Gene, 3rd edn.* Menlo Park, CA: W. A. Benjamin.

Willsky, G. R. (1979). Characterization of the plasma membrane $Mg^{2+}$-ATPase from the yeast, *Saccharomyces cerevisiae*. *J. Biol. Chem.*, **254**, 3326–32.

Wilson, T. H. & Lin, E. C. C. (1980). Evolution of membrane bioenergetics. *J. Supramol. Struct.*, **13**, 421–46.

Yoshimura, F. & Brodie, A. F. (1981). Interaction of vanadate with membrane-bound ATPase from *Mycobacterium phlei*. *J. Biol. Chem.*, **256**, 12239–42.

# Index

A-band, 7, 8
actin (*see also* Actin, F-; Actin, G-)
  evolution, 109
  primary structure, 45–6
  reactive Cys residues, 46
actin, F- (general features), 44–8 (*see also* actin, non-muscle)
  algal, 102
  arrowhead form, 47
  distribution, 44
  electron microscopy, 47, 49
  myosin head binding, effects, 52–3 (*see also* actin, F-, myosin head binding)
  non-muscle, 90–1
  rigidity, 47–8
  structural changes, 63–4
actin, F-, myosin head binding, 34, 62–4
  dissociation, *see* actomyosin
  effect of $Ca^{2+}$, 63, 65, 67, 69
  ionic strength effects, 49
  kinetics, 48
  on-off states, 66–7, 68
  tension levels, transition between states of, 75
  tertiary structure, 48–9
actin, G-, 44–5
  ATP binding, 45; site, 46
  DNase-1 binding, 46
  polymerization to F-actin, 44–5; mechanism, 46–7
  primary structure, 45–6
  tertiary structure, 46
actin, non muscle, 89–110
  binding proteins, *see* actin binding proteins
  cellular movement, 89–90
  F-, 90–1
  fibrils, 105
  intracellular distribution, 92–4
  organization, changes, 94–5; bundles, 94–5
  polymorphic forms, 95–6

structural features, 90–2
taxonomic distribution, 89
types, 91–2
actin binding (contractile network)
  proteins, 95–6, 98–100
  characteristics, 90–2
  crosslinking, 99
actinin, $\alpha$-, 92, 94
actin–myosin system, evolution, 108–10, 283
actinotanglin, 107
active transport, 184 (*see also specific ion*)
actomyosin, 5 (*see also* actin, F-, myosin head binding)
  ATPase, *see* actomyosin ATPase
  dissociation, nucleotide induced, 34, 36–7, 49–52; by analogs, 49–51; by ATP, 51–2; rate, 51; time course, 51
  formation, 48–9
actomyosin-ATP, dissociation, 54–5
actomyosin ATPase activity, 52–7
  burst heads, *see* myosin heads
  $Ca^{2+}$ regulation, 57–71; troponin inhibition, 61 (*see also* tropomyosin-tropinin complex)
  coupling to muscle contraction, 76–84; ATP hydrolysis, routes, 76–7
  free energy change, 78, 80
  in high salt and/or low temp., 53–4; actomyosin dissociation, 53, 55, 56
  in low salt/room temp., 54–5; actomyosin non-dissociation, 54, 55, 56; rate dependence, 54
  main routes, 53, 54, 55–7, 76–7
  molluscan, 69
  in near normal conditions, 55–7
  in other conditions, 56
  substrate inhibition, 65–7; by ATP, 65–7; $Ca^{2+}$ control, 65–6, 67, 68
actomyosin-P-ADP dissociation, 63, 73–4
acto-S1 complex, 47
  binding, 64